D0709516

Essays in Humanistic Mathematics

NATIONAL UNIVERSITY
LIBRARY   SAN DIEGO

Essays in Humanistic Mathematics

Alvin M. White, Editor

Published by
The Mathematical Association of America

NATIONAL UNIVERSITY
LIBRARY   SAN DIEGO

MAA Notes and Reports Series

The MAA Notes and Reports Series, started in 1982, addresses a broad range of topics and themes of interest to all who are involved with undergraduate mathematics. The volumes in this series are readable, informative, and useful, and help the mathematical community keep up with developments of importance to mathematics.

Editorial Board

Warren Page, Chair

Donald W. Bushaw   Vera S. Pless
Melvin Henriksen   David A. Smith
Joan Hutchinson   Tina Straley
John Neff

MAA Notes

1. Problem Solving in the Mathematics Curriculum, *Committee on the Teaching of Undergraduate Mathematics*, a subcommittee of the Committee on the Undergraduate Program in Mathematics, *Alan H. Schoenfeld*, Editor

2. Recommendations on the Mathematical Preparation of Teachers, *Committee on the Undergraduate Program in Mathematics, Panel on Teacher Training.*

3. Undergraduate Mathematics Education in the People's Republic of China, *Lynn A. Steen*, Editor.

5. American Perspectives on the Fifth International Congress on Mathematical Education, *Warren Page*, Editor.

6. Toward a Lean and Lively Calculus, *Ronald G. Douglas*, Editor.

8. Calculus for a New Century, *Lynn A. Steen*, Editor.

9. Computers and Mathematics: The Use of Computers in Undergraduate Instruction, *Committee on Computers in Mathematics Education, D. A. Smith, G. J. Porter, L. C. Leinbach, and R. H. Wenger*, Editors.

10. Guidelines for the Continuing Mathematical Education of Teachers, *Committee on the Mathematical Education of Teachers.*

11. Keys to Improved Instruction by Teaching Assistants and Part-Time Instructors, *Committee on Teaching Assistants and Part-Time Instructors, Bettye Anne Case*, Editor.

13. Reshaping College Mathematics, *Committee on the Undergraduate Program in Mathematics, Lynn A. Steen*, Editor.

14. Mathematical Writing, by *Donald E. Knuth, Tracy Larrabee, and Paul M. Roberts.*

15. Discrete Mathematics in the First Two Years, *Anthony Ralston*, Editor.

16. Using Writing to Teach Mathematics, *Andrew Sterrett*, Editor.

17. Priming the Calculus Pump: Innovations and Resources, *Committee on Calculus Reform and the First Two Years*, a subcomittee of the Committee on the Undergraduate Program in Mathematics, *Thomas W. Tucker*, Editor.

18. Models for Undergraduate Research in Mathematics, *Lester Senechal*, Editor.

19. Visualization in Teaching and Learning Mathematics, *Committee on Computers in Mathematics Education, Steve Cunningham and Walter S. Zimmermann*, Editors.

20. The Laboratory Approach to Teaching Calculus, *L. Carl Leinbach et al.*, Editors.

21. Perspectives on Contemporary Statistics, *David C. Hoaglin and David S. Moore*, Editors.

22. Heeding the Call for Change: Suggestions for Curricular Action, *Lynn A. Steen*, Editor.

23. Statistical Abstract of Undergraduate Programs in the Mathematical Sciences and Computer Science in the United States: 1990–91 CBMS Survey, *Donald J. Albers, Don O. Loftsgaarden, Donald C. Rung, and Ann E. Watkins.*

24. Symbolic Computation in Undergraduate Mathematics Education, *Zaven A. Karian*, Editor.

25. The Concept of Function: Aspects of Epistemology and Pedagogy, *Guershon Harel and Ed Dubinsky*, Editors.

26. Statistics for the Twenty-First Century, *Florence and Sheldon Gordon*, Editors.

27. Resources for Calculus Collection, Volume 1: Learning by Discovery: A Lab Manual for Calculus, *Anita E. Solow*, Editor.

28. Resources for Calculus Collection, Volume 2: Calculus Problems for a New Century, *Robert Fraga*, Editor.

29. Resources for Calculus Collection, Volume 3: Applications of Calculus, *Philip Straffin*, Editor.

MAA Reports

These volumes may be ordered from:
The Mathematical Association of America
1529 Eighteenth Street, NW
Washington, DC 20036
800-331-1MAA   FAX 202-265-2384

©1993 by the Mathematical Association of America

ISBN 0-88385-089-3

Library of Congress Catalog Number 93-78960

Printed in the United States of America

Current Printing
10 9 8 7 6 5 4 3 2 1

Preface

At their most vivid, the [humanities] are like the arts as well as the sciences.. The humanities are that form of knowledge in which the knower is revealed. All knowledge becomes humanistic when this effect takes place, when we are asked to contemplate not only the proposition, but the proposer, when we hear the human voice behind what is being said.

—Charles Frankel

In the preface to his book on semigroups, Einar Hille tells us that once the definition was formalized he noticed semigroups everywhere, in a variety of contexts. The career of humanistic mathematics is similar. Although the concept has not been formalized there have been many essays and much activity under its banner. Several conferences, symposia, contributed paper sessions, the Humanistic Mathematics Network Journal and this collection of essays indicate the vitality of the concept.

Humanistic mathematics carries with it an awareness of and a sensitivity to those things mathematics shares with the humanities such as literature, art, and music. Two themes have emerged: **1)** Teaching mathematics humanistically and **2)** Teaching humanistic mathematics. The first theme seeks to place the student more centrally in the position of inquirer than is generally the case, while at the same time acknowledging the emotional climate of the activity of learning mathematics. Included in this first theme is the encouragement of students to learn from each other and to better understand mathematics as meaningful, socially constructed knowledge, rather than as an arbitrary discipline.

The second theme focuses less upon the nature of the teaching and learning environment and more upon the need to reconstruct the curriculum and the discipline of mathematics itself. This reconstruction relates mathematical discoveries to personal courage, discovery to verification, mathematics to science, truth to utility, and mathematics to the culture in which it is embedded.

Humanistic dimensions of mathematics include:

An appreciation of the role of intuition in understanding and in creating concepts that appear in their finished versions to be "merely technical."

An understanding of the value judgments implied in the growth of any discipline. Logic alone never completely accounts for *what* is investigated, *how* it is investigated, or *why* it is investigated.

The essays in this volume illustrate and help to define humanistic mathematics. The variety and scope indicate the richness and fruitfulness of the concept. Although each essay is independent, a sense of unity emerges. Mathematics is common to all, of course; it is not the mathematics of facts, formulas, and algorithms with which most students are familiar. It is not the mathematics of anxiety and boredom. It is the mathematics of confidence and optimism. There is excitement, adventure, erudition. We hear not only the voices of the teachers, but the voices of the students as well.

There is another voice identified with the humanistic mathematics movement; a voice whose message is hope and liberation from routine. Many who have heard of humanistic mathematics quickly recognize this movement as a source of colleagues who feel as they do about mathematics in their lives and in their students' lives. The movement is also a source of courage—to seek new approaches that will make teaching more humanistic and to include humanistic dimensions in the perception of mathematics.

The movement, which began as the personal vision of a few, has now become a major part of mathematical culture. What was viewed with skepticism is now accepted and expected. Humanistic Mathematics is not a new discovery. It is a recent rediscovery of an idea that goes back to Plato, and later to D'Alembert, Condorcet, Bacon, and more recently to Wilder, Bronowski, and Cassirer. Almost all contemporary mathematicians view humanistic mathematics as a natural idea. Indeed, humanistic mathematics has provided a new vocabulary for previously unarticulated concepts and approaches.

Many nonmathematicians, however, consider humanistic mathematics an oxymoron, something beyond their comprehension. For too many, learning mathematics is an alienating experience. They will never understand E.O. Wilson's remark on National Public Radio:

Good science is play disguised as hard work.

Humanistic mathematics opens up a mathematical world of excitement, adventure and satisfaction.

Alvin White
Harvey Mudd College

Acknowledgements

I am grateful to the Exxon Education Foundation, especially Robert Witte, Michael Dooley, and Scott Miller, for being the midwives for the Humanistic Mathematics Network in 1986 and for sustaining and supporting the network and our newsletter and journal since then.

The MAA has welcomed our movement and has made humanistic mathematics a regular part of the annual national meetings and many sectional meetings. Hundreds of mathematicians have found their voices and a responsive audience at the contributed paper sessions.

Harvey Mudd College has recognized the scholarly content of humanistic mathematics, providing a home and administrative support to the network and its publications.

Michelle Ivey (HMC '95) is a master of desktop publishing on the Macintosh and associated software, and has brought her good humor and judgment to the task.

Special thanks to the MAA Notes Editorial Board Members Warren Page, Donald Bushaw, and Joan Hutchinson. Donald Bushaw was especially helpful with extensive editorial assistance and suggestions.

CONTENTS

Part IV: Teaching and Learning Experiences

Part V: Contemporary Views of Old Mathematics

Introduction

Part I: Introduction to Humanistic Mathematics

Philip Davis discusses the humanistic aspects of mathematics. Mathematics, like literature, has metaphor and that mathematics, like poetry, has ambiguity and cannot be totally formalized.

Thomas Tymoczko concludes that pure mathematics is ultimately one of the humanities because it is an intellectual discipline with a human perspective and a history that matters. Humanistic mathematics is not just "friendly math" or "touchy-feely" math. It is mathematics with a human face. Without it educators may teach students to compute and to solve, just as they can teach students to read and to write. But without it educators can't teach students to love or even like, to appreciate or even understand mathematics.

Reuben Hersh says that humanistic mathematics challenges dogmatic teaching styles that expect students to parrot the lecturer. It demands creativity of both the teacher and student. Many successful mathematics programs have not been sponsored by a mainstream national organization. Let us maintain that wild and woolly unrestricted arena for free experimentation and argumentation.

Part II: Mathematics in the World

Hardy Grant sketches the cultural evolution of mathematics. One may surmise that in the days of Euclid and Archimedes a science that unveiled apparently immutable truth about the ultimate realities was challenge and exhilaration enough, but this role of describing the existentially given, marked also the boundary of Greek mathematics.

Galileo represented, with a vividness probably unparalleled before or since, the several intellectual stances that together defined in his time the unique cultural influence of mathematics. Man's understanding, said Galileo, where mathematics can be brought to bear, rises to the level even of God's. An alarmed Church duly included this breathtaking assertion of human pride among the eight counts of heresy that brought the rash Florentine to his famous trial.

Baruch Spinoza's *Ethics* remains the most Euclidean treatise ever written on a humanistic subject, a resolute deduction of theorem after theorem from initial definitions and axioms.

God forbid, said William Blake, that truth should be confined to mathematical demonstration.

The younger Bolyai drew from his non-Euclidean geometry a more heartening promise. "From nothing I have created an entirely new world."

Most researchers still ascribe to their theorems a timeless, unconditional truth; Euclid may be dethroned, but Plato survives. Mathematics remains for many a self-contained world of enchantment, an inexhaustible realm of the strange, the diverting, the beautiful, the intellectually challenging.

Jack Wales describes the poetics of mathematics. The coincidence of pattern is a cause for delight. There is one thing of which geometry is not a picture, and that is the so-called real world. The poet uses these two crude, primitive, archaic forms of thought in the most uninhibited way, because his job is not to describe nature but to show you a world completely absorbed and possessed by the human mind. Mathematics, along with literature and music, is a language of the imagination.

The essence of science is the right to repeat an experiment, while the essence of mathematics is the right to understand. The aesthetics, properties of ceremony, formula and natural processes are intimations of complex and profound intellectual objects.

That mathematics, in common with the other humanities, can lead us beyond ordinary existence, and can show us something of the structure in which all creation hangs together, is no new idea. The structures with which mathematics deal are more like lace, the leaves of trees, and the play of light and shadow on a meadow or a human face, than they are like buildings and machines, the least of their representations.

We should let mathematics be, just as other disciplines are the pursuit of ways of seeing, the pursuit of visions.

The relations among humanistic mathematics, ethnomathematics, literature, and education are discussed by *Ubiratan D'Ambrosio*. This is a proposal to shift from the quantitative aspects of the investment in education. Remedial curricula are examples of how mathematics is being oversold. College courses on basic arithmetic, taught to adult students as if they were children, are aggressive and decontextualized, and they exemplify the criticism of overselling mathematics.

Ernst Mach, referring to offering the same curriculum for all students, said that in such a case the function of the schools would be simply to select the persons best fitted for being drilled whilst the finest special talents that do not submit to indiscriminate discipline would be excluded. The specific problem a child is trying to solve creates a mental set that determines the features selected for attention.

Literacy is the development of capabilities of communication among human beings, and *matheracy* is the development of capabilities of understanding, explaining, and coping with the reality that we call mathematics.

Setting out from the uncertainty of knowledge and the inadequacy of every form of dogmatism, conscious of the fragility and brevity of human civilization, confronted with the old historical cosmologies of progress, many sought to reawaken the positive value of the unconscious and dream life.

Confronted with powerful tools for developing the knowledge of self and of nature, capable of producing at paces hitherto unthinkable, mankind sees the breakdown of culture in competing ideologies and the incoherence of not only these ideologies and society as a whole, but of knowledge and the resulting voiding of the self.

Harald Ness celebrates mathematics in our culture. It is little known that mathematics has determined the direction and content of philosophical thought and has destroyed and rebuilt religious doctrine. New mathematical ideas, often developed with no thought of application, have repeatedly turned out to have immense scientific, technological, and economic benefits.

In ancient times music was considered a part of mathematics. Mathematics is the music of the mind; music is the mathematics of the soul. Mathematics is perhaps the purest of intellectual activities; to the outsider it seems like a private game—the manipulation of symbols to uncertain and unworldly ends.

A student, writing in a news magazine, stressed that the humanities deal with ideas while the sciences merely deal with skills.

Mathematics is the natural home of both abstract thought and the laws of nature. It is at once pure logic and creative art.

Robert Osserman traces and explains the development of musical scales from the time of Pythagoras. On the way, the contributions of Ptolemy, Galileo, Mersenne, Bach, and Brahms are considered. The necessity of introducing irrational numbers is part of the story.

Galileo, in *Two New Sciences* was one of the first to explain the physical origins of the mathematical ratios occurring in music. He points out that too much consonance strikes us as too bland and lacks fire. What we seek in our music is just the right amount of dissonance. "This produces a tickling and teasing of the cartilage of the eardrums, so that the sweetness is tempered by a sprinkling of sharpness, giving the impression of being simultaneously sweetly kissed and bitten."

Part III: Inner Life of Mathematics

Nelson Goodman modernizes the philosophy of mathematics. For Plato (*The Republic* VI, 510) the objects that mathematicians study are "ideals which can be seen only by the mind."

If right triangles exist only in our mind, asks Frege, must we speak of my Pythagorean theorem about my right triangles, and your Pythagorean theorem about your right triangles?

Mathematics consists of eternal truths about eternal objects. This was surely Plato's view. The mystery is how we could come to know such truths. Indeed, the increasing tendency of mathematicians to rely on computers clearly undermines the a priori character of their discipline. We learn mathematics by doing computations and other, less routine constructions and by being surprised by the results of those computations.

When we describe a physical situation mathematically, where does the vagueness go? It cannot just disappear since that would involve giving an exact description of a real situation which is almost never possible. The answer is that the vagueness goes into the word "approximately."

In most respects, therefore, mathematical theories resemble other scientific theories. These are constructed to solve particular problems and then develop a life of their own.

The interesting case for mathematics as an art, writes *Thomas Tymoczko*, is the possibility of regarding at least some of its products as objects of aesthetic enjoyment. Mathematics, rightly viewed, possesses not only truth, but supreme beauty.

Borel wants to argue that transfinite set theory can be justified better in aesthetic terms— because it is a beautiful theory with many elegant mathematical proofs and constructions.

Criticism is an essential feature of art. A good critic provides a way of seeing or hearing a work of art, and good art is subject to multiple analysis. A great work has a richness and complexity that a sensitive audience can be brought to acknowledge, and it is a function of criticism to bring them to this stage.

Formal proofs do not function as templates; they are rarely, if ever, written down or spoken aloud. In fact, really formal proofs are usually merely assumed to exist to validate the actually presented proof.

The Two Culture Controversy appeared as the latest in a long series of intellectual conflicts that have shaken the Western World, according to *Philip Davis*. There was the conflict between the Word and the Deed around 50 B.C. with Cicero as hero.

Art exhibited both redemptive and destructive features. Finally it was tempted by science, in that it proposed to bring in scientific methodological modes and modes of interpretation. This was perceived to be fatal for the humanities.

In technology the major event has been the widespread computerization which is entering every aspect of human life and changing it.

Theoretical physics has moved substantially away from the empirical, being inspired (and tempted) more and more by the formalisms of mathematics. Literature has split into the banal, the esoteric, and the sci-fi. The very act of reading has itself been threatened by the development of high-tech modalities of information transfer.

To be funded, what is experimental must often be backed up by "scientific" evaluations that display standard deviations of this and that computed routinely and most often thoughtlessly and pointlessly. The ritualistic display of such numbers gives the process a cachet of the scientific and objective.

Gian-Carlo Rota discusses the concept of mathematical truth. The truth of the theorems is seldom seen by staring at the axioms. Although the theorems may be tautological consequences of the axioms, at least "in principle", nonetheless such tautologies are not immediate or evident. The theorems themselves, whose truth had previously been made intuitively evident, are finally made inevitable by formal proof which comes almost as an afterthought.

All formalist theories of truth are reductionist. They derive from an unwarranted identification of mathematics with the axiomatic method of presentation of mathematics.

Thanks to high speed computers, we know today what Gauss could only guess—that conjectures in number theory may fail for integers so large as to be beyond the reach of even the best of today's computers, and, thus, that formal proof is today more indispensable than ever.

Contemporary mathematics, with its lack of a unifying trend is a further step on the way to the end of two embarrassing Victorian heritages: the idea of progress and the myth of definiteness.

Every mathematical theorem is eventually proved trivial. Every mathematical proof is a form of debunking. Science may be defined as the transformation of synthetic facts of nature into analytic statements of reason.

Rota in a second essay compares philosophy and mathematics and the misunderstanding between them. The assertions of philosophy are less reliable than the assertions of mathematics, but they run more deeply into the roots of our existence. Today's philosophers have rewritten Galileo's famous sentence to read "The great book of philosophy is written in the language of mathematics."

Mathematical logic no longer claims to give a foundation to mathematics. Our everyday reasoning is not precise, yet it is effective. Nature itself, from the cosmos to the gene, is approximate and inaccurate. The facts of mathematics are verified and presented by the axiomatic method. One must guard, however, against confusing the *presentation* of mathematics with the *context* of mathematics. A good definition is "justified" by the theorems one can prove with it.

Part IV: Teaching and Learning Experiences

Stephen I. Brown observes that unity, coherence, and clarity are illusions that appear to be conveyed not only through television but in most bureaucratized settings. The expert teacher, in contrast to the novice teacher in two recent reports, is a model of clarity. Students of expert teachers are never bored or confused. "Good teachers explain to students the problem, the solution, and when to use a specific process. Most expert teachers cover at least 40 problems a day. Novice teachers, on the other hand, may cover only six. Students with expert teachers are rarely lost during a math lesson. They know what is happening, what is going to happen and what they are supposed to do." The premises of these reports are shown to be unrealistic and against actual experience of problem solving or critical thinking. Brown contrasts the reports with the views of Polya and Dewey and shows how shallow and superficial these reports are.

A problem is, to some extent, a project for the future to which we commit ourselves by an act of will. Problems require attention and courage, and they involve a significant act of self-surrender. A problem is a hopeful enterprise that involves an act of faith.

We run a greater risk of misrepresenting the lived experience of inquiry when we discourage students from honoring their own doubts, ambivalence and disharmonies.

The first concern of teacher education is less with input of information about teaching and more with the creation of a climate for unearthing what teachers already believe about teaching.

Raffaela Borasi shows how students can be brought to personally experience the humanistic dimensions existing within technical mathematics. Students can thus be helped to appreciate how definitions are really created by us. Contrary to general expectations, confusion is often a positive achievement. It is not as if we were born and recognized a tree that has stood forever. It is that we invented forms from our own minds.

Larry Copes starts with a situation and generates questions. We don't stop with an answer. We reflect. We return to our original enjoyment of the problem's tug on our intuition. We want our students to be able to distinguish between intellectual and ethical correctness. For our students to be able to do mathematics, they must move beyond thinking of mathematics as a collection of formulas to be memorized.

A person's "Perry position" is closely tied with that person's identity. Changing the way of interpreting experience means changing who that person is.

As Clarence Stephens has said, one of the impediments of implementing a high-quality mathematics program such as that at Potsdam is "excessive faculty concern about the subject matter to be covered."

Like mathematicians, students may focus on answers; but our focus as teachers is no longer on the body of knowledge, but rather on the mathematical process: on students' acquiring the reasoning abilities, the critical thinking attitudes, and the maturity to look at things from a variety of perspectives.

Dorothy Buerk describes the public image of mathematics as an elegant, polished, finished product which obscures its human roots. It has a private life of human joy, challenge, reflection, puzzlement, intuition, struggle and excitement. Mathematics is a humanistic discipline, but the humanistic dimension is often limited to its private world. This chapter presents the story of Jackie, who knows mathematics' public image and was "successful" in high school, including a calculus course. She felt disconnected from mathematical ideas, however, and considered herself "math phobic".

Jackie tells the story of her experience in the Writing Seminar in Mathematics. Insights from Peter Elbow and William Perry illuminate the processes of the course and the student's growth and development.

When we ask whether a course of study is *humanistic*, writes *Robert Davis*, we ask a very different kind of question from those that are usually posed, and the answers can point us in quite different directions. A course is *humanistic* if its main goals include: helping students develop an appreciation of some of the great achievements of the human mind, learning to help others understand a mathematical situation or problem more deeply; being able to talk about and write about mathematical situations; developing a lively intellectual curiosity.

"Twelve years of being told how to perform largely meaningless operations on largely meaningless symbols is not humanistic education....It is not...education at all."

This chapter is about a calculus course at University High School of the University of Illinois and the effect on the students and the teacher. The interaction of the students with the intellectual challenges and the resulting excitement affords special attention to the humanistic dimensions.

Part V: Contemporary Views of Old Mathematics

William Dunham describes the first NEH summer seminar for high school mathematics teachers. The subject is the great mathematical treasures of the past.

"Even the terms mathematicians use to discuss their favorite theorems—terms like beautiful, elegant, powerful—are precisely those used to describe a Mozart concerto or an Eliot poem...

Never have I seen a discussion where this kind of fervor could be generated by mathematical ideas."

Elena Marchisotto shows how the familiar Pythagorean Theorem can be the basis for an exciting course for liberal arts students. The historical and cultural contexts allow the students to explore the philosophy of mathematics. The discovery of irrational numbers and of non Euclidean geometries are events that lead the students to new insights about mathematics.

Peter Flusser also starts with the Pythagorean tradition but takes a different turn. He compares Euclid, Diophantus, and Euler. "Reading Euler is a special experience; besides being mathematical treasures, his works are entertaining and suspenseful."

"Euler's work surpasses that of Euclid and Diophantus in one important respect. Euler's imperfection (which turns an otherwise perfect piece of workmanship into a work of art) raises many new questions,... It would seem that the more humanistic pieces of mathematics are also the more fruitful."

Abe Shenitzer rehearses Huygens' proof that the cycloid is the only tautochrone with an appreciation of the brilliance of his inventive capacities.

PART I

Introduction to Humanistic Mathematics

The Humanistic Aspects of Mathematics and Their Importance

Philip J. Davis
Brown University
Providence, Rhode Island

(A talk given at a panel on humanistic mathematics, Louisville, KY, January 17, 1990.)

We are now living in a period in which at least four revolutions are working themselves out—(not by the way, independently). There are revolutions of ethnic, familial, and gender related values. There is the political/economic revolution of Eastern Europe, which, in a way, represents the failure of one great idea put forward by rational thinkers to model human realities. There is the media revolution which is already changing

Like poetry, mathematics cannot be totally formalized.

the basic patterns of communication, socialization and learning. And there is the computer revolution, bringing with it, and enormously facilitating, mathematizations of every aspect of our lives.

"Any serious fundamental change in the intellectual outlook of human society must necessarily be followed by an educational revolution. It may be delayed for a generation by vested interests or by the passionate attachment of some leaders of thought to the cycle of ideas within which they received their own mental stimulus at an impressionable age. But the law is inexorable that education to be living and effective must be directed to informing pupils with those ideas, and to creating for them those capacities which will enable them to appreciate the current thought of their epoch." (A.N. Whitehead, *The Aims of Education*, "The Mathematical Curriculum.")

Mathematics lives both in the technological and in the humanistic cultures. It exhibits features that are science-like and features that one normally associates with the humanities.

What are some of the features of mathematics that are humanistic? I seek parallels with literature, which I take as a paradigm of humanistic expression. (And I follow here Jacques Barzun in his 1972 Bollingen lectures.)

Mathematics, like literature, has metaphor. (Models)

Mathematics, like poetry, has ambiguity. (Will the **true** geometry please stand up).

Mathematics possesses an aesthetic component which is strong and which is immediately apparent to the practitioner at the higher levels of the subject.

I would assert that like poetry (which, according to T.S. Eliot, cannot be totally written down), mathematics cannot be totally formalized.

Mathematics has paradox.

Mathematics has mystery and can convey awe.

Mathematics has: a sense of outcome, a feeling of rightness, a sense of catharsis.

Mathematics is allied with and has contributed mightily to philosophy.

Mathematics has contributed to theology. It grasps for the transcendental and in so doing, can be a surrogate for traditional forms of religious expression.

Like literature, mathematics can be an avenue of mental escape from this world.

Like contemporary literature, mathematics is done in a vacuum of belief, if one interprets this vacuum to refer to a formalist philosophy which most (pure) mathematicians adopt when queried about the essential nature of the materials they work with.

Like literature, mathematics exhibits both redemptive and destructive features.

Mathematics has a history. I emphasize this point, not because it is something unique to mathematics (presumably human ideas of whatever sort all have histories) but because it is often asserted that the truths of mathematicians are atemporal, and hence, stand outside of history.

Like anthropology and literature, mathematics embodies mythologies. I use the term "myth" not to mean that which is false, but that which is accepted as normative. My friend and coauthor, Reuben Hersh, has written about the four myths of mathematics: its unity, objectivity, universality, and certainty.

Christopher Ormell, distinguished British philosopher of mathematical education, has written about the need for the **demythologizing** of mathematics.

Perhaps what one wants is not so much a demythification but a de-hocus-pocus-ization to counteract a prevalent feeling among 99% of humans that mathematics is nothing but another form of magic.

It would be an undertaking of the first importance to work out in detail the parallels I have just suggested.

If mathematics exhibits humanistic features, then we may reasonably expect it to promote humanistic values.

Humanistic values are those which foster the consciousness of full human responsibility. To me, the phrase "mathematics as one of the humanities" means nothing if not that.

Mathematics is partly a language. Certain things are communicated by it. Things can be described, predicted, and prescribed by it. **The ability of mathematics to provide frameworks of reality and of action, and its ability to change our perception of what is, is very great.**

Our world is rapidly becoming mathematized. Some of our best talent is spent in putting mathematizations in place, creating and moving around abstractions. Cosmologists do it. Inside traders do it.

David Berlinski, a mathematician, philosopher and polemicist, has written that once a mathematization has been put in place, it is all but impossible to remove:

> Mathematical descriptions... tend to drive out all the others. Mathematics is often a matter of bondage with *things* in thrall to *theories*. Strong theories make for weak objects.

Philosophically, it would be better if we ceased regarding mathematics as that great, objective, people-free, supra-moral, a-temporal reservoir of eternal truths, whose procedures we chant mindlessly and before whose inevitabilities we must bend the knee.

With mathematizations proliferating (whether we install them ourselves or import them), the principal role of the teacher should **not** be to expound formal aspects in

---

If mathematics is taught simply as the learning of procedures, then none of the humanistic elements can be grasped.

---

a way that may be better accomplished by other means, or to drill on topics that have been automated out of the marketplace.

If mathematics is taught simply as the learning of procedures, then none of the humanistic elements can be grasped. It is rare that the teaching of mathematics, elementary or advanced, conveys or fosters the humanistic aspects of the subject.

If there has been a shortfall of self examination, it is not in the examination of its own inner material by its own methodology, but it is in a steady refusal to examine how the characteristic features of mathematical thought operate on us and affect us.

The principal role of the human teacher should be to humanize the subject.

To teach mathematics as a humanity means nothing less than to teach that it possesses the awesome power to influence and change our lives, and to teach that we who use and foster it, must subject it to constant study and scrutiny.

Very Short Bibliography

Jacques Barzun, *Bollingen Lectures*, Library of Congress, 1972.

Philip J. Davis and Reuben Hersh, *Descartes' Dream*, Harcourt Brace Jovanovich, 1986.

Alfred North Whitehead, *The Aims of Education and Other Essays*, Free Press, 1967.

Humanistic and Utilitarian Aspects of Mathematics

Thomas Tymoczko
Smith College
Northampton, Mass.

Excerpt from a talk at the International Congress of Mathematics Education VII, Quebec, Canada, August 1992.

Prof. Tymoczko began by arguing that, contrary to common opinion, applied mathematics was logically prior to pure mathematics and not vice-versa. By assimilating mathematics to science and common sense, Tymoczko claimed to have eliminated any special philosophical doubts about mathematical existence. Nevertheless, he went on to claim that pure mathematics could not be explained simply as the objective study of mathematical objects. We pick up his paper as he discusses what is missing from such an explanation.

So we must abandon the trivial answer to the question: what is pure mathematics? It's not just the study of arbitrary mathematical objects of the production of arbitrary proofs and computations. I suggest we can understand what pure mathematics is only if we abandon

---

Pure mathematics is ultimately humanistic mathematics because it is an intellectual discipline with a humanistic perspective and a history that matters.

---

the claim that mathematics is simply the study of the mathematical universe and embrace the thesis that what counts in mathematics is only what counts to mathematicians (eg., humans) at a given time in mathematical history. Mathematics is not just a universe of mathematical objects or formalisms or constructions; at the very least it includes a point of view on that universe. This is the essence of humanistic mathematics—mathematics requires a perspective, and a human perspective is the only perspective we can get.[1]

Several years ago, Alvin White of Harvey Mudd College in the USA began a campaign for what he called "humanistic mathematics."[2] While I admired the pedagogical reforms that issued from White's campaign—he wanted mathematics taught in a humane way—I failed to appreciate the significance of humanistic mathematics. To be sure, it was interesting to consider teaching mathematics as if it were one of the humanities, but what made mathematics one of the humanities? Certainly not the mere fact that humans did it; humans do science too. In writing this essay, I've rediscovered White's point. Pure mathematics is ultimately humanistic mathematics, one of the humanities, because it is an intellectual discipline with a human perspective and a history that matters. There is no final answer to the question: what is important in mathematics? We can only ask what is important in mathematics to human beings, with given abilities and limitations at a given point in their mathematical development. The discipline of pure mathematics is much more like geography than it is like physics. That is why I want to rename it "humanistic mathematics".

If, for the sake of argument, you grant my conclusions so far, then we can turn to the topic of how mathematics might be taught in a way that reveals its humanistic side or its relevant interest.

Earlier, I accused philosophers and mathematicians of making reciprocal mistakes. Philosophers have ignored utilitarian mathematics and thereby created for themselves the problem of mathematical existence. But educators, I claim, are prone to make the opposite mistake. In teaching mathematics, they insist on stressing its utility, even when it has none. As a result, they often hide from their students the excitement and intrinsic interest of mathematics: they hide it behind a facade of supposed utility. It's rather like trying to awaken people to the joy of mountain climbing by trying to convince them that someday they might need to climb a mountain (as if cars and busses wouldn't be able to satisfy any practical need).

Let me try to explain my view of humanistic mathematics in two ways. The first is by means of a concrete example concerning the teaching of quadratic equations in secondary schools in the USA. The second is with a more global metaphor for humanistic mathematics.

In my experience there is a "standard" way of teaching quadratic equations. It is organized according to utilitarian mathematics. The motivation is supplied, supposedly, by practical needs expressed in word problems. For example, suppose you have a rectangular plot of land and you want to build a sidewalk one meter wide around it, etc., etc. The student is led through hundreds of exercises involving various techniques of factorization. Finally the quadratic formula is derived (thus rendering otiose the effort that the student put into earlier attempts to factor or complete the square). This project easily consumes half of a school year.

Now, if the aim is to teach applied or utilitarian mathematics, perhaps this approach is all right—although it's unethical, from a utilitarian point of view, to delay the quadratic formula for so long. But, before committing yourself to the utilitarian viewpoint, you might try to remember the last time that you needed to solve a quadratic outside of a classroom, and you might try to explain why a computer program (or calculator) isn't a better way to solve such problems. Be that as it may, the standard approach does not introduce students to the discipline of mathematics or to humanistic mathematics as I conceive it.

To introduce students to humanistic mathematics is to introduce them to a human adventure, an adventure of which humans have actually partaken in history. The story of quadratics is part of a more general story of investigating equations—linear, quadratic, cubic, biquadratic, etc. These form "a natural class" of problems to us humans, and the quadratic equations are a piece of this richer puzzle. This puzzle is challenging to human mathematicians for the same reason that mountains are challenging to human mountain climbers: because the puzzles and the mountains are there for us.

By the time that they approach quadratics, students will find linear equations easy. But do students realize what a significant thing it is to find linear equations easy? The Greeks did not recognize negative solutions to linear equations, and even 16th century mathematicians classified quadratics into various subclasses because of their suspicion of negative numbers. It took human beings thousands of years to progress to the mathematical level of today's high school students, and perhaps teachers should mention this to students.

I was saying that humanistic mathematics tells the human story of mathematics. It puts the discussion of quadratics into the human-mathematical context that gives the mathematical topic its sense and its beauty.

The general story of quadratics provides an opportunity to discuss Arabian mathematics and the mathematician al-Khowârizmî, who preserved and developed the partial Greek solutions to the quadratic. (Of course, the words "algebra" and "algorithm" are derived from him and his work). Moreover, and this is the surprise, the investigation of quadratics could be put into perspective by spending just a week or two on cubic equations.

For starters, one could use the story of the cubic to expose students to the very different mathematical culture of Renaissance Italy—where mathematicians challenged each other like gun fighters in modern spaghetti Westerns. According to William Dunham's book "Journey Through Genius", from which I get my story, one Antonio Fior was bequeathed the solution to so-called "depressed cubics" by his teacher.[3] Fior immediately challenged Niccolo Fontana, known as Tartaglia the Stammerer, to a mathematical contest. Fontana proposed 30 problems, each asking for the solution to a depressed cubic equation! Tartaglia knew what was going on and, by working night and day, found the general solution in time to thoroughly humiliate Fontana, who did not know much besides the

> **To introduce students to humanistic mathematics is to introduce them to a human adventure, an adventure that humans have actually partaken of in history.**

formula for the depressed cubic. In the next twist of fate, that most bizarre, if not lunatic, mathematicianm Cardano extracted from Tartaglia his solution of a particular form of cubic equation. The price he paid was a solemn oath to Tartaglia "by the Sacred Gospel, and on my faith as a gentleman, never to publish your discoveries if you tell them to me..." Cardano went on to use Tartaglia's discovery to solve the general cubic and his student Ferrari exploited it to solve the biquadratic.[4]

My idea is that Cardano's analysis is well within the reach of secondary students; essentially it applies the quadratic formula to cleverly contrived cases of the cubic. And my suggestion is that the teaching of quadratic equation could be far more exciting if teachers used the quadratic solution to derive the solution of the cubic—as opposed to those endless and boring word problems. By highlighting the similarities and differences between the quadratic solution and the cubic solution, the comparison can give the student a deeper appreciation and understanding of mathematics for its own sake. Moreover, even less than bright students would rather listen to the story of Cardano, no doubt, than do a hundred so-called practical problems about quadratics.

Thus I suggest that some discussion of the cubic should be an essential part of the teaching of quadratic equations—not because it is useful, but because it makes sense. It puts the quadratic in its proper mathematical perspective. By comparing and contrasting the quadratic and the cubic, a student can begin to see the overall shape of the forest instead of just hundreds of trees.[5]

Moreover, if I were inculcating the discipline of mathematics—humanistic mathematics—I would not finish quadratics without mention of the work of the Norwegian Abel who showed that quintic equations are not solvable by radicals. This should generate an interesting class discussion. How can a mathematician show that a mathematical problem is unsolvable as opposed to merely failing to solve it? And, of course, it would be sinful not to mention the Frenchman Galois, who explained why equations to the fourth degree were solvable and why none higher were. This might even provide an opportunity to mention the concept of "group"—as well as ending the story where it began, with a "dueling mathematician"!

Let me briefly summarize. Standard approaches to the quadratic formula embed the quadratic formula in purely utilitarian mathematics. They suppress the aesthetical, the historical and the purely mathematical aspects of this mathematical problem in favor of touting the practical significance of answering various canned word problems. Students spend half a year mastering a variety of techniques leading up to a general solution which eliminates the need for their mastery of those techniques. But they are never told why anyone would think a general solution was intrinsically interesting for its own sake.

Humanistic mathematics can give quadratic equations their rightful mathematical significance by placing them in a context of pure mathematics—more particularly, by placing them in the context of historical progress toward answering a natural mathematical question. This is a history of approaches and conquests that stretches millennia from the halting efforts of the Greeks to the final summation of Galois. The general solution to quadratic equations is but one piece of this history. Humanistic mathematics is not just "friendly" math or "touchy feely" math. It is mathematics with a human face because there is no mathematical discipline without a human face. Stories of mathematicians are "color". It is interesting that Tartaglia was a stammerer who extracted a promise from Cardano. But stories about what historical individuals saw when they looked on the

mathematical universe at historical points of time are not color. They are mathematics. No one can learn mathematics without being inculcated into this tradition.

In conclusion, I want to sketch an analogy between humanistic mathematics and another human endeavor, the practice of mountain climbing.

Neither humanistic mathematics nor mountain climbing are practical human concerns—but both of them are rooted in practical concerns, for example, both have a foot in business and trade.

Neither humanistic mathematics nor mountain climbing are sciences—but both are bounded by objective constraints, mathematical facts and geological facts.

Both humanistic mathematics and mountain climbing fail as sciences for the same reason—each depends on the contingencies of the human condition. Mountain climbing is what it is because human beings are what we are; we have such and such size, such and such natural abilities, can do this easily and that with practice. Exactly the same applies to humanistic mathematics. It is shaped by human abilities and limitations, because we can do some thing easily, others only with difficulty. God's mathematics would be very different from ours—as would a beetle's conception of mountain climbing differ from ours.

Moreover, as with other humanistic disciplines, mountain climbing and humanistic mathematics both have a history. What is difficult at one period, becomes easy at another. The historical context of a given period

---

Humanistic mathematics is not just "friendly" math or "touchy feely" math. It is mathematics with a human face.

---

sets the goals of that period. If no one has solved the general cubic or climbed that particular mountain, then those are the goals of the day. Later, such goals might become exercises for apprentices. Furthermore, technology is especially important; it alters what can be done and our evaluations of various achievements. (Solving particular quadratic equations is not too impressive to one who has seen the formula for general solutions.)

In the end, humanistic mathematics and mountain climbing are both driven by a fundamental human characteristic: the ability to take joy in complex endeavors. In both cases we find activities or processes driven by goals or achievements. Without results, the theorems or the mountains climbed, we would not have the activity, but it is the journey to the results—the actual doing of mathematics and the actual climbing of mountains—that provides the day-to-day gratification that keeps these practices alive.

The point of the analogy between mathematics and mountain climbing is to exhibit a critical human, or subjective, component of mathematics. This human component is not a frill that might make learning mathematics more enjoyable for the mathematically handicapped; this human component is a *sine qua non*

But without it, educators can't teach students to love or even like, to appreciate or even understand, mathematics.

---

of a separate discipline of pure mathematics. In a nutshell, the human component imposes sense or intelligibility on mathematics; it imposes a human perspective on the arbitrary complexities of the mathematical universe—exactly as our human perspective shapes a coherent practice of mountain climbing on otherwise unwieldy mountains.

Educators ignore humanistic mathematics at their peril. Without it, educators may teach students to compute and to solve, just as they can teach students to read and to write. But without it, educators can't teach students to love or even like, to appreciate or even understand, mathematics.

Endnotes

1.   A case in point might be the rise of complexity theory and the resurgence of interest in discrete mathematics. My intuition is that in recursion theory all finite sets are trivially recursive and so uninteresting. But the development of computer technology has enabled us to raise interesting questions about distinctions in the finite realm: eg., the $P = NP$ problem.

2.   Further information on White's project is available in the Humanistic Mathematics Journal, published by White at Harvey Mudd College, Claremont, CA, 91711.

3.   Depressed cubics are cubic equations lacking a term involving the square of the unknown.

4.   For an interesting interpretation of the dispute between Cardano and Tartaglia, and of the practical difficulties that beset Abel and Galois, see Collins and Restivo.

5.   By the way, a natural human interest story arises here: how could Cardano and Ferrari reconcile the oath to Tartaglia with their desire to publish perhaps the most important mathematical discovery of the 16th century? Since we are interested in pure mathematics, I won't distract you by discussing their solution, but Dunham explains it in his lovely book.

Bibliography

Collins, Randall and Sal Restivo, "Robber Barons and Politicians in Mathematics: a Conflict Model of Science", *The Canadian Journal of Sociology*, 8, 1983, 199-227.

Dunham, William, *Journey Through Genius*, New York: John Wiley & Sons, Inc. 1990.

Edwards, C.H. Jr., *The Historical Development of the Calculus*, New York: Springer-Verlag. 1979.

Friedman, Michael, "Kant's Theory of Geometry", *Philosophical Review*, 94, 1985, pp. 455-506.

Kant, Immanuel, *Metaphysical Founders of Natural Science*, Indianapolis: Hackett, 1985.

Monk, Ray, *Ludwig Wittgenstein: the Duty of Genius*, New Your: the Free Press, 1990.

Quine, W. V. O, *From a Logical Point of View*, Cambridge: Harvard University Press, 1961.

Tymoczko, Thomas, "Mathematics, Science and Ontology", *Synthese*, 88, 1991, pp. 201-228.

"Making Room for Mathematicians in the Philosophy of Mathematics", *Mathematical Intelligencer*, 8, 1986, pp. 44-50.

New Directions in the Philosophy of Mathematics, Boston: Birkhäuser, 1985.

Wittgenstein, Ludwig, *Philosophical Investigations*, Oxford: Blackwell, 1953.

Remarks on the Foundations of Mathematics, Oxford: Blackwell, 1967.

Humanistic Mathematics and the Real World

Reuben Hersh
University of New Mexico
Albuquerque, New Mexico

Introduction

In the first part of this article, I draw the distinction between two interpretations of "humanistic mathematics," which I call Humanistic Mathematics #1 and Humanistic Mathematics #2. I discuss the relationship between them, pedagogically and intellectually. I argue that awareness of the two trends should enable each to make use of contributions offered by the other. The two views do not compete, but complement one another.

In the second part, I consider how our Humanistic Mathematics Network should relate to other groups concerned with the reform of American mathematics education.

Part One

"Humanistic mathematics." What could it mean? If we look at the titles of talks given at meetings on humanistic mathematics, we find two distinct trends. These correspond to two of the meanings of "humanistic", and also to two of the meanings of "mathematics."

The first view of "humanistic mathematics" takes humanistic to mean "concerned with the humanities." These are the liberal or humane studies: history, literature, philosophy. Then, there are, peripherally, law, psychology, anthropology, sociology. The study of the works, the creations, thoughts and deeds, of man and woman. To connect this understanding of "humanistic" with "mathematics" the word "mathematics" must be taken in the sense of "field of study," "a domain of knowledge." In other words, "humanistic mathematics" has to do with course titles, syllabi, content of lectures and courses. In this direction, history of mathematics is a long recognized, though much neglected, topic in the mathematics curriculum. Philosophy of mathematics also has a long pedigree. We can "humanize" the content of our courses by bringing in historical and philosophical issues, and explain how mathematical applications affected people. Tell what Euler and Cauchy were like as human beings. This is one direction that we can detect in our humanistic mathematics movement. It is expressed in talks and proposals with such names as "Mathematics and Poetry," or "Let's Teach Philosophy of Math!"

The second view of "humanistic mathematics" takes a different interpretation for both "humanistic" and "mathematics." Humanistic now means human or people-oriented. Mathematics means mathematical activity: what we actually do when we study or teach mathematics. This tack in humanistic mathematics challenges the lecture or lecture-recitation system. It challenges dogmatic teaching styles that expect students to parrot the lecturer. It challenges drill by copious exercises that demand only a mechanical mastery of explicitly given "rules" and "methods." It demands instead student initiative, student independence, indeed creativity of both teacher and student in the mathematics

---

It challenges dogmatic teaching styles that expect students to parrot the lecturer.

---

classroom. It is expressed in talks and proposals with such names as "A student-centered calculus course" or "Encouraging Trigonometry Students to Fight Back."

I hope these examples make it clear that these two trends are conceptually distinct, even though both have every right to call themselves "humanistic mathematics."

Of the two, the first is much less difficult and less controversial. We may not be able to get our colleagues to join us, but, if we ourselves want to offer courses on "Mathematics and Society" or "Mathematics and Human Thought," most likely we can do so with little opposition. Such activity may not help greatly in our yearly departmental evaluation. The big problem is that, most likely, the course runs only while we keep pushing it. It's hard to pass on the torch.

The other kind of humanistic math—teaching math in a humanistic manner—is a much harder hill to climb.

We are fighting all the habits of subservience and thoughtlessness that students bring from twelve years of miseducation. Truthfully, I suspect, we ourselves don't

---

It demands instead student initiative, student independence, indeed creativity of both teacher and student in the mathematics classroom.

---

really know how we *should* teach, if we gave up lectures and drill, while still having the same old students and the same prerequisites demanded by the next course in the sequence, or by the physicists and engineers. Though we may not be sure how to teach humanistically, lots of promising ideas float around: teaching by projects; writing across the curriculum; working in groups. Which way *should* we go? It's all humanistic.

It seems to me that the two approaches to humanistic mathematics have important help to offer each other.

For approach one, which seeks to "humanize" the syllabus without seriously criticizing the pedagogy, there is an obvious need to look at teaching styles as well as content. The second approach has developed many proposals for enlivening the classroom, which are just as valid (or even more valid) in a "Math and consciousness" course as in a calculus course.

Still, those whose primary focus is on pedagogic reform ought not to avoid questions of course content even while moving around student's desks and the podium. Material prepared and presented by those primarily concerned with curricular refrom fits well with the aims of those stressing changes in teaching methodology.

I find it remarkable that two such different approaches work in the same organization without the proponents' of either approach being aware of their own differences in perspective, perhaps because all of us are driven by the same despair about traditional mathematics teaching. In any case, it seems to me that, if two separate organizations had somehow come into being, we would want to merge them into one.

One thing I believe both Humanistic Math #1 and Humanistic Math #2 have in common is that they both focus on the means, and take their goals for granted. After all, neither a pedagogy nor a curriculum is an end in itself. They are tools in the hoped-for transformation of the student. That transformation is the goal for both of them.

What, then, do we hope to accomplish for our students, apart from mere transfer of information? I think the question is not hard to answer, if we recognize that humanistic math education is a subset of humanistic education in general.

The human values we seek, it seems to me, are: independence, skepticism, intellectual honesty, self-reliance, and recognizing and valuing some great achievements of the human mind—NOT JUST TEST SCORES.

Today, there is a ferment in American education. There is a consensus that our students know much less than those in many other countries, and that somehow our educational systems, including our teaching methods, must be partly at fault. Mathematics has been late to feel the effect of this dissatisfaction.

There is no reason to think this stirring will fade away soon. The impulses that brought the humanistic mathematics network into being will be here for a long time. We can expect mathematics teachers to continue to look for ways out of the present dead end of most typical mathematics teaching.

Part Two

This leads to the second question I want to talk about. The relation of Humanistic Math #1 to Humanistic Math #2 is an internal question, having to do with the inner life of our organization. My second topic deals with our outer life—how we relate to the rest of the world.

It is clear that, in the business of trying to reform mathematics education in America today, we are far from alone. There is MSEB—the Math Sciences Education Board, producer of the widely circulated report, "Everybody Counts." There is CBMS—the Conference Board of Mathematical Science, and its "Issues in Mathematics Education." There is MER, the Mathematicians Education Reform Network, which had two sessions on the program of the national AMS meeting at Louisville in January, 1990. It is sponsored by the National Science Foundation. There are other important NSF activities, such as the Calculus Curriculum Reform Project. There is UME Trends, edited at Purdue by Ed Dubinsky. There are the various local Math Projects, whose prototype, BAMP, the Bay Area Math Project, was inspired by Leon Henkin. And, of course, the AMS, the MAA, the NCTM (National Council of Teachers of Mathematics), the Ford Foundation, the Rockefeller Foundation, the Carnegie Foundation and the MacArthur Foundation. In a recent issue of Focus, Leonard Gilman described four successful programs: Clarence Stephens' at Potsdam; Uri Treisman's at Berkeley; Manuel Berriozabal's at San

Antonio and statewide in Texas; and in Las Cruces, NM. He ended his report by pointing out that none of these had received any significant attention or inspiration from the MAA.

In the face of all this sudden commotion about the reform of math teaching, we have to ask ourselves: how do we fit in? Are we really necessary?

The major difference I see when I compare our network with these other groups is that we are a bottom-up organization, and that many of them are, to a great extent, top-down.

There is a typical sequence by which a social problem in America "surfaces." First, it attracts media attention, and is spoken of by political candidates. Eventually "funding agencies"—private or public—put it on their agendas. With publicity (by open bid) or privately (by direct negotiation with a contractor) they set up projects

---

The human values we seek are: independence, skepticism, intellectual honesty, self reliance, and recognizing and valuing some great achievement of the human mind—NOT JUST TEST SCORES.

---

to work on the problem. Some people bid on the job in direct response to the agencies request, and the agencies fund them perhaps without realizing that others have already been working on it, with little or no funding. This has been the sequence in our social problem: the failure of mathematics education.

The Humanistic Mathematics Network is quite a different story. Let me hasten to acknowledge the very valuable generous important financial support granted by the Exxon foundation, and by the AMS-MAA Joint National Meeting in granting us a place on their official program. Nevertheless, the impetus, the motivation, the goals, the values for our group came and still come from a scattered, heterogeneous, almost disconnected set of people who were or are now already finding their own way to transform mathematics teaching—to humanize it. We never waited for a panel of nationally visible experts to give us an agenda. We plunged in, to see what can be done, and to do it, now, where we are.

Our need was to know what each other are doing, to find other like-minded people teaching college math and let them know what we are doing, and finally, to let the profession as a whole know what can be done, and what

has been done to move out of the traditional rut and find a new way, a more human way.

In some respects, we have accomplished very little in comparison with other reform movements. For instance, there is visibility. Only a small handful of departments are receiving support from NSF for calculus curriculum reform projects. But the *existence* of such an activity is widely known, simply because NSF announced a million dollars would be spent on it. That fact alone was enough to attract the attention of most department chairs, and to result in memos to all faculty urging a response.

On the other hand, even though we have been an official part of the last two AMS-MAA national meetings, we are still almost unheard of in the American mathematical community. That's not a tragedy. Once we have great accomplishments, recognition will follow.

That we are a "grass roots" group and not an "establishment-oriented" group is neither good nor bad? Visibility, money, and status are often thought to be assets in getting things done. If that is true, then our "do-it-yourself" status is a disadvantage. On the other hand, I think that there is an integrity, a truth to experience, a dedication, if you will, that comes with self-motivated and self-determined work. That is our strength.

The truth is we are growing, we are being noticed by the national organizations and foundations. It will require a very delicate balance to take advantage of the good possibilities that their attention offers us, without losing our unique special character that makes us worthy of such attention. The crucial thing is to keep an eye on the classroom and on the students. Visibility, funding, and other conseiderations sometimes distract people from their primary goals. I trust that will not happen to us.

What should be the relationship between our work and that of the more "mainstream" reform groups?

Let's consciously seek to foster interchange of ideas and information with all other groups interested in reform of math education. At the same time, let's maintain and defend our status as a grass-roots movement. We are a place where math teachers and math professors are free to yell about what's wrong, and to tell each other what we have done or what we propose to do about it. We don't have to check with anybody else's guidelines, policies, agendas, deadlines, criteria, or evaluations. Hopefully, we do take account of the comments and criticisms which we can give each other.

Since, by all accounts, this is a time of great ferment,

flux, experimentation, in math education, it is a time when the ideas and experience of teachers can be offered and heard. Our kind of uninhibited testing ground is something that government agencies and large private foundations find difficult to provide. We can do it; we are doing it. Let us continue to maintain that character of wild and woolly unrestricted arena for free experimentation and argumentation.

PART II

Mathematics in the World

A Sketch of the Cultural Career of Mathematics

Hardy Grant
York University
North York, Ontario, Canada

For most of Western cultural history, mathematics enjoyed a unique "image", and a consequent prestige. Those perceptions were shaped, for centuries to come, by the achievement and outlook of ancient Greece, which saw in mathematics both a particular insight into the substance of "reality" and an unparalleled certainty of reasoning and conclusions. In the 16th and 17th centuries the founders of modern science fused with this classic legacy the further conviction—denied by the Greeks—of the enormous potential of mathematics for the description and control of the sensible world of our everyday experience. The spectacular success of the ensuing "Scientific Revolution" sent the cultural status of mathematics to unprecedented heights, whence its precision and its methodology offered inspiration and example to such diverse spheres as politics, ethics, philosophy and the arts. In the last two centuries this pre-eminence of mathematics has been threatened by internal developments—non-Euclidean geometry, Gödel's "Incompleteness" Theorem—that have cast grave doubt on its claims to unshakable sureness and absolute truth. But these same two centuries have also opened to the ancient science new and exciting vistas, which in our time embody its humanistic values, its potential for cultural enrichment, as never before. Such is the story which, in brief and superficial compass, I propose to tell.

I

The primary source and symbol of mathematics's long influence was Euclid's *Elements* (c. 300 B.C.). Later centuries found in this book a paradigm of the acquisition and organization of a body of knowledge—the foundations explicit and clear, the sequence of theorems unfolding with inexorable logic, the whole brought together in a masterpiece of arrangement and exposition. These geometrical propositions carried, of course, the stamp of an absolute certainty. But this—it may go without saying—was far from being the consequence merely of their logical form, the vacuous irrefutability possessed by statements (for example, any of the form "A or not A") that are eternally true but void of all significance. Euclid's theorems, in contrast, had for the Greeks certainty *and* content: a proposition like "the angle sum in a triangle is 180°" was held *both* to be logically necessary—in the sense that to deny it would be to end in contradiction—*and* to state a fact

about the world. Of course the truth of such theorems hinged on the truth of the underlying postulates, but these—with one nagging exception—were *manifestly* true, the self-evident products of innate intuition. The one source of unease was the "parallel" postulate, which asserted (in effect) that through a point P not on a line l there passes exactly one line parallel to l; this, though plausible enough, seemed much less obviously true than the other axioms. But this minor flaw (if such it was) proved scant deterrent to Euclid's admirers. His system's astonishing double strength, its attainment of

Plato and Aristotle agreed that mathematics and physics do not fit, and differed only over which was at fault.

genuine knowledge with absolute sureness, dazzled the centuries. The ensuing chorus of tribute, the many attempts at emulation, would fill an anthology of diverse voices. Even thinkers not generally enamored of mathematics sometimes paid their respects; thus—to take two especially striking cases—Montesquieu, whose *Persian Letters* (1720) smile genially at the sometime excesses of *l'esprit géométrique*, confided twice to his notebooks that he took nothing as certain save the pages of Euclid,[1] while Goethe, eager advocate of a holistic philosophy of nature as against the linear thinking of geometry, nevertheless urged that all scientists strive for the great Alexandrian's clarity and rigor of argument.[2] Such examples could, of course, be much multiplied. Except only the Bible, no book in the Western heritage has had an impact so lasting and so wide.

A second aspect of the Greek mathematical bequest ran deeper. Much ancient thought assumed, and transmitted to posterity, a kind of epistemological optimism, a confidence that the world presents an objective reality that human beings can know. At one level this is perhaps only the everyday assumption of the unreflecting; but Greek philosophy articulated it with self-conscious precision, as the belief in a fundamental congruence of our minds with the transcendent order of creation. "That which it is possible to think," said Parmenides (c. 475 B.C.) "is identical with that which can Be."[3] Some in antiquity saw the bridge between the two in the miracle of language—or rather in a primal, universal language that once mirrored perfectly the world in the word, a tongue whose loss was told in the story of Babel. But others looked instead to mathematics as the key to—because a reflection of—the ultimate cosmic order. Why mathematics? Perhaps the wide consensus enjoyed by its first principles, and the logical necessity of its conclusions, hinted at truths independent of the fleeting and fallible perceptions of individuals. But more: mathematics could be seen to reach insights into the form, the structure, the relations of things, rather than into their physical matter, and hence to grasp the enduring amid the perishable, the essential as against the contingent. Was this not the obvious lesson in the familiar distinction between (say) the triangle drawn by the geometer in the sand as an aid to his reasoning and the mental triangle that the reasoning truly contemplates? Is not the number 2 more lasting, more universal, in a word more "real", than any mortal married couple, any material pair of shoes? The tradition born of such speculation passed down the centuries.

The first explicit claim of mathematics's privileged ontological status was the mystic utterance of the Pythagoreans (6th century B.C.) that all things are numbers. This surely stemmed, at least in part, from an observation long familiar to musicians, that harmonious chords are produced by the vibration of strings whose lengths are in whole-number ratios. Its lasting impact, fittingly, would be centuries of belief, in classical and then in Christian culture, that such ratios are our minds' intimations of harmonies built deep in the nature of things, governing alike the cosmos, the state, and the relationships of individuals.[4] The assigning of true "reality" and significance to mathematical entities was given a new and historic direction by Plato. He it was who brought into philosophy, and endowed with classic statement and lasting importance, that suggestive contrast (cited above) between the geometer's idealized mental triangle and its crude, temporary physical incarnations. Mathematical objects became for Plato prime examples of the "Ideas" or "Forms" at the core of his theories of existence and of knowledge: entities apprehended by human minds but (according to him) persisting independently, and mirrored dimly by the objects our senses perceive. The famous educational program set out in the *Republic* prescribes long immersion of the student in mathematics, whose objects Plato ranked as only just below—and as offering an essential path toward—the Idea of the Good, the summit of all being and morality alike.[5]

Of course, not all contemporary opinion followed Plato in this lofty vision of mathematics. His pupil Aristotle—his only rival at the pinnacle of Greek philosophy—located true being in the physical, and held that Plato's Ideas, including the objects of mathematics, are merely human concepts, abstracted from experience but having no existence outside our minds. Yet thinkers on both sides of this divide shared crucial common ground. All could agree that mathematical objects, whatever their "reality", come first into our consciousness through the impressions of our senses. But also—and here lies a deep paradox—the Greeks were

Galileo represented, with a vividness probably unparalleled before or since, the several intellectual stances that together defined in his time the unique cultural influence of mathematics.

equally unanimous in holding that mathematics, despite these physical roots, is no tool for describing or manipulating the changing panorama of everyday experience. For their mathematics portrayed, after all, the static and the ideal, a mode of being removed and, so to say, insulated from the mutability of all things physical—perhaps by independent existence (as for Plato), perhaps by an act of abstraction (as for Aristotle). Why (the Greeks might have asked) should the necessary, permanent relations among mathematical entities be *expected* to apply to a sensible world whose only constancy was change? Plato and Aristotle "agreed that mathematics and physics do not fit, and differed only over which was at fault"[6]—the one looking down on the physical as existentially inferior, the other seeing mathematics as barred from the analysis of change by the very fixity of its abstractions. Archimedes, much the most modern mind in Greek mathematics, came close in spirit to bridging the gap, but ended by reaching only results governing the static, like the quasi-geometrical "law" of the lever. Nor was astronomy's mathematical modeling of celestial movements, by Hipparchus, Ptolemy and others, an exception; for—apart from the fact that these epicycles and eccentrics were in many minds mere mathematical fictions, designed only to "save" the phenomena—the heavens were by universal agreement precisely the region where no secular change (as opposed to the eternal, uniform whirling of celestial bodies) ever occurs.

This inability to cope with change was, in easy hindsight, one profound limitation of the cultural role of Greek mathematics. There was another. One may well surmise that in the days of Euclid and Archimedes a science which unveiled apparently immutable truth about the ultimate realities was challenge and exhilaration enough; but however that may be, this role of describing the existentially given marked also the *boundary* of Greek mathematics. The geometers dealt of course in abstractions from the sensible world, but they went beyond experience in no other way. To the Greeks, wrote Carl Boyer, "mathematics, instead of being the science of possible relations, was ... the study of situations thought to subsist in nature."[7] Greek mathematics thus declined always to invent, to define new concepts lacking physical reference, to fashion new worlds from pure imagining. To Euclid the product of two line segments was an area, the product of three was a volume, the product of four was—meaningless. In the Pythagorean discovery of incommensurability[A] the real drama, the real significance was that here the mathematical mind searched the *given* stock of numbers for the measure of (say) the diagonal of a unit square, found none, and accepted the finding—leaving to later generations the creation of irrational numbers. The genius even of Archimedes, greatest of the Greek mathematicians, shone forth in sheer technical mastery and in a marvelous methodological ingenuity, but remained bound to the objects of his experience; even Archimedes imagined no new realms.

II

The Greek mathematical achievement, only partly known and poorly understood during the Middle Ages, was fully recovered by the end of the Renaissance. That inheritance included the twin sources here suggested of the subject's perceived uniqueness, the supreme certainty claimed by geometry and the Pythagorean-Platonic vision of ultimate reality as embodied in mathematical objects and relations. From one point of

---

Man's understanding, said Galileo, where mathematics can be brought to bear, rises to the level even of God's. An alarmed Church duly included this breathtaking assertion of human pride among the eight counts of heresy that brought the rash Florentine to his famous trial.

---

view the ensuing "Scientific Revolution" consisted precisely of the union of these two Greek convictions with the radical overturning of a third, in the insistence by the "moderns" that an abstract mathematics can lead

to understanding and control of the physical world. The origins of that historic reversal resist easy analysis; many would contrast the leisured, aristocratic Greek philosopher contemplating a static realm of eternal Ideas with the urgent, Faustian drive of modern Europe for mastery of nature, and this hackneyed dichotomy remains deeply suggestive. In any case practical problems that never challenged antiquity now spurred the cleverness of mathematicians: the path of a light ray refracted by a lens, the instantaneous velocity and maximum range of a projectile fired from a cannon. The resulting development wrought a strange alteration in the way some mathematical entities themselves were conceived; an ellipse, for example, which for the Greeks had been the result of the *completed* slicing of a cone by a plane, was now seen as the path traced—without completion, in a kind of "timeless time"[8]—by a moving point. Mathematics, hitherto the science of the static, was preparing to describe a world of change.

Of course, the vital instrument of this mathematizing of nature was and remains the calculus. But the great archetypal figure of the Scientific revolution did his work independently of—indeed, largely before—the full development of that powerful algorithmic machinery. Galileo represented, with a vividness probably unparalleled before or since, the several intellectual stances that together defined in his time the unique cultural influence of mathematics. The sureness of geometrical reasoning, the ontological primacy of mathematical objects and relations, the applicability of mathematics to the physical world—all of these were for him central convictions passionately urged. The "grand book" of the universe, he declared, "is written in the language of mathematics, and its characters are triangles, circles and other geometric figures without which it is humanly impossible to understand a single word of it."[9] The final statement of his physics (*Two New Sciences,* 1638) begins its celebrated discussion of free fall with definitions and axioms rendered indubitable (so Galileo believed) by experiments, then deduces a long sequence of further results *by pure geometry;* the inquiry thus mimics exactly its Euclidean and Archimedean models, and carried for its author the same conviction. Mathematical deductions from sure premises, he felt, made the conclusions so unassailable that no experimental verification was necessary, unless to win over the obtuse.[10] More dramatically still, the truths he had reached seemed to him so absolute that no other *kind* of knowledge or explanation of the phenomena under study seemed either possible or necessary; in this sense, said Galileo, man's understanding, where mathematics can be brought to bear, rises to the level even of God's.[11] An alarmed Church duly included this breathtaking assertion of human pride among the eight counts of heresy that brought the rash Florentine to his famous trial.[12]

While Galileo's physical treatises retained the form and techniques of Greek geometry, there unfolded around him a time of rapid and revolutionary progress in pure mathematics. Pierre de Fermat touched so many aspects of this advance as almost to seem its central protagonist: founder of modern number theory, successor to François Viète in the development of symbolic algebra, co-creator with René Descartes of analytic geometry, co-founder with Blaise Pascal of the theory of probability, important contributor to the early history of the calculus. It was, of course, Isaac Newton and Gottfried Leibniz who brought this last evolution to its first great synthesis, opening the way to a century and more of truly explosive elaboration and to a corresponding surge of mathematical physics. Together, this growth of pure mathematics and (even more) the attendant examples of its triumphant application to nature, gave definitive shape to a revolution already stirring in European thought and sensibility. Here, too, Newton marks a milestone, with his *Principia* (1687); his work, though forbiddingly difficult, was better calculated than Galileo's to inspire

---

Baruch Spinoza's *Ethics* remains the most Euclidean treatise ever written in a humanistic subject, a resolute deduction of theorem after theorem from initial definitions and axioms.

---

the lay imagination, for even the mathematically unlettered could grasp and marvel at the reduction of universal gravitation to a simple formula describing equally the fall of an apple from a tree and the motions of the stars in their courses. So Newton became the symbol of—and eventually gave his name to—the complex of diverse reactions which together form the high-water mark in all of history for the cultural influence of mathematics.

That story has been much told,[13] but a quick survey may be forgiveable. Some aspects of this "Newtonianism" predated the *Principia,* and must be ascribed instead to the earlier progress in mathematized science, to the perennial appeal of Euclid, and to Descartes's influential urging[14] of mathematics as the key to all philosophical methodology. Baruch Spinoza's *Ethics* (1660s) remains the most Euclidean treatise ever written on a "humanistic" subject, a resolute deduction of theorem after theorem from initial definitions and axioms. The rationale is profoundly characteristic: he would discuss people's actions and appetites, said Spinoza, as if they were points or lines, for they belong to the single, uniform order of nature,

and must be studied by the methods applicable in other enquiries.[15] In England Thomas Hobbes laid it down (1650s) that "Geometry is the only science which it hath pleased God to bestow on mankind," urged universal use of its techniques, and set himself the goal of inferring the immutable laws of civil society from appropriate postulates.[16] Bernard de Fontenelle (1657-1757), the pioneering popularizer of science who arguably ranks second only to Voltaire in influence on the French Enlightenment, simultaneously proclaimed the goal and hailed its realization:

> Works in ethics, politics, criticism, even eloquence, all things otherwise being equal, will be better if they bear the mark of the geometer. The order, clarity, precision and exactitude that have been prevalent in good books for some time could very well have had their source in this geometric spirit, which is being more widely spread than ever...[17]

System-builders in many spheres sought axioms that might rival geometry's in sureness; the American Declaration's "We hold these truths to be self-evident" thus breathes the spirit of its age. Arbiters of literary taste championed a mathematical plainness of discourse, whose adoption promised (in their view) to end the verbal ambiguities, the semantic fogs, with which the muddled or unscrupulous contrive to veil the face of truth. Others pursued the goal of clarity in the precision of numbers, as when Immanuel Kant decreed that no branch of learning is truly rigorous until quantified.[18] Others again saw in the new mathematics of probability the key to a "calculus" of moral and political behavior, that would bring civilized agreement to these chronically contentious realms. Indeed we should remember—or risk crucially distorting our picture of the age—that for many this espousal of mathematics and science had a fervent emotional side, the hope and conviction that the use of rational strategies, and the spread of "enlightenment" through education[B], would topple the oppressive institutions (the monarchy, the feudal order, the Church) that throve on the ignorance of the people, promoting in their stead the triumph of reason in social affairs and the indefinite perfectibility of mankind.

So strong a movement inevitably bred a backlash, as if in illustration of Newton's Third Law. Against an excessive rationalism that would make mathematical exactness and logical demonstration the sole criteria of value, dissenters proclaimed the rights of the heart over the mind, of feeling and instinct over reason, of impulse and passion over calculation. At the very dawn of the Age of Reason, Pascal—most telling of critics, by virtue of his own great mathematical gifts—set the geometric and "intuitive" minds in vivid contrast, underscoring the strengths and limitations of each.[20] As time passed such voices grew in number and vehemence, until, by the beginning of the nineteenth century, mathematics had become a prime target—and Newton the particular *bête noire*—of the Romantic protest against all that the Age of Reason stood for. This familiar indictment pictured mathematics and mathematized science as cold, dessicating abstractions that rob nature and life of all beauty, poetry and joy. John Keats phrased it in lines so lovely as almost to persuade. "Philosophy," he wrote, meaning the science of his day,

> will clip an Angel's wings,
> Conquer all mysteries by rule and line,
> Empty the haunted air, and gnomed mine,
> Unweave a rainbow.[21]

"God forbid," said William Blake, more starkly, "that Truth should be Confined to Mathematical Demonstration."[22] But if this kind of thing was the commonest of contemporary reactions there was another variety of protest, less passionate but possibly more compelling. Its most original statement came (1725) near the very height, but far from the geographic center, of the French Enlightenment, from the lonely figure of Giambattista Vico, an obscure professor of rhetoric at Naples. Vico drew on a tradition reaching back to the Middle Ages, which held that only he who makes something can truly know and understand it; on this view, for example, only God can have perfect insight into the physical world, human science (pace Galileo) reaching only intimations. Similarly—said Vico—the supposed certainty that we attain in and through mathematics merely reflects the fact that this science is our own creation; its deepest theorems are only the spinning out of our own assumptions.[23] Vico was echoed (unknowingly) in mid-century by Georges Louis Leclerc, Comte de Buffon. Doubtless this great writer on natural history felt deeply the sense, shared by many biologists[C], that the tidy categories of mathematics can never do justice to the teeming variety and vast complexity of living things. In any case he held that mathematics begins with abstractions and remains confined to them, for "there is nothing in this science that we have not put there and the truths that we draw from it can only be different expressions of the same suppositions that we have employed."[25]

III

In its time this was very much a minority view. Far more commonly, mathematics was still seen in, say, 1800 A.D., as the Greeks had seen it, as a science grounded in self-evident perceptions of the physical world, attaining a certainty beyond the merely tautological, and allowing—though confined to—an exact description of an objective reality. But now an epochal change was on the horizon, the result not of external criticism but of mathematics's own inherent progress. The great watershed was the creation of non-Euclidean geometry. After hesitant beginnings in other hands, mature formulations were reached independently by Carl Friedrich Gauss in Germany (from 1792), Wolfgang and Johann Bolyai, father and son, in Hungary (1815), and Nikolai Lobachevsky in Russia (1826). All made a single change in the axiomatic base of Euclidcean geometry, in effect replacing the troublesome "parallel" postulate by an alternative which denied it, and all concluded that the resulting geometry

> "Philosophy will clip an Angel's wings, Conquer all mysteries by rule and line, Empty the haunted air, and gnomed mine, Unweave a rainbow." —John Keats

is mathematically just as valid, because just as consistent (free from internal contradiction) as Euclid's. To those who could and would heed, the lesson was clear. The proud claim of mathematics to absolute truth had been a delusion. Vico and Buffon had been right: the validity of any mathematical theory hinges only on its underlying assumptions. And because such assumptions are arbitrary, there is no reason to credit the resulting theories with any relevance to our understanding of nature. In particular the old idea that Euclidean geometry describes physical space, so far from being self-evident, might not even be true; the new geometry might serve as well or better, and (as Gauss already saw) the choice between the two had become empirical, to be made by such observation and measurement as might avail. The Queen of the Sciences tottered precariously on her pedestal.

So understood, this historic watershed would indeed seem a harbinger of unrelieved gloom. But already the younger Bolyai drew from his non-Euclidean geometry a more heartening promise. "From nothing," he exulted, "I have created an entirely new world."[26] Succeeding decades would echo that perspective, and that enthusiasm, many times. No doubt the links of mathematics to the world of experience, though far from actually severed—for the 18th century's splendid development of rational mechanics seemed, after all, as valid as ever—had been rendered more mysterious and problematic than before.[27] But now beckoned, as if in compensation, a widened picture of mathematics as outstripping the *limits* of mundane reality, a creative endeavor subject always to the requirement of consistency but otherwise knowing no boundaries save those of imagination. Of course, there had been some earlier hints of this passing beyond physical experience, this leveling of the barrier that had always confined the Greeks; but full acceptance and exploitation had tended to lag behind. Thus the 16th century introduced "imaginary" entities as roots of algebraic equations, but mathematicians at first viewed these with deep unease, and they waited nearly three hundred years for recognition as legitimate numbers. Likewise the analytic geometry of Descartes and Fermat inherently invited extrapolation to dimensions beyond the third. Yet here, too, the real breakthrough came only in the early 19th century. These cases are typical: only in that age, coincident with the rise of non-Euclidean geometry, did mathematicians begin to feel and pursue a real sense of creative freedom. Various developments reflected or reinforced the trend: the growing emphasis on abstract structures as opposed to their concrete interpretation, the apparent mathematical taming by Georg Cantor (1870s) of a realm of thought palpably without physical counterpart: the infinite. There was—there still is—a darker side: notoriously, the tendency to build systems on arbitrary postulates has led in some hands to an arid and sterile formalism. But, at the other extreme, minds within and outside mathematical ranks were now stretched and beguiled by such novel fantastications as the Möbius (one-sided) strip and the Koch "snowflake" (whose infinite perimeter encloses a finite area); here were delights capable, perhaps, of disarming even the hostile Romantics. Nor did the new boldness of imagination compromise the power of mathematics in the study of nature. On the contrary—some theories born of no motive save their intrinsic mathematical interest later proved almost magically fruitful in outside

---

God forbid, said William Blake, more starkly, that Truth should be Confined to Mathematical Demonstration.

---

applications: a version of non-Euclidean geometry due to Bernhard Riemann (1854) awaited Einstein's general relativity, and Arthur Cayley's matrix algebra (1855) lay to hand when, seventy years later, Werner Heisenberg needed it in the development of wave mechanics. A hundred years after the severe foundational crisis threatened by non-Euclidean geometry, mathematics seemed more vigorous than ever, in internal richness and external relevance alike.

The 20th century produced another famous result with the potential to shatter confidence. Kurt Gödel showed (1931) that *any* attempt to axiomatize even so simple a mathematical system as elementary arithmetic leaves "undecidable" propositions (that is, propositions that can be neither proved nor disproved within the system), and moreover that the consistency of such a system cannot *in principle* be proved without the use of methods whose own consistency is equally (or more) in need of justification. Taken together, these findings imply fatal limitations on any hope of deducing a body

of knowledge completely and certainly by (finitary) formal methods. And yet even this apparently stunning setback has in practice proved survivable. Some working mathematicians have merely responded with a "bland indifference"[28]; but others have urged, even as the Greeks urged long ago, that intuition guarantees both the objective validity and the logical coherence of the foundations. Gödel himself wrote:

> Despite their remoteness from sense experience, we do have something like a perception of the objects of set theory, as is seen from the fact that axioms force themselves on us as being true. I don't see any reason why we should have less confidence in this kind of perception, i.e., in mathematical intuition, than in sense perception... They, too, may represent an aspect of objective reality.[29]

Perhaps there is in such pronouncements an almost Kierkegaardian leap of faith—another aspect of modern mathematics that would have appealed to the Romantics. In any case it seems that, despite all undermining of foundations, most researchers still ascribe to their theorems a timeless, unconditioned truth; Euclid may be dethroned, but Plato survives. Nor, manifestly, do foundational qualms slow the attempted use of mathematics in other spheres. Economics, psychology, sociology, anthropology, history, biology, medicine—all constantly seek in numbers and formulas the precision long enjoyed by the physical sciences. These, in turn, continue to draw applications from even the most abstract or exotic of mathematical creations; so infinite-dimensional spaces play a vital role in modern physics, and the captivating new study of "fractals" offers a tool for the fruitful modeling of mountains and coastlines. At deeper levels, many of our most central formulations of scientific inquiry—quantum theory, relativity theory, cosmology—have come to rely on mathematics as the language of nature in a sense transcending any urged even by Galileo, as a unique, necessary, *probably untranslatable* encoding of such perception and understanding of physical reality as we may hope to attain.

But beyond all this, beyond all considerations of external reference, mathematics remains for many a self-contained world of enchantment, an inexhaustible realm of the strange, the diverting, the beautiful, the intellectually challenging. And hence comes a capacity for cultural impact and enrichment very different from, but not less than, the Age of Reason's exuberant embrace. To be sure, the subject's sheer proliferation, the technical vocabulary, the subtlety of reasoning, will always be formidable barriers to lay understanding. But much nontrivial diffusion occurs, and rewards enormously. "Touch on even its more abstruse regions," writes George Steiner—he is thinking of a wide area comprising both pure and applied mathematics—"and a deep elegance, a quickness and merriment of the spirit come through." He gives as example the Banach-Tarski Paradox, in its common popularization to the effect that a spherical pea can be finitely divided into pieces rearrangeable by rigid motions into another sphere the size of the sun; "what surrealist fantasy yields a more precise wonder?"[30] Through such fascination the new realms explored by mathematics have helped to shape many significant works of literary and visual art. The "fourth dimension"

Most researchers still ascribe to their theorems a timeless, unconditioned truth; Euclid may be dethroned, but Plato survives.

---

has intrigued novelists of the stature of Dostoevsky, Conrad and Proust;[31] non-Euclidean geometry played a role in the revolutionary visions of Cubist painting.[32] The borrowing goes on: let the cunningly interwoven, subtly varied "tessellations" (repeated patterns) in Maurits Escher's famously intricate designs, and the absorbed play with post-Cantorian infinities in Jorge Luis Borges' delectable *ficciones,* stand as two especially distinguished recent examples. That such cases are not even more numerous—in particular the echoes of mathematics in modern science fiction are surprisingly faint[33]—suggests an ongoing task of dissemination for the research community and for educators. But the potential is surely very great. Stripped of its ancient certitude but still a prodigiously vital and growing enterprise, illumining the world of physical experience but no longer confined there, modern mathematics expresses the human spirit in three distinct but intertwining ways, as a "man-made universe"[34] to be cultivated and cherished for its own sake, as an indispensable instrument for our understanding of nature, and as a limitless source and playground of delighted imagination.

A.   That is, the discovery that there exist pairs of geometrical "magnitudes"—line segments, or areas, or volumes—with the property that no unit magnitude of their type measures both of them an integer number of times. For us, but *not* for the Greeks, two magnitudes are incommensurable in this sense if and only if the ratio of their measures is an irrational number.

B.   Some filtering down of the intellectuals' passion for mathematics is hinted by anecdotal evidence, as of the young ladies in late-17th-century France who allegedly refused otherwise eligible suitors

for having no new ideas on the squaring of the circle.[19]

C   Aristotle's opinions on mathematics can usefully be viewed from this perspective, and Ernst Mayr is its eloquent advocate in our time.[24]

NOTES

1.   Montesquieu, *Oeuvres Complètes* (Paris, 1966), pp. 856, 959; cf. *Lettres Persanes,* Letter 129.

2.   H.B. Nisbet, *Goethe and the Scientific Tradition* (London, 1972), p. 50.

3.   Parmenides, frag. B3, in Kathleen Freeman, *Ancilla to the Pre-Socratic Philosophers* (Cambridge, Mass., 1983), p. 42, note 2.

4.   Cf. the fascinating book of Leo Spitzer, *Classical and Christian Ideas of World Harmony* (Baltimore, 1963).

5.   Plato, *Republic,* Bk. VII.

6.   C.C. Gillispie, *The Edge of Objectivity* (Princeton, N.J., 1960), p. 15.

7.   C.B. Boyer, *The History of the Calculus and its Conceptual Development* (New York, 1959), p. 25.

8.   A. Koyré, "The Significance of the Newtonian Synthesis", in his *Newtonian Studies* (Chicago, 1965), pp. 10-11.

9.   Galileo, "The Assayer", in S. Drake, ed., *Discoveries and Opinions of Galileo* (Garden City, N.Y., 1957), pp. 237-38.

10.   Galileo, letter to Carcavi, in *Opere,* Vol. 17, pp. 90-91.

11.   Galileo, *Dialogue Concerning the Two Chief World Systems,* tr. S. Drake (Berkeley and Los Angeles, 1970), pp. 103.

12.   Ibid., p. 474.

13.   Notably by M. Kline, *Mathematics: A Cultural Approach* (Reading, Mass., 1962), Chaps. 20-22; idem., *Mathematics in Western Culture.*

14.   E.g. Discourse on the Method, pt. 2, in J. Cottingham et al., eds., *The Philosophical Writings of Descartes,* Vol I, p. 120.

15 .   Spinoza, *Ethics*, Part III, preface.

16.   Hobbes, *Leviathan*, Pt. 1, Chap. IV (Penguin ed. p. 105); H. Grant, "Geometry and Politics: Mathematics in the Thought of Thomas Hobbes", *Mathematics Magazine*, 63, No. 3 (1990), 147-54.

17.   Fontenelle, *Preface on the Utility of Mathematics and Natural Science*, tr. L.M. Marsak in *French Philosophers from Descartes to Sartre* (Cleveland and New York, 1961), p. 89.

18.   Cited in Kline, *Mathematics: A Cultural Approach* (n. 13), p. 449.

19.   *Journal des savants*, 4 March 1686, quoted in P. Hazard, *The European Mind 1680-1715* tr. J.L. May (Penguin ed., 1964), p. 351.

20.   Pascal, *Pensées*, tr. A.J. Krailsheimer (London, 1966), pp. 211-12 (Brunschvicg ed. nos. 1,4).

21.   Keats, *Lamia*, II, 234-37.

22.   Blake, "Annotations to Sir Joshua Reynolds' *Discourses*", in *Poetry and Prose of William Blake*, ed. G. Keynes (Bloomsbury, 1927), p. 1009.

23.   Vico, *Opere*, ed. R. Parenti (Naples, 1972), I, 83. Cf. the superb exposition by I. Berlin, "Vico's Concept of Knowledge", in his *Against the Current* (New York, 1980), pp. 111-19.

24.   E. Mayr, *The Growth of Biological Thought: Diversity, Evolution and Inheritance* (Cambridge, Mass., 1982), pp. 37-43, 54-55.

25.   Quoted in T.L. Hankins, *Jean D'Alembert: Science and the Enlightenment* (Oxford, 1970), p. 76.

26.   J. Bolyai, letter to W. Bolyai (1823), quoted in M.J. Greenberg, *Euclidean and non-Euclidean Geometries* (San Francisco, 1972), p. 143.

27.   Thus Jean Dieudonné in 1961, quoted in Greenberg (n. 26), p. 252.

28.   C. Davis, "Materialist Mathematics", in R.S. Cohen et al, eds., *For Dirk Struik* (Dordrecht, 1974), p . 43 .

29.   Quoted in P.J. Davis and R. Hersh, *The Mathematical Experience* (Boston, 1981), p. 319.

30.   G. Steiner, *In Bluebeard's Castle: Some Notes toward the Re-definition of Culture* (London, 1976), p. 98.

31.   Dostoevsky, The Brothers Karamazov, Bk. V, Chap. 3; Conrad and F.M. Heuffes, *The Inheritors*; Proust, *Remembrance of Things Past*, tr. C.K. Scott Moncrieff and T. Kilmartin (Penguin ed., 1983), I, 66.

32.   L.D. Henderson, *The Fourth Dimension and Non-Euclidean Geometry in Modern Art* (Princeton, 1983).

33.   P. Versins, *Encyclopédie de l'Utopie, des Voyages Extraordinaires et de la Science Fiction* (Lausanne, 1972), art, "Dimensions".

34.   From the title of a well-known book by Sherman Stein, *Mathematics, the Man-Made Universe*.

Mathematics and Its Application

Jack V. Wales, Jr.
The Thacher School
Ojai, California

Rather than teaching mathematics on the basis of its practical applicability, it would be better to teach that mathematics is the study of an independent, extant reality and that any application of mathematics is the postulating of an analogy between a set of worldly circumstances and a set of mathematical circumstances. Seen that way, the coincidence of pattern is a cause for delight, and the limitations are expected, even if they are not seen.

But, instead, mathematics, as it is generally taught, justifies itself on the basis of its applicability in the worldly circumstances which are the focus of the application at hand. Such a belief does not encourage the student to investigate the limitations of mathematics to the situation being examined. Further, such an emphasis directs students into seeing worldly reality as a mathematical edifice to be conquered and oneself as feeble before it.

I wish to argue that mathematics teachers should not feel that it is necessary to justify whatever mathematics we teach by its applicability, and further that even the most applicable mathematics should be presented as being worthy of study independent of its applicability, not least because its application will be done best when

---

"There is one thing...of which a geometry is *not* a picture, and...that is the so-called real world,"

---

its independence is accepted. By "application" I mean not only the use of mathematical patterns to increase control or predictive power in practical situations, but also the use of mathematical patterns to increase understanding in practical or impractical or even fantastical situations.

When I was learning to fly, my instructors told me, "The throttle controls your altitude, the elevator controls your speed, not the other way around." I now believe that is a distortion, but it was a very useful one.

When one begins flying an airplane, he has a natural tendency to believe that the throttle controls the airplane's speed and the elevator makes the plane go up or down. In fact, the situation is much more complicated than that; and my instructors' dictum was instrumental in my learning to use the two controls effectively. I would use it today if I were to take the controls of an airplane again.

I don't think the phenomenon of an idea's being of good effect even if it is a distortion is a particularly unusual one. Another example might be the dictum, "There is no important difference between men and women." Or, "When acting in a play by Shakespeare, one should never give emphasis to a personal pronoun." If those sentences are not true, acting as if they were might nevertheless have good effects because they counteract some deeply ingrained erroneous biases.

The idea I want to offer here is, I suspect, correct. Even if it is not, however, acting as if it were would have good effects. And so I want to present the idea here, examine some of its consequences, and defend that complex consisting of the idea and its consequences.

The idea is this: first, that mathematics is an account of an extant reality that is independent of physical reality or social reality; and, second, that it is therefore appropriate to understand applications of mathematics in various arenas as analogies—analogies which often appear as metaphors.

I don't know, nor do I care, just what philosophical position I am taking when I say that mathematics is an account of an extant reality. The important point is that mathematical truth is independent of human judgment. 29 is prime. 27 is not prime. Those sentences are true not because I or any expert says so, but because 29 *is* prime and 27 is *not* prime. That is just the way they are. I don't know offhand whether forty-three times sixty-seven is equal to fifty-one times sixty-one, but there is no doubt that it either is or it isn't, independent of anything anyone does or says. Mathematical truth is about the least contingent truth around. Mathematical reality is there to be discovered or observed.

At the same time, mathematics gives an account of logical inevitability, not physical inevitability. "There is one thing...of which a geometry is *not* a picture, and...that is the so-called real world," says the famous mathematician, G. H. Hardy. (l) No one can mathematically prove that the sun will rise tomorrow or that it won't, that the stone will fall when released, that closing the circuit will cause the light bulb to come on, or whatever; physical reality may more or less coincide with mathematical pattern, but it is not constrained by it. This fact gives substance to the word 'unreasonable' in the title of Eugene Wigner's essay, "The Unreasonable Effectiveness of Mathematics in the Natural Sciences" (2).

What, then, is the relationship between mathematics and, say, physics? Is mathematics a language in which physics is expressed, as is suggested by the title of Tobias Dantzig's book, *Number: The Language of Science?* (3) If I am right that mathematics is an account of an extant reality, then it is certainly more than a language. Perhaps mathematical symbolism is a language, but the mathematical ideas, the mathematical

---

"The structure that most emphatically exhibits the power of mind nevertheless leads to the denigration of the human mind."

---

facts, expressed by the symbols are not mere language. Rather, physics is a science which, among other things, makes analogies between mathematical circumstances and physical circumstances. If extent of spatial separation is like number, and this particular spatial separation has an extent which we take to be like the number 5, then that one has an extent which is like the number 7, and, furthermore, the two of them together constitute a spatial separation the extent of which is like the number 12. This is not a particularly unusual idea; mathematicians and scientists often refer to mathematical "models" of worldly circumstances, by which they don't mean anything too different from what I mean when I use the word "analogy".

Why, then, do I choose to use that somewhat unusual word? Because the word "analogy" tends to remind us of certain things that the word "model" does not. The important points are these: any analogy has two parts or sides, and the extent of validity of an analogy is always in question. (Note Wendell Berry's words, quoted later in this paper.)

There are, roughly speaking, two kinds of analogy: explicit and implicit, simile and metaphor. The literary critic Northrop Frye observes,

In descriptive writing you have to be careful of associative language. You'll find that [simile], or likeness to something else, is very tricky to handle in description, because the differences are as important as the resemblances. As for metaphor, where you're really saying "this *is* that," you're turning your back on logic and reason completely, because logically two things can never be the same thing and still remain two things....The motive for metaphor, according to Wallace Stevens, is a desire to associate, and finally to identify, the human mind with what goes on outside it, because the only genuine joy you can have is in those rare moments when you feel that although we may know in part, as Paul says, we are also a part of what we know. (4)

If application of mathematics is the making of analogies, then in much of human knowledge, mathematics is (ironically, for if mathematics is not logical, then nothing is) not a simile—it is a metaphor. It is, in fact an unconscious metaphor, a culturally subconscious metaphor, for other sorts of reality.

The words left out by the ellipsis in the quote by Northrop Frye are,

The poet, however, uses these two crude, primitive, archaic forms of thought in the most uninhibited way, because his job is not to describe nature, but to show you a world completely absorbed and possessed by the human mind. So he produces what Baudelaire called a 'suggestive magic including at the same time object and subject, the world outside the artist and the artist himself.' (5)

And when Frye mentions Wallace Stevens, he is referring to the following poem by Stevens:

The Motive for Metaphor

You like it under the trees in autumn,
Because everything is half dead.
The wind moves like a cripple among the
    leaves
And repeats words without meaning.

In the same way, you were happy in spring,
With the half colors of quarter-things,
The slightly brighter sky, the melting clouds,
The single bird, the obscure moon—

The obscure moon lighting an obscure world
Of things that would never be quite expressed,
Where you yourself were never quite yourself
And did not want nor have to be,

Desiring the exhilarations of changes:
The motive for metaphor, shrinking from
The weight of primary noon,
The A B C of being,

The ruddy temper, the hammer
Of red and blue, the hard sound—
Steel against intimation—the sharp flash,
The vital, arrogant, fatal, dominant X.

Frye's contrast between "describing nature" and "showing a world possessed by the human mind," and Stevens's use of technical symbols (A B C, X, weight, primary noon, temper, steel) for what is shrunk from when one is motivated to metaphor, suggest that I am off base, that the whole point of what they are talking about is the contrast between poetry on the one hand and the "objective" worlds described by math and science on the other. But remember that Stevens's title is "The *Motive* for Metaphor." Is not science an (in some ways quite successful) attempt by the human mind to absorb and possess a world?

It is one of the ironies of Man's present condition that a motive of identification should have been a driving force behind what seems to so many to be the source of so much alienation, i.e., the application of mathematics to worldly circumstances. As I put the finishing touches on this essay, I am starting to read a new book by the philosopher William Barrett, in which I found this:

> We note the extraordinary power and constructivity of the human mind in producing the great edifice of modern science. And yet, precisely here occurs one of the supreme ironies of modern history: The structure that most emphatically exhibits the power of mind nevertheless leads to the denigration of the human mind. The success of the physical sciences leads to the attitude of scientific materialism, according to which the mind becomes, in one way or another, merely the passive plaything of material forces. The offspring turns against its parent. We forget what we should have learned from Kant: that the imprint of mind is everywhere on the body of this science, and without the founding power of mind it would not exist.

The irony here is not one that we can merely sit back and enjoy aesthetically. This doubt of the mind, in its actual consequences, in the lives of individuals and societies, provides one of the ordeals that modern civilization will have to go through. (6)

Frye suggests earlier in his essay that, as science moves from data towards laws, it "moves toward imagination" and it tends to invoke mathematics, which, along with literature and music, is (he says) a language of the imagination. "A highly developed science and a highly developed art are very close together, psychologically and otherwise." (7) Pursuit of mathematics is the pursuit of understanding. (Henry Pollack has said that the essence of science is the right to repeat an experiment, while the essence of mathematics is the right to understand (8)). Surely understanding is, in Frye's terms, the identification of the human mind with

Mathematics, along with literature and music, is a language of the imagination.

something outside it, the attribution of meaning to the coincidence of pattern. In the case of applications of mathematics, meaning is often attributed to the coincidence of mathematical patterns with patterns of worldly circumstances. Carried along by the logical inevitability of mathematics, undaunted by obscurities in the coincidence of the patterns, users of mathematical applications can come to believe that the mathematics is "in" the worldly circumstances (9). Understanding becomes "knowledge," and that knowledge is passed from one human being to another, often as subject matter in academic courses. And so we can see how mathematical circumstances, which are intrinsically logical, become illogically (but perhaps more or less appropriately) identified with things they are not, namely worldly circumstances.

Scott Buchanan gives a more thorough account of the passage from understanding to "knowledge" in his book *Poetry and Mathematics*. Here are some relevant excerpts:

> Belief is the natural attitude of a thwarted mind. It arises from fatigue and confusion.... For the most part confusion is of two sorts, one involving symbols, and the other metaphysical nostalgia, the tendency of thought toward the absolute. (10)

For Buchanan, symbols are things (aesthetic objects) which point to ideas (intellectual objects). He explains the confusion involving symbols as follows:

> The aesthetic properties of ceremony, formula, natural processes are intimations of complex and profound intellectual objects, but the difficulties of intellectual clarification and discrimination leave the mind in various attitudes of belief. For every intellectual

object, half-comprehended, there is an aesthetic object before which we bow in more or less deep reverence. Pure aesthetic contemplation and complete intellectual clarity are seldom found in human beings, and any middle ground is touched with credulity and idolatry. (11)

As to the metaphysical nostalgia,

> [a stage of mathematical discovery] results when we can see the *relations* holding between *qualities*.... Mathematical functions find elementary values in qualities. Qualities find their relations in the functions of mathematics. Whenever this happens, a system is recognized, and it takes on a quasi-independence and reality. Often the effect on the thinker is a conviction. Belief attaches itself only to such systems. The further expansions and the wider assumptions are ignored and there is a resting point for thought in a mathematico-poetic allegory. (12)

Of course, we are dealing here not with a disorder or aberration in human thought processes, but with the very nature of thought itself. Buchanan's account has some similarities with this description by Ernst Cassirer:

> [For the religious genius,] the power of his belief first proves itself in being made public. He must communicate his belief to others, he must fill them with his own religious passion and fervor, in order to be certain of his belief. This is possible only by means of religious constructs—constructs which begin as symbols and end as dogmas. Thus, even here, every initial expression of feeling is already the beginning of an alienation. It is the destiny and, in a sense, the immanent tragedy of every spiritual form that it can never overcome this inner tension; to extinguish it is to extinguish the life of the spirit. For the life of the spirit consists in this very act of severing what is whole in order that what has been severed may be even more securely united. (13)

And, in Cassirer's view, this pattern runs deep. Later, but talking about the same example, he says,

> And so it is that here, too, we find the same oscillation which sets in within all forms of culture as they begin to take shape[,]...the ceaseless and irresistible rhythm of life itself. (14)

Mathematics is often a tremendously effective metaphor, "unreasonably effective," in Eugene Wigner's words. But there is danger here. Essayist Wendell Berry, in his volume on culture and agriculture in the United States, refers in passing to

> ...the model of the scientists and planners:... an exclusive, narrowly defined ideal which affects destructively whatever it does not include. (15)

Joseph Weizenbaum, a computer scientist and teacher of computer science, speaks this message to his fellow computer science teachers:

> I...affirm that the computer is a powerful new metaphor for helping us to understand many aspects of the world, but that it enslaves the mind that has no other metaphors and few other resources to call on. The world is many things, and no single framework is large enough to contain them all, neither that of man's science nor that of his poetry, neither that of calculating reason nor that of pure intuition. And just as love of music does not suffice to enable one to play the violin—one must also master the craft of the instrument and of music itself—so is it not enough to love humanity in order to help it survive. The teacher's calling to teach his craft is therefore an honorable one. But he must do more than that: he must teach more than one metaphor, and he must teach more by the example of his conduct than by what he writes on the blackboard. He must teach the limitations of his tools as well as their power. (16)

Mathematics is, like the computer, a powerful metaphor. Unlike the computer, it is not a new metaphor. Any metaphor can take on excessive significance in a human imagination. Perhaps because mathematics is not a new metaphor, we are inured to it a bit; but that may just mean that we have lost the nervous uncertainty that comes with venturing into new territory, but have not seen enough failure of the metaphor to undermine our credulity. Metaphor is a wonderful thing, not at all to be scorned. Even if it were not wonderful, it would probably be inevitable (17). The motive for metaphor is a glorious aspect of human nature. G. Spencer-Brown observes, "That mathematics, in common with other art forms, can lead us beyond ordinary existence, and can show us something of the structure in which all creation hangs together, is no new idea." (18) But we need not take away from the value of the metaphor, or from the metaphorical experience, indeed we will enhance them, if we keep in mind the question, "What are the limits of this metaphor?"

If the application of mathematics is the construction of analogy, whether simile or metaphor; and if we are to effectively understand the limits of the analogy; then we must have some understanding of each of the two things being compared, and that understanding must be wider than what is immediately relevant to the analogy. Hearing Macbeth say of Duncan,

> After life's fitful fever he sleeps well, (19)

we bring to bear all our experience of life, fever, sleep, and death; and if we did not know enough of sleep and death, of life and fever, to know many ways in which they are not alike, as well as ways in which they are, then the metaphors would not be as rich and effective, the sentence not as beautiful, as they are. Of course, at the same time, analogies broaden our experience, deepen our understanding, give us new insights into the things being compared; indeed, that is their very function.

What does all this mean for the teaching and learning of mathematics? It means we should teach and learn mathematics beyond that which is "relevant", that which appears explicitly in applications of practical importance. The perennial students' question, "When will we ever use this?" is a misguided question, one to which we should not succumb. That is not to say that we should refuse to answer it; but we should deny that the question is determinative of what is important in a

---

"That mathematics, in common with other art forms, can lead us beyond ordinary existence, and can show us something of the structure in which all creation hangs together, is no new idea."

---

person's study of mathematics. Mathematics is one side of a myriad of important analogies; if we are to understand that side, then we must understand, we must teach and learn, mathematics itself. (And, of course, the study of mathematics itself has its own rewards, quite apart from applicability. Alfred North Whitehead describes the pursuit of mathematics as "a divine madness of the human spirit, a refuge from the goading urgency of contingent happenings." (20) Hardy says, "Real mathematics...must be justified as art if it can be justified at all." (21) Scott Buchanan observes, "The structures with which mathematics deals are more like lace, the leaves of trees, and the play of light and shadow on a meadow or a human face, than they are like buildings and machines, the least of their representatives." (22)) Knowledge of calculus has vocational value for engineers and others, but its value is increased by deep understanding of the mathematical theory of calculus; knowledge of statistics is perhaps necessary for informed citizenship; knowledge of arithmetic has survival value; knowledge of number theory may have none of these, but it strengthens one's understanding of mathematics as an independent reality which is in some of its facets analogous to some facets of other kinds of reality. (23)

It also means that we should resist, and teach our

students to resist, any tendency to neglect those aspects of other sorts of reality that do not fit into those analogies with mathematical reality that we call applications. We should in fact look for them. We should seek to understand the limitations of our analogies, and we will understand them better if we know what is beyond them, on both sides. We will be the richer for knowing what of mathematical reality does not fit the physical circumstances, and what of physical reality does not fit the mathematical circumstances, of whatever mathematical application with which we are dealing.

We should, in short, let mathematics be, just as other disciplines are, the pursuit of ways of seeing, the pursuit of visions. We should teach our students to look for mathematical analogies, to delight in them when they find them, to stretch them and test them and savor them, but not to be consumed by them lest they—we—suffer the fate of all tragic heroes. (24) We will lay the proper foundation if we teach them mathematics itself, as independent, extant reality, whose "applications" are in fact analogies which often appear as metaphors.

Notes

(1)　　G. H. Hardy, "What is Geometry?," Presidential Address to the Mathematical Association, 1925.

(2)　　Eugene P. Wigner, "The Unreasonable Effectiveness of Mathematics in the Natural Sciences," *Communications on Pure and Applied Mathematics*, Vol. XIII, 001-14 (1960).

(3)　　Tobias Dantzig, *Number: The Language of Science* (Garden City, NY: Doubleday & Company, Inc., 1954)

(4)　　Northrop Frye, *The Educated Imagination* (Bloomington: Indiana University Press, 1971), pp. 32-3.

(5)　　Ibid, pp. 32-3.

(6)　　William Barrett, *Death of the Soul* (Garden City, NY: Anchor Press/Doubleday, 1986), p. 75.

(7)　　Frye, *The Educated Imagination*, pp. 23-4. Perhaps one of the more important differences has to do with belief. See the discussion of the passage from understanding to "knowledge" on pp 31-32.

(8)　　This was in a talk that Dr. Pollack gave on "The History of an Application" at the 1986 Woodrow

Wilson Summer Institute on High School Mathematics in Princeton, New Jersey.

(9)   As part of its promotion of "Math Education Month" (April, 1987), the National Council of Teachers of Mathematics is offering for sale a bumper sticker which reads, "Math Keeps the World in Motion." Of course, as is proper for a bumper sticker, the sentence is ambiguous, suggestive, a play on words. It points to the central role of mathematics in modern technology, government, economics, etc. But it also suggests that math *causes* the earth to rotate on its axis and revolve around the sun, the automobiles and telephones to work, and so on. It can be read as ironic (in a couple of subtly different ways), but for one who cannot see the irony, it could be an intimidating and depressing, an unfortunate, message.

(10)   Scott Buchanan, *Poetry and Mathematics* (Chicago: The University of Chicago Press, 1975), p. 135.

(11)   Ibid, p. 140.

(12)   Ibid, pp. 146-7.

(13)   Ernst Cassirer, *The Logic of the Humanities* (New Haven: Yale University Press, 1974), pp. 115-6.

(14)   Ibid, p. 214 .

(15)   Wendell Berry, *The Unsettling of America: Culture and Agriculture* (San Francisco: Sierra Club Books, 1986), p. 112.

(16)   Joseph Weizenbaum, *Computer Power and Human Reason* (San Francisco: W. H. Freeman and Company, 1976), p. 277.

(17)   When looking up the sentence about belief and the thwarted mind, I discovered how much my thinking in this essay had been affected by Buchanan's book. For example, the last sentence in one of his chapters is, "Any history of thought might begin and end with the statement that man is an analogical animal." Buchanan, *Poetry and Mathematics*, p. 141.

(18)   G. Spencer-Brown, *Laws of Form* (New York: E. P. Dutton, 1979), p. xxix.

(19)   This example will be recognized as borrowed from Hardy.

(20)   Alfred North Whitehead, *Science and the Modern World* (New York: The New American Library of World Literature, 1954), p. 22.

(21)   G. H. Hardy, *A Mathematician's Apology* (New York: Cambridge University Press, 1977), p. 139.

(22)   Buchanan, *Poetry and Mathematics*, p. 36.

(23)   Buchanan makes several comments relevant to mathematical pedagogy. Here is one that relates to the use of mathematical applications in the classroom:

> Mathematics is not a compendium of memorizable formula and magically manipulated figures. Sometimes it uses formulae and manipulates figures, but it does this because it is concerned with ideas already familiar to the ordinary mind, but needing special sets of words or symbols for the sake of precise expression and efficient communication. Further, the abstraction thus signalized, which most people from bad emotional habits fear, is actually much more familiar to the untrained mind than any observed facts could possibly be. Abstract ideas are of the very tissue of the human mind. For this reason and for many others, illustration of mathematics by concrete event, fact, or object is never as effective as illustration by equally abstract analogous ideas.

Ibid, pp. 35-6. Of course, in this passage Buchanan is speaking of the use of the concrete to illustrate mathematics, rather than the use of mathematics to explain the concrete.

(24)   I have heard attributed to Mark Twain the remark, "It ain't what people don't know that causes all the trouble, it's what they do know that ain't so."

Mathematics and Literature

Ubiratan D'Ambrosio
Univ Estadual de Campinas
Campinas, Brazil

MATHEMATICS Dries up the heart. Gustave Flaubert's *Dictionary of Received Ideas.* [8; p. 316]

Introduction

The first reaction we get when mentioning "humanistic mathematics" to someone is a question: "What does this mean?" Either our interlocutor does not see any relation between mathematics and the humanities or he takes for granted that mathematics is an human endeavour, so why qualify mathematics?

The origin of the movement for "humanistic mathematics" can be traced in two directions. One comes as a sort of desperate effort to keep mathematical hegemony in the school curricula. Research mathematicians, teachers and mathematics educators— all concerned professionals coming from different origins, which have traditionally been throwing barbs at each other and would hardly mix—now appeal together to the dangers of the mathematical unpreparedness of the current school generation. They call for less rote learning and for stronger emphasis on problem solving and on the understanding of the fundamental concepts, with the objective of helping students to understand and to apply mathematics to practical situations. The proponents of reforms hope for better achievement by setting standards and by calling upon teachers to become more concerned with "reality", as perceived by themselves, adults and research mathematicians, teachers and mathematics educators. But mathematics remains untouched; it is "the cake that I like, so you have to swallow it". I have no doubts that the results of this late millenium effort will be equally disastrous as have been the results of reform movements in the last fifty years. The subsequent assessments of school achievement in mathematics will show even poorer results. We will be nostalgic for the good results of the SIMS(Second International Math Study)!

The other direction in the movement for humanistic mathematics comes as a sort of natural reaction to over-specialization in doing and teaching mathematics. Currently, mathematics is taught as if it were a universe closed in itself, conceived, and for some even received, in the supposedly final form in which it is known now. Merely looking into the etymology of mathematics, we recognize a much broader and dynamical concept. It means the technique or art (*techne* or *tics*) of understanding, knowing, explaining (*mathema*). Indeed, the earliest recorded use of the word by Pythagoras was in this sense, and it thus prevailed throughout Plato and practically all philosophers of antiquity. Calculating, numbering, measuring are different techniques which, when combined, lead the human being to survive with less fear and more security in this environment.

The history of every individual and every society reveals an effort towards understanding, explaining, managing and coping with reality. This is present, as organized forms of knowledge, in every culture, in every civilization. Different manifestations of techniques with the same purpose are known to all of us: magic, divination, religion, art, science. By achieving self-domination, domination of his ambiance and ascendence

Ours is a proposal to shift from the quantitative to the qualitative aspects of the investment in education.

over the others, in other terms by gaining power, man gets close to creation, and is driven into the search for omniscience, omnipotence, omnipresence. He thus gets closer to a superior being, the Creator himself, who necessarily must be omniscient, omnipotent and omnipresence. Already in the earliest manifestations of our species we can identify those objectives both in man's drive to command and understand nature and reality as a whole. In the associated intellectual search throughout times and places, different conceptions of a superior being have generated different forms of behavior. This is beautifully illustrated in Arthur Clarke's "2001: An Odyssey in Space". This conception underlies religious practices, living habits, dressing modes, food and drinking tastes, painting and sculpture, music and dancing, theater and literature, and obviously mathematics and science. These different manifestations are part of the same culture and period. None of these manifestations can be regarded outside of the proper context, and their transmission will be either by violent imposition or by contextualization.

This paper is motivated by the general concern with the teaching of mathematics, particularly with ever lowering achievement in mathematics throughout the school systems. Efforts to improve academic achievement in mathematics are not producing the desired results. Large financial and human resources committed to this seem to be failing in producing the expected results. Of course, claims that these resources are insufficient are always made. Ours is a proposal to shift from the quantitative to the qualitative aspects of the investment in education. Of course, such a shift affects mainly minorities and the sectors of society which have been in disfavor. But we have to face the issue that for the minorities no investment in education

---

Remedial curricula is an example of how mathematics is being oversold. College courses on basic elementary arithmetic, taught to adult students as if they were children, are aggressive, decontextualized and exemplify my criticism of overselling mathematics.

---

is meaningful without corresponding actions in housing, in job opportunities and in health. It is absolutely false to think that school hours are independent of the rest of the daily life of a student. Individuals live twenty-four hours a day, seven days a week, fifty-two weeks a year. Their school year is made of more than the 1,200 hours in classroom. This is the essential meaning of sharing experiences, sharing a full year of life, sharing perceptions about the past, sharing each one's personal history, sharing one's visions and expectations about the future.

I hope to show in the course of this paper that my views on mathematics will put me on the side of those for whom humanistic mathematics is a redundancy. To avoid redundancy, instead of calling this a movement in favor of humanistic mathematics, I would rather call it a movement against the prevalence of Alienated (and Alienating) Mathematics. Current school mathematics

is alienated in several senses: socially, politically, culturally, and as a branch of knowledge.

In this paper I will try to show how mathematics is woven into the fabric of literature and how much light this can throw upon the needed critical analysis of mathematics and science in general. The pulse of every cultural stage in the history of mankind has always been sensed by global thinkers, as in literature and the arts. To take into account their views is a must for a "healthier" teaching of Mathematics.

Overall Goals of Education.

In the context above I base my remarks on a philosophy of education centered on the student and reflecting societal behavior as a whole, with a concept of curriculum and all the rest that is associated with schooling. My basic idea rests on the principle of sharing experiences. Without that sharing of personal experience, the student's own history has to be put aside and the student must accept the personal history and expectations of a teacher whose own experiences do not reflect the student's own history, assumptions, or aspirations. When teachers presume a set of assumptions which are drawn from the dominating groups of society, they will fail to engage students whose experiences separate them from those assumptions. How can the white teacher develop rapport with the black student without sharing experiences? Schooling without sharing is not teaching—it is indoctrination. Therefore, teachers should not only talk—they should listen. In fact, listening is more important than talking!

Listening to students is the major methodological component of this approach. Of course, listening methodologies lead to ethnographical methods, to participatory teaching, and may suggest several different ways of conducting the classroom in a dynamical way. In research this leads to naturalistic or qualitative research. This leads naturally to Ethnomathematics, and I refer the reader to my recent book on the subject [5] or to [4]. Later in this paper we will discuss this.

I dare to say that what is intrinsically wrong in our school systems is *the mathematics* we are teaching. The main point I want to make is that no changes in curriculum can bring much improvement.

In an earlier essay [1] I state that by curriculum I understand the the three-dimensional strategy of educational action which considers objectives, contents and methods, where all three are interdependent. In that short paper I analyzed what went wrong with 20 years invested in the Modern Math movement. My conclusion was that there are two basic reasons why the teaching of mathematics defies our efforts and why the results we are achieving are not commensurate with the resources society has been pouring into it. First, there is a mistaken view of the concept of the curriculum, which has caused a trichotomy of it into objectives, contents and methods, clearly reflected in the programs for teacher training. But even more serious is the intrinsic obsolescence of the mathematics taught when we consider modern thought as a whole. Somehow we seem to be aiming in the wrong direction. Even when students become mathematicians, in many cases they are the least creative professionals in the scientific fields. A critical view of the mathematical advances in

---

Ernst Mach referred to offering the same curriculum for all the students saying that "In such a case the function of the schools would be simply to select the persons best fitted for being drilled, whilst precisely the finest special talents, which do not submit to indiscriminate discipline, would be excluded."

---

the 20th century as compared with other fields is enough to convince anyone about this. With remarkable exceptions, mathematical thought seems to go diametrically opposed to most of modern behavior, which stresses action, fantasy, freedom, egalitarianism. Mathematics seems to forget that it is but one of the multiple manifestations of human intellectual endeavour, and that, like all the other such manifestations, it results from the natural drive of mankind towards understanding, explaining and managing reality. It is as if losing track of the *Zeitgeist*, mathematics has created an existence of its own and, in the way it is taught, passes this on to the new generation, disregarding the complexity of its own search for explaining, understanding and managing reality. Normally, the focus of our search for improvement goes from the student to the teacher and vice-versa. It is taken for granted that mathematics is to be taught and learned. What seems to be wrong is the idea that mathematics is to be taught in a way which is boring, uninteresting and above all arrogant, claiming too much, overbearing and dogmatic. Maybe this is intentionally done in order to have "useful" scientists, well trained but obedient and acritical of the implications of their scientific work to society as a whole. More than ever, the curriculum decisions, based on the structural arrangement of the disciplines themselves, should be reconsidered and mathematics teachers should learn how to get input from the students on what should be the focus of attention and interests in schools.

Mathematics and Literacy.

As I have said in the introduction, there is an enormous concern with the lowering of achievement in mathematics in the school systems. Indeed, these worries are not restricted to mathematics. Literacy is similarly imperiled. In the case of mathematics, one component remains untouched: mathematics itself, the content. Some successful professional parents claim that they never did well in mathematics, but that their children are doomed to failure if they do not learn this or that content.

I dare to suggest that the idea of a universal mathematics, supposedly to be part of the intellectual baggage of every citizen, merits some reflection. I see many points in common with my position and the criticism raised by Frank Smith in his beautiful article about how, by overselling literacy, school systems may be doing more damage than benefit to our children. The following remarks by him easily apply to mathematics. "Let me stress at the outset that I am in favor of literacy. I think that people who don't read and write miss something in their lives. But I think the same about anyone who doesn't appreciate some form of music. Nevertheless, people who aren't musical aren't usually regarded as failures or social outcasts....Literacy doesn't make anyone a better person. Some of the greatest tyrants in the world have been voracious readers and compelling writers." [3; p. 354]. I obviously endorse these statements and they apply to our case if we read "Mathematics" where Frank Smith refers to "Literacy".

The ultimate importance given to mathematics, mainly by creating an aura of genius for those who do well in mathematics, may have resulted in a major barrier in bringing mathematics to a better reception and appreciation in schools. Without any doubt this aura has been responsible for failures and anxieties. Overselling mathematics may lead to the radical attitude of giving up the universality of mathematics in school systems. Remedial curricula is an example of how mathematics is being oversold. College courses on basic elementary arithmetic, taught to adult students as if they were children, are aggressive, decontextualized and exemplify my criticism of overselling mathematics. They fulfill no purpose at all and it would be better not to require any mathematics at all. The overall practical results would not be different. People would continue to perform mathematically as poorly—or as well—in their everyday life, as they have been performing, in spite of the system's telling them they are failures in mathematics. Frank Smith has a point!

It is quite possible for someone doing poorly, or close to nothing, in school mathematics to become an achiever even in professions that require mathematical ability.   So, why all the fuss about universal mathematics in school systems and expectations that everybody perform equally well according to some standardized tests?   Already in the end of the last century, during the big controversy about abolishing Latin and Greek in general education, Ernst Mach referred to offering the same curriculum for all the students saying that "In such a case the function of the schools would be...simply to select the persons best fitted for being drilled, whilst precisely the finest special talents, which do not submit to indiscriminate discipline, would be excluded from the

---

A life that would be always and everywhere nothing but rational is not liveable.

---

contest." [8; p. 370].   Aren't universal mathematics and standards of achievement a somewhat more sophisticated form of the aptitude and the IQ tests and other testing devices so "efficiently" used for discriminatory practices?

Humanistic Mathematics and Ethnomathematics.

Humanistic mathematics, in conjunction with ethnomathematics, may succeed where other movements with similar goals (Back to Basics, Standards and Problem Solving) are, in my perception, doomed to failure.   The main reason why I believe the combination humanistic mathematics + ethnomathematics is a good option is the fact that Humanistic Mathematics provides the component of cultural background—that *plus* which is recognized when we say "He/She is a cultured person", while ethnomathematics provides what one needs to perform critically and consciously in his/her professional surroundings.   It is a matter of not only living, but of also giving quality to one's life.

To manage and cope with reality is basic, but is not all. We want to understand and to explain. Regrettably, the modern world and pressures for mere survival tend to remove quality from one's life. Human beings ask for more than satisfaction of their basic needs for survival, they ask for freedom, for fearlessness, for space to be creative, for explanations, for knowledge.   The combination   ethnomathematics + humanistic mathematics aims at bringing these dimensions of quality to life.

Essentially, I see mathematics as this combination, as a

mode of thought in a sort of duality, the result of a complementary drive of the human species to survive and to manage its daily life and at the same time to raise its thoughts to understand and to explain the meaning of this survival and of existence itself. In other words, the theoretical reflection upon one's practices guides and improves these practices and establishes the dialectics of living and reflecting upon living. This dual aspect of mathematics, clearly spelled out in Plato, is seen in the development of a technique or art (*techne*) of coping and managing reality (the environment), and at the same time reflecting upon it, trying to understand and explain it (*mathema*). These correspond respectively to what we have been calling *Ethnomathematics* and *Humanistic Mathematics.* For a more detailed discussion of this see [4] and [5].

The movements to bring humanistic mathematics and Ethnomathematics to the school systems come as a sort of natural reaction to over-specialization and formalism in doing and teaching mathematics. Mathematics is usually taught, and regrettably also done, as if it were a universe closed in itself, conceived, and received, even by some "successful" graduate students, in the final form in which it is now. Techniques, results and theories are presented as if they were conceived in the final form they are presented, without any reference to the history of the ideas behind them. By merely looking into the etymology of mathematics we recognize a very broad concept. As we have said in the Introduction, the word mathematics is composed of *mathema* and *tics*, which mean the technique or art (*techne*) of understanding, knowing, explaining, managing and coping with reality. These techniques are developed according to the specificities of the natural and cultural environments, thus the importance of bringing the (*ethno*) component into our considerations and the talking about ethnomathematics.

But everyone claims, and I do not disagree, that school should go further than ethnomathematics, that school should care also about "real math". By this somewhat aggressive way of putting it, the critics of Ethnomathematics mean academic mathematics, with the agreed symbols and names and codes, equivalent to what socio-linguists, particularly Basil Bernstein [15], call the cultured norm of the language. Indeed, school should be concerned with teaching mathematics as the most central of the cultural achievements of western civilization. Like it or not, we are under the imprint of western civilization, with its sets of values, with its styles of behavior, with its modes of production and property.   But we can hardly agree that this is permanent. As current history tells us, the dreams to build empires that would last for a thousand years were nothing but nightmares!

Western civilization, as we know it, is barely 400 years

old. This is a short period in the history of ideas and much improvement is expected in the coming decades and we hope, (this is indeed my dream) that mankind will eventually speak of a really Universal Civilization. Of course, this is an evolutionary process. But the only chance of positive change comes from western civilization itself, which is the dominating one, as a result of internal criticism. This is the only way of smooth socio-political changes, as basic political science shows us. Thus, there is nothing wrong with teaching mathematics (academic mathematics, which is the imprint of western civilization) in schools, provided it opens space for facilitating this internal criticism of society as a whole. This can be achieved through studies on the history and sociology of mathematics, which are inherent to both humanistic mathematics and ethnomathematics, and through the perception that mathematics (the imprint of western civilization) is indeed a result of cultural dynamics, with contributions from so many different civilizations, and not the absolute mode of thought of individuals or societal classes predestined to mastery. Thus the improvement in academic mathematics will eventually come out of criticism which is inherent to this approach. The way mathematics is currently taught and sometimes learned can be not only useless, but also negative. It may lead to arrogance and build up tensions, anxiety and servility. It may easily destroy self image and block creativity.

The Generation of Knowledge.

The drive towards understanding, explaining, managing and coping with reality, present in every individual, in every society, in every culture and civilization, was discussed in the introduction. This drive, in its different

---

The specific problem a child is trying to solve creates a mental set that determines the features selected for attention.

---

manifestations, is part of the same culture and period. As such, this is registered, reported and transmitted to fellow human beings by practices and rituals, by artifacts and by tradition, both oral and, later on, written tradition. These we call literature. In it we find a large amount of experiences, of facts and fantasies, of predictions and dreams, of attempts and errors, of successful and misleading explanations, of correct and false understandings, which have been accumulated throughout the entire existence of our species.

The absolute preponderance of the exact, the rigorous, the rational tend to dismiss this immense reserve of

intelligence. As Michel Serres put it, "a life that would be always and everywhere nothing but rational is not liveable. It is therefore reasonable to be reasonably wary of the rational. We are not judging the merits (acknowledged), the grandeur (admitted), the effectiveness (a thousand times proved), the value, in sum, of science; we are worried about its victories for fear of one day being under its complete domination. No matter how correct a theory may be, it becomes odious when it reigns alone." [6; p. 4].

The wish to recover the sense of individuality, of freedom of choice, of free will for mankind as a whole is, without any doubt, a philosophical posture, close to what might be considered a "value statement" or even a "metaphysical statement". Of course, dealing with statements of this kind invokes a branch of knowledge, and this has been the pursuit of sages and heroes, saints and mystics, philosophers and moralists. All of them have followers, some in disagreement, some even killing each other in the name of something they call "truth". Other branches of knowledge or modes of thought, are those relating to "imperative and attitude statements", "empirical statements" and "analytic statements". These steps correspond to the process of generating, transmitting and diffusing knowledge. Of course, this goes in parallel with the meaning of truth. To say that a statement is true, one must respond to the three following steps, all essential: (i) to know what the statement means; (ii) to know the right way to verify it; (iii) to have good evidence for believing it. It is clear that even when the first two are verified, the third one remains very difficult and brings about the need for some form of agreement.

Mathematics is the result of a string of agreements that can be traced back centuries. These agreements have been coherently used in scientific hypotheses whose result seem to have "worked". Through the centuries we have been collecting the evidence we want from observations which we know are relevant to the "proof" of our truth. This is how Science has gone and goes. But how about the other branches of knowledge, as, for example philosophy, psychology, sociology, politics, economics. John Wilson, from whom I borrowed the classification of statements, claims that: "Our problem in dealing with value and metaphysical statements, then, is to see whether we do actually share experiences of a kind which would make it useful for us to agree upon an established meaning and verification for at least some of such statement. Here it is evident that we must begin to distinguish different *types* of experience. Empirical statements are verified, in the last analysis, by sense-experience. Analytic statements are verified by our experience and knowledge of the rules of logic. Can we find different types of experience which can be usefully employed to help us build up a framework of communications for value and metaphysical statements?" [7; p. 87].

Remarks of the same nature underlie the most modern approaches to human mental development. To understand human experience, infancy, and cognitive development our attention must be directed towards the context. "The specific problem a child is trying to solve creates a mental set that determines the features selected for attention" [19; p. 209]. Thus the need for a pedagogy which relies on our understanding of the processes behind the generation of knowledge properly contextualized to personal history and to socio-cultural contours. Most of the times these conditioners are better dealt with in literature and the arts in general than in the writings of education theorists.

Literacy and Matheracy.

In a somewhat different context, the ideas above were discussed in [2]. There we assumed as desirable the ideal of developing in each individual a capability which I have called *matheracy* and which is closely related to the magnificent cultural achievements throughout history which we call mathematics, whose etymology we have discussed above. Matheracy is essentially the development of cognitive capabilities which allow the individual to understand, to explain, to cope with and to manage its reality. The term was borrowed from Professor Tadasu Takeushi, but I have been using it in a different and broader sense. The meaning of matheracy is clear when we think of literacy. Literacy is understood as the use of language, which is a structured set of words, so that people can establish communications (ie, *action in common*) with each other, thus building up social behavior. Oral language is particularity effective when actors are present. The introduction of written language leads to an important improvement by allowing action in common by remote (in time and in space) actors, thus building up cultural behavior. For a more detailed discussion of this see [7]. Essentially, literacy is the development of capabilities of communication among human beings, which is the art or technique which we call language, and matheracy is the development of capabilities of understanding, explaining, coping with and managing reality, which is the art or technique which we call mathematics. Similar to what happens when we oversell literacy by stressing the importance of dominating formal aspects of language, I dare to say that the ways in which we are teaching and promoting mathematics destroy the matheracy that I see among our main goals for school systems. The situation is even more serious when we recognize matheracy as a spontaneous attribute, the same as creativity, of every human being.

Let me bring into discussion an important distinction

between matheracy and literacy which has implications for schooling. Both are based on systems of codes. Matheracy allows one to describe reality (the natural, the artificial, and the social environment) through representations (or models) upon which the individual builds models, i.e. "codified approximations" of reality which he/she utilizes in the explanation, understanding, coping and management of reality itself. The matherate individual works upon representations in communicating with—and we had better say, in this case, in dealing with—reality. The system of codes is built by the individual; there is no need for "agreement" with the other part (reality). Matheracy is a one-way process. Literacy is essentially the same: representations, through codes, of ideas, but with the purpose of establishing a communicative link between individuals. Hence, it is a two-way process. Of course, schooling is a many-way process. Building upon the experience of individuals in matheracy is much more complex than building upon the experiences in literacy. The essential step of "agreement"—which is a step towards socialization—is intrinsic to literacy, while in matheracy what might be called the interlocutor (i.e., reality) is passive in the process. This is why in many cases we do not recognize the individual as matherate. In this respect, the diverse versions of *"enfant sauvage"* (in particular the one put in cinema by Werner Herzog with the little "Kasper Hauser" in the early seventies) are good illustrations.

Literature and the Sciences.

The renowned historian and philosopher of science and a leading figure in the new French critical philosophy Michel Serres begins his paper on *Literature and the Exact Sciences* by claiming: "that every body of literature, and more broadly, the sum of all academic, literary and popular traditions, anecdotes and songs, old

---

...literacy is the development of capabilities of communication among human beings, which is the art or technique which we call language, and matheracy is the development of capabilities of understanding, explaining, coping with and managing reality, which is the art or technique which we call mathematics.

---

wives' tales, sayings and cliches, beliefs, myths, languages, and even religions, form a sort of reserve, a stock, the local treasure of ethnic groups, of cultures, of diverse groups, and that these constitute, in addition, the global treasure of mankind, accumulated in a fluctuating and unstable way through history and time." Later on he complains: "Time is in danger... for a very new reason: the crushing of all cultures by the culture that is led astray by exact sciences. The exact sciences take all the places and soon all the space: money, professional openings, techniques, research, politics, power, the religious confidence of the population. ... Here the complete tyranny...does not come from a single locality

---

Wrong doings, inequities, oppression and infamous inhumanities, some of very recent memory, have been in most cases performed with the acritical use of the instruments indirectly provided by mathematics.

---

that would forcibly take the place of all others, from a single language that would impose itself, no matter how unique, as a universal language. The terror comes, if I dare to say it, not from the fact of power, but from rightness. The thing is that science is right—it is demonstrably right, factually right. It is thus right in asserting itself. It is thus right in destroying that which is not right." And Serres concludes, as a preamble for his beautiful paper, by saying: "I hope to die before I live in a world that will only speak reason. For a life that would be always and everywhere nothing but rational is not liveable." [6; pp. 3-4].

This citation of Serres touches many of the points we want to raise in this paper. Essentially, he puts Sciences on the side of the powerful, of the effective and of the legitimate and leaves literature as the legacy of the poor—emptied of power, of reason and of legitimacy. By leaving literature aside in trying to have a critical view of Mathematics, Science, Education and in essence the future of mankind, we are risking the killing of time, i.e. killing history and not learning from the past, thus allowing the future generations to repeat mistakes, errors, wrong-doings, injustices, brutalities and inhumanities. These are frequently spelled out in literature, but are absolutely disregarded in the mathematical discourse which becomes, at a rapid pace, the universal unique language of acceptable knowledge. Wrong-doings, inequities, oppression and infamous inhumanities, some of very recent memory, have been in most cases performed with the acritical use of the instruments indirectly provided by mathematics.

The kind of criticism raised above also comes from inside the profession, from inside the "mathematical club". There has been much criticism of the alienating factor intrinsic to Mathematical thought. To mention "mathematical club" is indeed a dangerous provocation! Mathematicians usually claim that the outsiders are those who were not granted admission for lack of the desired "intellectual resources" for admission! We see that this view is agreed upon by society as a whole when someone says "He is a mathematician. What a bright guy!". This classification of "outsiders" and "insiders" or "practitioners"—the bright part of the population as a whole—is at least educationally very damaging. We should better think of learned individuals, successful professionals who have passed the filters of school mathematics and yet claim their hatred of mathematics, while others reassert their love of mathematics. In all cases we are dealing with "achievers", in the sense that they succeeded in going through the filters of school systems, but did not become "practitioners".

As a remark, let me say that I frequently use the word "filter" when referring to the prevailing evaluation and degrees system in schools and society as whole. While I will not elaborate on this, I would like to refer to my basic sources in this matter: Alexander Grothendieck [16] and Pierre Samuel [17], unquestionably full members of the "mathematical club", and among the earliest to refer to mathematics as the "new religion" and to denounce the filtering character of mathematics in modern days. But indeed we find a very similar discourse two thousand years ago in Plato.

Views of Mathematics in World Literature.

How do mathematics and literature relate in this broad picture? How is mathematics perceived by non-professionals, by those who surely manage mathematics in their every day life, who probably have struggled with mathematics in their school days and who see society with the critical and romantic eyes of a novelist?

Of course, critical views do not imply negative views. It is a general perception—and we mathematicians agree—that there is an intrinsic beauty and elegance in mathematics, that mathematics helps in developing clearer ways of reasoning, that mathematics is useful in the modern world and was so in all the ancient worlds. Robert Musil was a major Austrian essayist and novelist (1880-1942). Ulrich, Robert Musil's "Man Without Qualities", is an individual "in whom all qualities merge spectrally into the whiteness of none" [12; p. iv], in a clear reference to Newton's chromatometer, and has only a first name during the entire novel. Ulrich, who stands for the intellectually successful John Doe, "could see that [he was not so much scientifically as humanly in love with science] in all the problems that came into its orbit, people in science thought differently from the way ordinary people thought. If for 'scientific attitude' one were to read 'attitude to life', for 'hypothesis', 'attempt' and for 'truth', 'action', then there would be no considerable natural scientist or mathematician whose life's work did not in courage and revolutionary power far outmatch the greatest deeds in history." [12; p. 41]. That was the opinion of Ulrich, who himself was a mathematician, commenting on the fact that "there were people who were prophesying the collapse of European civilization on the grounds that there was no longer any faith, any love, any simplicity or any goodness left in mankind" [12; p. 41]. These are people witnessing scientific and technological progress with a distorted view of mathematics and the sense of absolute truth associated with it. This view of mathematics disregards any relativism due to social, political and cultural factors. An example of such a distorted view—prevailing in society as a whole but regrettably shared by a few mathematicians—is that among school subjects there is no beauty or elegance outside mathematics, that no one

People in science thought differently from the way ordinary people thought.

else helps to develop clearer ways of reasoning, that no other subject is as useful as Mathematics. And as a consequence children are punished, are labeled "stupid" or told that they are doomed to failure in life if they do not do well in mathematics.

A certain contempt about mathematics, implicit in the citation of Flaubert in the caption of this paper, is highly contrasted by the writings of his contemporary, Isidore Ducasse / Lautreamont (1846-70): "O mathématiques sévères, je ne vous ai pas oubliées, depuis que vos savants leçons, plus douces que le miel, filtrèrent dans mon coeur, comme une onde refraîchissante. J'aspirais instinctivement, dès le berceau, à boire à votre source, plus ancienne que le soleil.... ...Arithmétique! algèbre! géométrie! trinité grandiose! triangle limineux! Celui qui ne vous a pas connues est un insensé!" [10; p. 106-7]. And praises for mathematics go on in the works of this important writer.

Of course, it is important to recognize what is behind the author(s) opinions and feelings about mathematics. In a sense, these remarks reflect much of the author's character, personality, feelings, posture, and memory. But this is no less true with the mathematical production itself. We can hardly grasp the full meaning of a piece of mathematics—a technique, an idea, a result or a theory—without placing it in the global context of

its conception and generation. Its transmission and its diffusion, and each of these steps, clearly can not be dissociated from the author's own history. This is the main reason why we study history, including the history of mathematics—in order to get a deeper and more effective understanding of mathematics. Otherwise mathematics is taken out of context, leaving it mostly devoid of inner and even fewer outer connections. The same is true of all other branches of science and of knowledge in general, including literature.

We must also refer to the important remarks about differential calculus, science and history in general by Leo Tolstoy in his monumental *War and Peace* [18]. We see there the views of mathematics as placed in the kernel of culture by someone who had in his life history an important educational experiment, the Yasnaya Polyana pedagogical project.

Perhaps the most impressive example of how mathematics is seen critically by writers comes from the same Robert Musil in another of his important novels. Witness to the late days of the Franz Joseph empire, Musil is considered among the most important writers of the German language. His critical view on the profound cultural changes of the turn of the century reflects his experience of living in that intellectually magnificent *fin de siècle* in Vienna, dominated by the new thinking of Dilthey, Freud, Klimt, Mach, Weber and many others. The generation in which Musil towers include such names as Hermann Hesse, Thomas Mann, Karl Jaspers, Franz Kafka, George Lukacs, Gustav Mahler, Paul Tillich, Ludwig Wittgenstein and many others. Although some of them were somewhat inhibited by their respect for reason and science, others, "Setting out from the uncertainty of knowledge and the inadequacy of every form of dogmatism, ... were conscious of the fragility and brevity of human civilization ... Confronted with the decline of the old historical cosmologies of progress, they sought to reawaken the positive value of the unconscious, sexuality, and dream life." [9; p. 17]. This same period of revolutionary and unconformable thought sees the consolidation of the foundations of mathematics which have been built upon over centuries. I see this generation as the precursor of new ideas in science, such as quantum physics and the later questioning of paradigms and the search for a new metaphysics and an associated modern ethics for the scientific and industrial world, which impregnates much of the scientific circles of nowadays, but with a conspicuous absence of mathematicians. As I remark above, there have been few new ideas in mathematics in the 20th century as compared with other sciences.

Robert Musil has a prominent position among a generation of critical writers. More than the others, he hits the crux of the problem of reconciliation of modern culture and industrial society, of intellect and feeling, of rationality and passion, of the material and the spiritual. He was born in a middle-class family and went to a military academy and to an engineering school, finishing his career with a Ph.D. dissertation presented in Berlin in 1908 on *Beitrag zur Beurteilung der Lehren Machs*. He opted for a career in writing soon after obtaining his degree. Indeed, his first novel *Die Verwirrunger des Zoeglings Toerless* (The Confusions of Young Toerless) was first published in 1906 to immediate acclaim. It is a classical novel, focusing on

---

And what can you imagine from being told that parallel lines intersect at infinity? It seems to me if one were to be over-conscientious there wouldn't be any such thing as mathematics at all.

---

rites of passage, set in an Austrian military academy at the turn of the century in which adolescents experience their emotional, sexual, social, and intellectual crises. The novel is unquestionably autobiographical. For us, of special interest is the passage below of a dialogue between Toerless and colleague, Beineberg, about mathematics and a later related dialogue between Toerless and his mathematics teacher. Although it is somewhat long, I feel it is appropriate to quote parts of it.

> During the mathematics period Toerless was suddenly struck by an idea.
>
> For some days past he had been following lessons with special interest, thinking to himself: "If this is really supposed to be preparation for life, as they say, it must surely contain some clue to what I am looking for, too."
>
> It was actually of mathematics that he had been thinking, and this even before he had those thoughts about infinity.
>
> And now, right in the middle of the lesson, it had shot into head with searing intensity. [13; p. 105]

When the class was dismissed, Toerless looked for his colleague who, in the novel, serves as a kind of mentor of Toerless for "the facts of life" in the initiation process which is the essence of the novel. Intellectually

strong, Beineberg echoes much of the reaction of society to the doubts and anguishes of our hero. The dialogue of Toerless with Beineberg is remarkable.

"I say, did you really understand all that stuff?"

"What stuff?"

"All that about imaginary numbers."

"Yes. It's not particularly difficult, is it? All you have to do is to remember that the square root of minus one is the basic unit to work with."

"But that's just it. I mean, there is no such thing. The square of every number, whether it's positive or negative, produces a positive quantity. So there *can't* be any real number that could be the square root of a minus quantity."

"Quite so. But why shouldn't one try to perform the operation of working out the square root of a minus quantity, all the same? Of course, it can't produce any real value, and so that's why one calls the result an imaginary one. It's as though one were to say: someone always used to sit here, so let us put a chair ready for him today, too, even if he has died in the meantime, we shall go on behaving as if he were coming."

"But how can you, when you know with certainty, with mathematical certainty, that it's impossible?"

"Well, you just go on behaving as if it weren't so, in spite of everything. It'll probably produce some sort of result. And after all, where is this so different from irrational numbers—division that is never finished, a fraction for which the value will never, never, never be finally arrived at, no matter how long you may go on calculating away at it? And what can you imagine from being told that parallel lines intersect at infinity? It seems to me if one were to be over-conscientious there wouldn't be any such thing as mathematics at all."

"You're quite right about that. If one pictures it that way, it's queer enough. But what is actually so odd is that you *can really* go through quite ordinary operations with imaginary or other impossible quantities, all the same, and come out at the end with a tangible result!"

"Well, yes, the imaginary factors must cancel each other in the course of the operation just so that does happen."

"Yes, yes, I know all that just as well as you do. But isn't there still something very odd indeed about the whole thing? I don't quite know how to put it. Look, think of it like this: in a calculation like that you begin with ordinary, solid numbers, representing measures of length or weight or something else that's quite tangible. At any rate, they are real numbers. And at the end you have real numbers. But these two lots of real numbers are connected by something that simply doesn't exist. Isn't that like a bridge where the piles are there only at the beginning and at the end, with none in the middle, and yet one crosses it just as surely and safely as if the whole of it were there? That sort of operation makes me feel a bit giddy, as if it led part of the way God knows where. But what I really feel is uncanny is the force that lies in a problem like that, which keeps such a firm hold of you that in the end you land safely on the other side." [13; p.106]

After some further discussions of common facts, they return to the main issue: mathematics. And Toerless reveals his doubts about the nature of mathematics.

"Why shouldn't it be impossible to explain? I'm inclined to think it's quite likely that in this case the inventors of mathematics have tripped over their own feet. Why, after all, shouldn't something that lies beyond the limits of our intellect have played a little joke on the intellect? But I'm not going to rack my brains about it: these things never get anyone anywhere."[13; p. 110]

Toerless is encouraged to have a conversation with the teacher of mathematics, and that same day he asks for an interview to discuss some points in the last lesson. The teacher is described as a fair, nervous young man of no more than thirty.

The arrival of Toerless to the teacher's apartment is remarkable. Even more is his first contact with him. After a long silence, and after being invited by the teacher to explain his doubts, Toerless tries to make his questions. The reaction of the teacher, consistently called "master" by Musil, is a mix of lack of any insight into the students psychology and at the same a show of arrogance mixed with insecurity. As Musil has said before, he was "quite a sound mathematician, who had already submitted several important papers to the academy." [13; p. 105]. Through many explanations

and doing his best to expose his doubts to a teacher who was listening or not, Toerless gets his reaction.

The master smiled, now and then gave a little fidgety cough, ... And finally he cut Toerless short. "I am delighted, my dear Toerles, yes, I am indeed delighted," he said, interrupting him, "your qualms are indications of a seriousness and a readiness to think for yourself, of a ...h'm...but it is not at all easy to give you the explanation you want...Now, you must not misunderstand what I am going to say.

"It is like this, you see—you have been speaking of the intervention of transcendent, h'm, yes—of what are called *transcendent* factors....

"Now, of course I don't know what you feel about this. It's always a very delicate matter dealing with the suprasensual and all this lies beyond the strict limits of reason. I am not really qualified to intervene there in any way. It doesn't come into my field. One may hold this view or that, and I should greatly wish to avoid entering into any sort of controversy with anyone...But as regards mathematics," and he stressed the word "mathematics" as though he were slamming some fateful door once and all, "yes, as regards mathematics, we can be quite definite that here the relationships also work out in a natural and purely mathematical way. Only I should really—in order to be strictly scientific—I should really have to begin by posing certain preliminary hypothesis that you would scarcely grasp, at your stage. And apart from that, we do not have the time.

"You know, I am quite prepared to admit that, for instance, these imaginary numbers, these quantities that have no existence whatsoever, ha-ha, are not easy for a young student to crack. You must accept the fact that such mathematical concepts are nothing more or less than concepts inherent in the nature of purely mathematical thought. You must bear in mind that to anyone at the elementary stage at which you still are it is very difficult to give the right explanation of many things that have to be touched upon.

"Fortunately, very few boys at your age feel this, but if one does really come along, as you have today—and of course, as I said before, I am delighted—really all one can say is: My dear young friend, you must simply take it on trust. Some day, when you know ten times as much mathematics as you do today, you will understand—but for the present: believe!

"There is nothing else for it, my dear Toerless. Mathematics is a world in itself and one has to have lived in it for quite a while in order to feel all that essentially pertains to it." [13; p. 111]

When preparing to leave the apartment, obviously disappointed with the interview, Toerless notices on a little table a volume of Kant, lying there for the sake of appearances. The teacher takes up the volume and holds it out to Toerless.

"You see this book. Here is philosophy. It treats of the grounds determining our actions. And if you could fathom this, if you could feel your way into the depths of such principles, which are inherent in the nature of thought and do in fact determine everything; they themselves cannot be understood immediately as with mathematics. And nevertheless we continually act on these principles. There you have the proof of how important these things are. But," he said, smiling as he saw Toerless actually opening the book and turning the pages, "that is something you may leave on one side for the present. I only wanted to give you an example which you may remember some day, later on. For the present I think it would still be a little beyond you." [13; pp. 112].

POOR TOERLESS! The book proceeds with many other interesting remarks on mathematics. We must insist that this comes from an author who became a mathematician and that the book is recognized as autobiographical by literary critics. Clearly Musil's criticism is directed to a certain kind of alienating mathematics and alienated mathematicians. This is evident in the first part of his masterpiece, *Der Mann ohne Eiqenschaften,* whose first volume was published in Berlin, 1930, Vol. II in Berlin, in 1933, and Vol. III was posthumously edited by Martha von Musil from unfinished notes in Lausanne, 1943. We have referred to this above when mentioning the main personage, Ulrich. Also of an autobiographical character, and we might even say biographical of an entire generation, the several characters of the book search for order in the chaos which characterizes the transition from the 19th to 20th centuries. Being provided with powerful tools for the knowledge of the self and of nature, capable of producing at paces hitherto unthinkable, mankind sees the breaking down of culture in competing ideologies and the incoherence of not only these ideologies and society as a whole, but of knowledge itself, and the resulting voiding of the self.

The *Man Without Qualities* is about Ulrich, in the sense that his qualities were so many that one could not distinguish one from another. Professionally he is a mathematician and the book is full of references by some of the characters on mathematics and mathematicians, always related to the socio-political atmosphere which permeates the entire novel.

There are numerous other examples. I limited references to the few contemporary writers above to keep this paper within a reasonable length. I was tempted to mention also important analysis of earlier authors such as Jonathan Swift, William Blake, Lewis Carrol, Bertrand Russel as a novelist, and the rich ground offered by the arts in general.

Somewhat related are references to mnemonics, drawing on literature (mainly popular literature), on music and on folklore. Of course, memory plays an important role in cognitive development, in the building-up of knowledge, and also in schooling. Everyone is aware of

---

Being provided with powerful tools for the knowledge of the self and of nature, capable of producing at paces hitherto unthinkable, mankind sees the breaking down of culture in competing ideologies and the incoherence of not only these ideologies and society as a whole, but of knowledge itself, and the resulting voiding of the self.

---

the exaggerations of rote learning in the teaching of Mathematics. We will not go through this here,. But I have to mention the important and rather complete Ph.D. thesis of Robert Hrees (Indiana University, Bloomington, 1986) where, in 14 volumes, nearly 4,000 pages, the author analyzes, throughout history, mnemonics in school mathematics. This thesis surveys world literature, both popular and cultured, as related to school mathematics. We insist on the fact that rote learning goes against our proposals, although we have to recognize that the development of memory plays an essential role in the generation of thought and cognitive development.

Conclusions and Suggestions for Classroom.

The main conclusions of this paper have already been mentioned in its introduction. We can make mathematics education, in all levels, including the training of future mathematicians, more vigorous, more attractive, less frustrating and less damaging, i.e., healthier, if we bring into it the overall views of society on mathematics. These are the signs of the times.

We should not forget the press, a most current and important form of literature. In papers and magazines, and, of course, radio and TV, we find much that is relevant to our social, political and cultural life. Many of these materials are loaded with Mathematics *alive*, in the form in which it is accepted, known and learned, and, above all, found interesting. We have to mention again the importance of popular or folk literature in this respect. The critical study, both in classroom practice and in teacher training, of mnemonics has very important payoffs in recognizing the views of society as a whole on mathematics. Mathematical appreciation and consequently mathematical understanding are enhanced. It is difficult to understand and above all learn what is not appreciated.

Some Mathematics texts are fortunately directed to this. The recent book by Marilyn Fankenstein on *Relearning Maths,*[20] draws on this. I must also mention a recent dissertation by Sergio Roberto Nobre (UNESP, Rio Claro, Brasil, 1989) with the same focus. A forthcoming book by Florence Fasanelli, V. Frederick Rickey, Richard Thorington and Caroline Thorington on *Mathematics on the Mall* is based on the Mathematics behind the Mall in Washington, DC. And, of course, we can not leave aside the enormous methodological potential for Mathematical Education of projects such as FOXFIRE, active under the direction of Eliot Willington since the late 60's.

Of course, this implies that a good part of our teaching should be devoted to talk *about mathematics* instead of talking mathematics. For example, some lectures devoted to listening and talking about Tom Lehrer's songs might be very attractive in all levels, but especially in teacher education. Mnemonics is especially useful in elementary education as a motivational element. Of course, resist the temptation of "back to basics"! Beware of testing and rote learning: they are twins.

In my teacher training classes I include as mandatory readings, with comments, pieces of literature and art in general. Students are supposed to write essays on literary works such as "Young Toerless", "Madame Bovary" and others, go to movies like "Kasper Hauser", and go to art galleries, concerts, theater, sports and several other "non-mathematical" activities, always reporting on them with emphasis on what this has to do with mathematics, on their own professional experiences and on professional experiences of relatives and friends. In every classroom we find expertise, at least amateurish expertise, in several different areas of human activity. To draw on these expertises has been for me a most successful methodology. This is true also at the secondary and elementary levels and has proven to be a successful methodology. These are the basic ideas which helped us in building up Ethnomathematics as a research and pedagogical program.

We have noticed that with such practices students like what they do, they are not merely "observers" of the process; they indeed are doing research, are drawing conclusions and have an opportunity to be creative. Summing up, they are intellectually alive—and, above all, they are treated as such!

Everyday life, life in general, is our best partner in Mathematics Education. Regrettably, many of our colleagues refuse this partnership.

> "It seems to me that the basic intuitional assumptions of any group of people must be sought among the things they take so much for granted that they are unaware of them."
> W.M. Ivins, Jr. [14; p.x]

Refererences

1. Ubiratan D'Ambrosio: Successes and failures of Mathematics curricula in the past two decades: A developing society viewpoint in a holistic framework, *Proceedings of the Fourth International Congress on Mathematical Education (Berkeley 1980)*, ed. Marilyn Zweng e et al., Birkhauser, Boston, 1983, pp. 362-364.

2. Ubiratan D'Ambrosio: Mathematics Education in a Cultural Setting, *International Journal of Mathematics Education in Science and Technology*, v.16, n.4, 1985, pp. 469-477.

3. Frank Smith: Overselling Literacy, *Phi Delta KAPPAN*, vol. 70, n. 5, January 1989, pp. 352-359.

4. Ubiratan D'Ambrosio: On Ethnomathematics, *Philosophia Mathematica*, Second Series, vol. 4, n. 1, 1989, pp. 3-14.

5. Ubiratan D'Ambrosio: *Etnomatematica. Arte ou tecnica de explicar e conhecer.*, Editora Atica, São Paulo, 1990.

6. Michel Serres: Literature and the Exact Sciences, *Substance*, n. 59, 1989, pp. 3- 34.

7. John Wilson: *Language and the Pursuit of Truth,* Cambridge University Press, Cambridge, 1979 (orig. edn. 1956)

8. Ernst Mach: On the Classics and the Sciences, in *Popular Scientific Lectures,* Open Court Pub. Co., La Salle, 1986 (orig. edn. 1893).

9. Gustave Flaubert: *Bouvard et Pecuchet with the Dictionary of Received Ideas,* Penguin Books, London, 1987.

10. Isidore Ducasse Comte de Lautreamont: Les chants de Maldoror. Chant deuxieme, in *Deuvress completes,* Flammarion, Paris, 1969 (orig.edn. 1869).

11. David S. Luft: *Robert Musil and the Crisis of European Culture 1880-1942,* University of California Press, Berkeley, 1984.

12. Robert Musil: *The Man Without Qualities,* Vol. I, translation and foreword by Eithne Wilkins & Ernst Kaiser, The Putnam Pub. Co., New York, 1980.

13. Robert Musil: *Young Toerless,* translation by Eithne Wilkins & Ernst Kaiser, Pantheon Books, New York, 1955.

14. Willian M. Ivins, Jr.: *Art & Geometry. A Study in Space Intuitions,* Dover Publications, Inc., New York, 1964.

15. Basil Bernstein: *Class, Codes and Controls,* vol. 1, Routledge and Kegan Paul, London, 1971.

16. Alexander Grothendieck: La Nouvelle Eglise Universelle, *Pouquoi la Mathematique?,* Robert Jaulin ed., 10I18, Union Generale d'Editions, Paris, 1974, pp. 11-25.

17. Pierre Samuel: Mathematiques, Latin et Selection des Elites, *Pourquoi la Mathematique?,* Robert Jaulin ed., 10I18, Union Generale d'Editions, Paris, 1974, pp. 147-171.

18. Leo Tolstoy: *War and Peace,* ed. George Gibian, W.W. Norton and Company, Inc., New York, 1966, pp. 1348-1351.

19. Jerome Kagan: *The Nature of the Child,* Basic Books Inc., New York, 1984.

20. Marilyn Frankenstein: *Relearning Mathematics,* Free Association Books, London, 1989.

Mathematics; an Integral Part of Our Culture

Harald M. Ness, Jr.
University of Wisconsin Centers—Fond du Lac
Fond du Lac, Wisconsin

"Mathematics was born and nurtured in a cultural environment. Without the perspective which the cultural background affords, a proper appreciation of the context and state of present-day mathematics is hardly possible."

Raymond L. Wilder

"Science is not a mechanism but a human progress, and not a set of findings but a search for them." Thus said Jacob Bronowski in *Science and Human Values*. The theme that ran throughout that book, the main concept that Bronowski was trying to communicate, was that "science is as integral a part of the culture of our age as

---

It is even less widely known that mathematics has determined the direction and content of philosophical thought, has destroyed and rebuilt religious doctrine...

---

the arts are". The same could be said of mathematics in particular, which Bronowski never separated from science. Unfortunately, this view of mathematics as an integral part of our culture, and as a significant force in the development of that culture, is not the view held by the general public, the well educated people outside of mathematics, and, we must admit, even some mathematicians. This was not always the case. Probably the most elegant statement of the problem was made by Morris Kline [8]: "after an unbroken tradition of many centuries, mathematics has ceased to be generally considered as an integral part of culture in our era of mass education. Almost everyone knows that mathematics serves the very practical purpose of dictating engineering design. Fewer people seem to be aware that mathematics carries the main burden of scientific reasoning and is the core of the major theories of physical science. It is even less widely known that mathematics has determined the direction and content of philosophical thought, has destroyed and rebuilt religious doctrine, has supplied substance to economics and political theories, has fashioned major painting,

musical, architectural, and literary styles, has fathered our logic, and has furnished the best answers we have to fundamental questions about the nature of man and his universe. Despite these by no means modest contributions to our life and thought, educated people almost universally reject mathematics as an intellectual interest."

C. P. Snow, who was a scientist by training and a writer by vocation, expressed in *The Two Cultures* great concern that his scientific colleagues and his literary colleagues, although "comparable in intelligence" and "not grossly different in social origin", had almost ceased to communicate with each other at all. He saw the intellectual life of Western society increasingly being split into two polar groups, and mentioned literary intellectuals "who incidentally while no one was looking took to referring to themselves as 'intellectuals' as if there were no others." Referring to the literary intellectuals' lack of understanding of the "other" culture, he wrote: "They still like to pretend that the traditional culture is the whole of 'culture', as though the natural order didn't exist. As though the exploration of the natural order was of no interest either in its own value or its consequence. As though the scientific edifice of the physical world was not, in its intellectual depth, complexity and articulation, the most beautiful and wonderful collective work of the mind of man. Yet most non-scientists have no conception of that edifice at all." Four years after he wrote about the two cultures, in what he called a "second look", Snow realized that his idea was not original or new. Many people had been thinking about this, and he mentioned Jacob Bronowski, Merle Kling, and A. D. C. Peterson.

Raymond Wilder [12] discusses mathematics as an integral part of our culture to a great extent. Mathematics is, as indicated in Kline's statement, a significant force in the development of our culture, and Wilder conversely describes the effect of the culture on the development of mathematics. Wilder categorizes mathematical development as being motivated either by "environmental stress" or "hereditary stress". Environmental stress, which I prefer to call cultural stress, provides motivation for the development of mathematics from that part of the culture external to mathematics. Take, for example, the development of numeration systems to represent natural numbers for purposes of commerce, the development of trigonometry to meet the needs of the astronomer, and the development of differential calculus to meet the needs of the astronomers and physicists to communicate and study quantities which changed with time. Hereditary stress, which could very well be called intellectual curiosity of the mathematician, provides motivation from within mathematics itself. Much of mathematics developed from intellectual curiosity, without any practical application in mind, as, for example, the so-called "imaginary" numbers, which came into being in order to make sense out of the process of finding roots of cubic equations. The name indicates that it was understood that these numbers had no real meaning nor practical use whatsoever. The extension of the concept of imaginary numbers in William R. Hamilton's invention of quaternions to satisfy an intellectual need is described by Hamilton himself [7], "...because I felt a problem to have been at

---

New mathematical ideas, often developed with no thought of application, have repeatedly turned out to have immense scientific, technological, and economic benefits.

---

that moment solved, an intellectual want relieved, which had haunted me for fifteen years before." The "intellectual want" was a method for multiplying vectors. Other well-known examples are the non-Euclidean geometries which evolved out of the frustration of mathematicians in their inability to prove the parallel postulate of Euclid and were developed by Gauss, Bolyai, Lobachevski, and Riemann to satisfy the curiosity as to what would be the consequences of replacing it with a contrary statement. Incidentally, the culturalogical basis for the independent discoveries of mathematical concepts such as this (the geometry of Gauss, Bolyai, and Lobachevski) is well described by Wilder [12].

This characteristic of mathematicians that compels them to follow their intellectual curiosity regardless of the usefulness of the outcome and the fact that the "useless" outcomes may become useful, sometimes centuries

later, is well known to mathematicians, and should be communicated to students and the general public. Edward E. David, Jr. stated [5], "New mathematical ideas, often developed with no thought of application, have repeatedly turned out to have immense scientific, technological, and economic benefits." He refers to G. Harold Hardy's statement, "I have never done anything useful. No discovery of mine has made, or is likely to make, directly or indirectly, for good or ill, the least difference to the amenity of the world." Bronowski [4] refers to this activity as play: "Abstract thought is the neoteny of the intellect, by which man is able to carry out activities which have no immediate goal (other animals play only while young) in order to prepare himself for long term strategies and plans." The fact that mathematical concepts developed to satisfy intellectual curiosity sometimes become useful can be illustrated by Arthur M. Jaffe's response to Hardy's statement relative to number theory, which was Hardy's specialty. David further states, "Forty five years later (after Hardy's statement), as Jaffe points out, number theory is at the heart of issues of national security: it has become the basis of several new schemes for constructing secret codes." Also, the previously mentioned non-Euclidean geometry and the algebra that evolved from the discovery of quaternions helped make possible the development of Einstein's relativity theory and Heisenberg's wave mechanics.

The role that mathematics plays in the culture is not only important but dynamic. This ever changing role of mathematics is nicely illustrated in Jacob Bronowski's *Music of the Spheres*, one of the films in his *Ascent of Man* series. Bronowski discusses the development of mathematics and culturally related aspects from the Pythagoreans where Bronowski says *mathematics* began with the use of the power of reasoning to discover and communicate relationships between numbers. He then traces it through the great change from a static to a dynamic description of nature with the invention of the calculus. The relationship between music and mathematics was discovered by the Pythagoreans (6th century, B.C.). In ancient times, music was considered a part of mathematics. Mathematics is the music of the mind; music is the mathematics of the soul. Bronowski traces the flow of culture from the Greek domination to the Islamic domination, where, he says, the mathematician and the artist were one (and says he means that literally). The Islamic period art was very mathematical and made extensive use of symmetries in its patterns. Another concept that relates mathematics to art is perspective (Albrecht Durer, 15th century), Which was the foundation for projective geometry. Also, the grid that Durer and other artists used as an aid became the graph paper that is an important tool of the mathematician.

Bronowski traces the cultural development from Islam

via Spain into Europe and the development of the calculus as a response to the needs of the astronomer and physicist. We see here the changing role of mathematics—commerce, astronomy (and astrology), engineering, music, art, and physics. Certainly, the relativity theory of Einstein and the quantum theory of Heisenberg and Schrodinger are mathematical constructs. More recently, we see the mathematics of Morgenstern and von Neumann laying the foundation for modern economic theory. We also see more and more use of mathematical models for communication and insight into the concepts of biology. As Edward E. David said [5], "Mathematical modeling of natural phenomena is hardly new. Nevertheless, advances in numerical analysis and the development of the computer have made it possible in ways that are much more complex and more realistic than ever before. Mathematical modeling in partnership with the computer is rapidly becoming a third element of the scientific method, coequal with the more traditional elements of theory and experiment." In many intellectual endeavors, the extent of mathematization of a discipline is a gauge of the maturity of that discipline.

Unless we want mathematics to continue to be viewed as something distinct and separate from the mainstream of culture and consisting of a bag of clever tricks or skills, we must change the way we relate to the general public and the way we teach mathematics. George A. W. Boehm stated, [2] "Popular knowledge of mathematics is now appallingly scanty. Newspapers and magazines conscientiously report the latest developments in medicine, chemistry, physics, biology, and technology. Yet few publications ever discuss mathematics. Late in 1956, for example, Chen Ning Yang and Tsung Dao Lee demolished the principle of parity. The press did a commendable job of describing the abstruse physical ideas and implications of Yang's and Lee's work. Significantly, however, nothing was said about the mathematics of group theory on which the physicists based their conclusions. The reason is that there is no established popular terminology for talking about mathematics. Where physicists, with profound reservations, talk about atoms and subatomic particles as if they were tiny marbles, mathematicians talk only in mathematical terms. The public cannot understand and it does not listen." Our dilemma is that we are well aware that speaking or writing with a popularizing language destroys the precise, unambiguous nature for which the mathematical language was devised. What do we do, keep the public unaware, or make them aware in the imprecise and ambiguous way that they seem to demand? Lewis Wolpert, a biology professor, stated in a recent letter in *The Chronicle of Higher Education*, "It (mathematics) is, perhaps the purest and most rigorous of intellectual activities, and is often thought of as queen of the sciences. To the outsider, however, it seems like a

private game, the manipulation of symbols to uncertain and unworldly ends." Although we all agree with Professor Wolpert (except perhaps that the "perhaps" should be eliminated), we might wonder where the outsiders referred to get that idea. I maintain that this is a reflection of the way we relate to the public and, even more so, the way we teach our subject. John Lucas, in his article, "Heuristic Thinking and Mathematics", lists common myths about mathematics. (1) Mathematics is a collection of isolated facts and tricky techniques that must be memorized. (2) Mathematical truth is absolute. (3) Mathematics is an exact science. (4) Mathematics deals primarily with symbolic representation and manipulation, so ordinary writing and speaking skills are not necessary for communicating mathematics. (5) Mathematics is something one does alone. A student, writing in one of the news magazines, stressed that those in the humanities deal with ideas, while those in the sciences deal merely with

---

In ancient times, music was considered a part of mathematics. Mathematics is the music of the mind; music is the mathematics of the soul.

---

skills. He was a college student, completely unaware that many of the most important and powerful ideas in our culture are mathematical. If we want to eliminate the misunderstanding of what mathematics is, we must change the way we teach mathematics. We must teach from a culturalogical perspective. We must view mathematics as an integral part of our culture and a significant force in our culture, and we must communicate this to our students.

What, then, is culture? I think Samuel Taylor Coleridge said it best when he described culture as the "qualities and faculties which characterize our humanity". Certainly mathematics, with its language, modes of thought, and techniques of analysis and problem solving is an important part of what Coleridge described. What do we want to communicate to our students that mathematics is? The Lawrence University catalogue describes mathematics this way: "Born of man's primitive urge to seek order in his world, mathematics is an ever-evolving language for the study of structure and pattern. Grounded in and renewed by physical reality, mathematics rises through sheer intellectual curiosity to levels of abstraction and generality where unexpected, beautiful, and often extremely useful connections and patterns emerge. Mathematics is the natural home of both abstract thought and the laws of nature. It is at once pure logic and creative art."

While we must communicate the true nature of mathematics in all our classes, it is of prime importance in our math courses for liberal arts students and education majors. I was told a story of a visit by John von Neumann to a well-known university where he lectured about his work to the mathematics students. He was scheduled to speak in the evening to the general

---

Mathematics is the natural home of both abstract thought and the laws of nature. It is at once pure logic and creative art.

---

public. One of the students, wondering whether he should attend, asked whether he was going to say anything important. "Young man", von Neumann replied, "What I am going to say this evening is far more important than anything I've said this afternoon." What we say to the general public and what we say to our liberal arts students is very important; it is likely to influence greatly their perception of what mathematics is and what its place is in our culture.

References:
1.   Albers, Donald J. & Alexanderson, G. L., eds.; *Mathematical People*; Birkhauser; 1985.

2.   Boehm, George A.W.; *The New World of Math*; Dial Press; 1958.

3.   Bronowski, Jacob; *The Ascent of Man*; Little Brown & Co.; 1973.

4.   Bronowski, Jacob; *Science and Human Values*, Revised Edition; Harper & Row; 1965.

5.   David, Edward E., Jr.; "The Federal Support of Mathematics"; *Scientific American*; May, 1985, V252, #5.

6.   Kenji, Qugimoto; *Albert Einstein; A Photographic Biography*; Schocken Books; 1989.

7.   Kline, Morris; *Mathematical Thought from Ancient to Modern Times*; Oxford Univ. Press, 1972.

8.   Kline, Morris; *Mathematics in Western Culture*; Oxford Univ. Press; 1953.

9.   Lucas, John F.; *Heuristic Thinking and Mathematics*; Contributed Paper; AMS-MAA Mtg. Phoenix, AZ; Jan. 1989.

10.   Newman, James R.; *The World of Mathematics*; Simon & Schuster; 1956.

11.   Snow, C. P.; *The Two Cultures: And a Second Look*; Cambridge Univ. Press; 1965.

12.   Wilder, Raymond L; *Evolution of Mathematical Concepts: An Elementary Study*; John Wiley & Sons; 1968.

Rational and Irrational: Music and Mathematics

Robert Osserman
Stanford University
Palo Alto, California

"...some specifically human tendency to create and notice organized patterns, hierarchies, and sequences"
John A. Sloboda, *The Musical Mind*

"...I still feel that mathematics, more than any other discipline, studies the fundamental, pervasive patterns of the universe....To me, the deepest and most mysterious of all patterns is music."
Douglas Hofstadter, Introduction to *Metamagical Themas*

The name of Pythagoras is most frequently associated with the Pythagorean Theorem, perhaps the best known and most often cited result in all of mathematics. But the appellation "Pythagorean" occurs in another context, of probably far greater importance for science as a whole: the Pythagorean theory of music. The basic tenet of the theory is that music has profound underpinnings in mathematics. The reason for the powerful impact of an apparently narrowly-focused theory, is that it was viewed as merely an illustration and confirmation of a broad doctrine, holding that mathematical relationships are at the heart of the physical world. That doctrine, in the hands of Kepler, Einstein, and many others, has led to major advances in our understanding of the universe.

Our goal here is to describe the Pythagorean theory, but with a kind of reverse twist: not to explain music in terms of mathematics, but, rather, to use music to motivate and clarify some elusive mathematical concepts. One might say that the goal is to prove that there really is rhythm in a logarithm.

Long before Pythagoras, it must have been observed that, if a string is held taut and either struck or plucked, it will produce a musical note: the shorter the string (assuming that it is maintained at the same tension), the higher the note. Furthermore, if two strings are struck simultaneously, the result will sometimes be perceived as discordant or dissonant, and sometimes as concordant or harmonious.

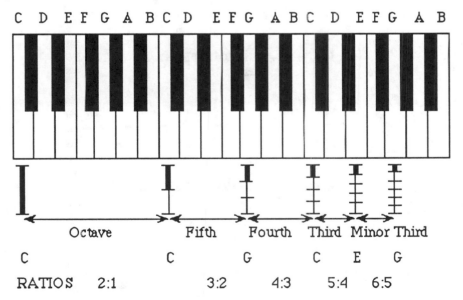

Figure 1.

What the Pythagorean theory[1] states is that, provided the strings are both of the same material and at equal tension, they will sound harmonious precisely when the lengths of the strings form a simple mathematical ratio such as 2:1, 3:2, 4:3—the more complicated the ratio, the more dissonant the sound. The simplest of these, the ratio 2:1, produces sounds that are so concordant that we often describe them as "the same note, an octave apart". What we hear in each case is a pair of notes that we perceive as being a certain "distance" apart. That distance is referred to as a *musical interval*. The ratio 3:2 corresponds to the interval called a *fifth*, the ratio 4:3 to a *fourth*. The terminology derives simply from the position of the corresponding notes on the scale. If we start the scale on the note corresponding to the longer of the two strings, then the shorter one will be the fifth note up on the scale if their ratio is 3:2. The word *octave* is simply Latin for "eighth", the position on the scale corresponding to the ratio 2:1.

In figure 1 we show how this process works when referred to a standard piano keyboard. The vertical line segment at the left represents a string length corresponding to the note "C" directly above it. As it is successively divided into 2, 3, 4, 5, and 6 equal parts, the resulting string lengths—the darkened part of each interval—correspond to the notes directly above them on the scale. The ratios of successive string lengths and the corresponding musical intervals are indicated below. For example, dividing the left-hand segment into two equal parts, we obtain a string half as long which will produce a note again designated "C", but an octave

above the original. Dividing the original string into three equal parts, we obtain a string of a third the original length. Compared to the previous one—half the original length—we obtain two strings in the ratio 3:2, producing the musical interval "a fifth", with the upper note designated by "G". Proceeding in the same fashion, we obtain the series of notes shown in Figure 1.

We now ask the following fundamental question: "How is the 'addition' of intervals expressed in the language of ratios?" The answer turns out to be as simple as it is fundamental: *adding musical intervals* translates to *multiplying the ratios*. Referring back to Figure 1, we see, for example, that if we wish to add a fifth and a fourth, we must first take a string 2/3 the length of the original one, and then a string 3/4 the length of the second one. The resulting string will be 2/3 times 3/4 the original length, or half as long, and the corresponding interval is an octave. Musically, we have first moved up from C to G, and then up from G to the following C, a total of one octave. (See Figures 1 and 2.) The same process is completely general: moving from one note to a higher one through a certain interval means going from one string to a second one whose length is a certain fraction of the original length. Moving up from there through another interval, we arrive at a third string whose length is some fraction of the length of the second string. To find the relation of the third string to the first, we must take a *fraction of a fraction*; that is, we must *multiply* the corresponding fractions.

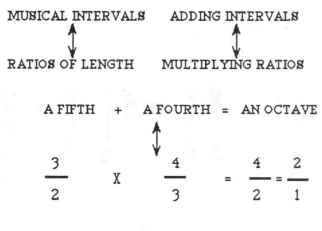

Figure 2.

To summarize, each simple ratio of string lengths corresponds to a musical interval, and *multiplying* two ratios corresponds to *adding* the musical intervals.

For someone with a mathematical background, such a description is bound to strike a responsive chord: the conversion of multiplication to addition is precisely the basic property of *logarithms*. Thus, the fact of logarithms is built into the physiology of sound perception, even though it was not until 1614 that the idea of logarithms was extracted and explicitly formulated by Napier.

One point that tends to obscure the relationship between numerical ratios and musical intervals is that there are two peculiarities in the *naming* of musical intervals. The first, and more superficial, is that an interval is named after the position on the scale of the upper of the two notes, assuming that the lower one is the first note on the scale. It would be much more natural to name an interval according to the space between the two notes, or the *difference* in position: that is, the number of steps on the scale needed to go from the first note to the second. For example, what we have called a "fifth", the interval from C to G, should rightly be assigned the number *four*, since there are four steps in between them: C to D, D to E, E to F, and F to G. Continuing from G up to the C above requires three more steps (see Figure 1), so that we should assign the number *three* to the interval we have called a "fourth". Then, adding the two musical intervals would correspond to simple numerical addition: 3 + 4 = 7; it takes *seven* steps to go from C to the C above, an interval of an "octave". Renaming all musical intervals in this fashion might be a worthwhile reform, but it is unlikely to happen; the names of musical intervals are as firmly entrenched as the many bizarre spellings in our language, and appear equally resistant to rationalization.

The second, and much deeper source of confusion, is that in naming musical intervals according to the position of notes on a scale, we are implicitly assuming that we already have an established scale. But upon reflection we realize that the intervals were there prior to, and independent of, any particular scale. In fact, the true sequence of events, both logically and historically, is the following:

    1. discovery of *musical intervals*: preferred pairs of notes,

    2. adoption of a certain set of related notes, called a *scale*,

    3. *naming* the original musical intervals according to the way the notes lie on the scale.

We are thus led to one of the most fundamental questions in music. How do we go about choosing, out of all the (infinite) possibilities for musical notes, the small number that will make up our "scale"?

The basic idea is simplicity itself. Out of all possible string lengths, the ones we wish to choose are those that sound most harmonious together. Following Pythagoras (or Ling Lun), we may start with some arbitrary length, and then choose the others, using the simplest possible ratios. To begin with, we may restrict ourselves to the ratios 2:1 and 3:2. If our original string length corresponds to the note "C" on the piano, then using the ratio 3:2 will bring us up a "fifth" to the note designated by G. Repeating the process, we start on "G" and find that the fifth note up (corresponding to the ratio 3:2 once more) is the note "D". (Check once more on figure 1.) We may now use our ratio 2:1 to move down an octave, and arrive at the note "D" just above our original "C".

Pythagoras tells us to continue in the same fashion, each time adding a new note to what is to become the privileged set of notes, called our "scale". If we carry out the process, we arrive successively at the notes C, G, D, A, E, B, then the five "black notes" lying between those with letters, then F, and then back to C. In other words, we obtain the *entire scale* on the piano, just by using the ratios 2:1 and 3:2. If you have a piano handy, try it out, starting at C near the bottom of the keyboard, and moving upwards in successive fifths. You will find that after twelve steps you will have struck each different note exactly once (where notes that differ by any number of octaves are counted as the same) and have come back to C, seven octaves above your starting point. We may express that fact as a musical equation:

    12 fifths = 7 octaves

or

    a fifth + a fifth + ... + a fifth (12 times)

    = an octave + ... + an octave (7 times).

Since adding intervals corresponds to multiplying ratios, that musical equation converts to the mathematical one:

    " $(3/2)^{12} = 2^7$ "

We have all at once arrived at a very suspicious-looking "equation", and, indeed, a quick check on our calculator reveals that

    $(3/2)^{12} = 129.746...$, while $2^7 = 128$.

So the two sides of our "equation" are close, but

definitely not equal. What has gone wrong?

Before we answer that question, let us make a brief digression, to examine more closely the offending "equation". We may ask whether there is some way to see why the two sides cannot be equal, other than by a mindless calculation. One way to approach the problem would be to try to eliminate the fractions. In fact,

$$(3/2)^{12} = 3^{12}/2^{12}$$

so that the "equation" takes the form

$$\text{``} 3^{12}/2^{12} = 2^7 \text{''}$$

and multiplying through by 2^{12} leads to

$$\text{``} 3^{12} = 2^7 \cdot 2^{12} = 2^{19} . \text{''}$$

Now each side is an integer, and the two sides are transparently not equal. One could give a variety of reasons, the simplest being that the left-hand side is odd, and the right is even. We therefore conclude that our original "equation" must also not be valid.

Is this more roundabout argument any better than a direct check with a calculator? That is clearly a matter of subjective judgment, but the abstract argument does yield more insight. In fact, the same argument holds much more broadly, and shows that *no* power of 3/2 can ever equal any power of 2. Converting back to music, we conclude that no successive number of "pure fifths" (corresponding to the ratio 3:2) can ever equal any exact number of octaves. That simple fact has profound musical implications, and it leads us back to the problem faced by Pythagoras: what can be done about the frustrating fact that twelve fifths is *almost* but not quite, equal to seven octaves? The answer is the highly unsatisfying but inevitable one: you have to stretch things a bit here and there, and hope that no one will notice. Over the centuries, many modifications of the Pythagorean approach have been proposed and hotly debated. One simple one was suggested by the famous Greek astronomer Ptolemy, who also devoted considerable energy to musical theory. He noted that the interval of a major third (from C to E) would correspond in the strict Pythagorean theory to the ratio
$$(3/2)^4/2^2 = 81/64$$

obtained by going up four successive fifths and down two octaves, but that a far simpler (and hence more harmonious) ratio would be

$$80/64 = 5/4,$$

which is the one we have used in figure 1. Ptolemy proposed in general using the simplest ratios that approximated the notes obtained from the Pythagorean scheme. The advantage of Ptolemy's approach was that it produced simpler ratios, and presumably, as a consequence, more pleasing sounds. The disadvantage was that what formerly were equal intervals, such as from C to D, and from D to E, were now different, corresponding to the ratios 10/9 and 9/8, respectively.

And so the debate continued from generation to generation, with new proposals and compromises. A leading figure in the sixteenth century was Vincenzo Galilei, who suggested a method for placing frets on a lute to divide the octave into twelve equal intervals. Those intervals are called *semitones* and correspond to the interval between adjacent white and black notes on the piano. Vincenzo's method amounted to a choice of the ratio 18/17 for each semitone, and his claim that twelve of those intervals made up an octave translates to the equation

$$(18/17)^{12} = 2 .$$

But that cannot be exactly correct, for much the same reasons we gave in the earlier case (or again, by a direct computation, as was rapidly pointed out by his contemporaries). In any case, Vincenzo's main claim to immortality is as the father of Galileo Galilei, who, among his other accomplishments, became in turn the father of Physics, with the publication in the Spring of 1638 of his book *Two New Sciences*, a document very much in the Pythagorean tradition, using precise mathematical reasoning to analyze the working of the physical world.

As time went on, the debate began to focus more and more on the conflict between simple ratios and equal intervals, with the gradual realization that the two were incompatible. Finally, in a book called *Harmonie Universelle*, published in 1636 (just two years before Galileo's), the French mathematician Marin Mersenne spelled out the details of what has since become known as *equal temperament*: the division of an octave into twelve exactly equal intervals. From our modern standpoint the solution is almost trivial. If r denotes the ratio of strings needed to produce an interval which repeated twelve times gives an octave, then, since the ratio for an octave is 2:1, our basic correspondence principle tells us that necessarily

$$r^{12} = 2,$$

and therefore,

$$r = \sqrt[12]{2} .$$

Mersenne also gave an approximate value for r. To four decimal places, it is

$$r \sim 1.0595.$$

For historical completeness we should mention that

once again the West was anticipated by a Chinese scholar, Prince Chu Tsai-yu, who in 1596 explained the principles of equal temperament, and gave the correct value of r to nine decimal places!

With these discoveries the terms of the debate shifted. Now the argument was between *just intonation*, where intervals were determined by simple ratios of whole numbers, and equal temperament, where simple ratios were abandoned, but all intervals were equal in size. After Bach took up the banner of equal temperament, displaying its great versatility in his collection of preludes and fugues called *The Well-Tempered Clavichord*, there has been virtually no contest, with almost universal acceptance (at least in the West) of the equal-tempered scale. However, there have always been, and continue to be holdouts. To understand the reasons for the objections, consider just one interval on the well-tempered scale—the one that divides the octave precisely in half. Since there are twelve semi-tones to the octave, the interval in question would consist of six semi-tones. Starting at C and moving up six semi-tones, we arrive at the black note between the F and G. That note is called F-sharp, and written F$^{\#}$. The corresponding interval from C to F$^{\#}$ is called an *augmented fourth*. (See Figure 3.) The interval from F$^{\#}$ to the C above is again an augmented fourth, as we may check by confirming that it consists also of six semi-tones.

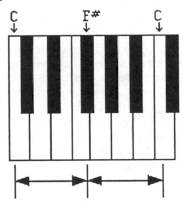

Figure 3

We now ask the question: What is the ratio of string lengths needed to produce the musical interval of an augmented fourth on the well-tempered scale? Since that interval added to itself gives an octave, we deduce immediately that the corresponding ratio, call it m/n, must satisfy the eqauation

$$(m/n)^2 = 2/1 .$$

But, as the Pythagoreans discovered, *this equation has no solution*: there are no whole numbers m and n satisfying it. In the terminology of the time, the strings needed to produce the interval of an augmented

fourth must be *incommensurable*. In modern notation, a number whose square is two would be written as $\sqrt{2}$, and we call it an *irrational* number, since it cannot be expressed as the ratio of two whole numbers.

How are we to interpret this situation? Looked at from the Pythagorean point of view, where the most harmonious intervals correspond to the simplest ratios, and more complicated ratios become more and more discordant, an interval corresponding to *no* ratio at all should be maximally discordant. Interestingly enough, this musical interval, also called the *tritone*, was considered terribly jarring, and referred to as *Diabolus in Musica* in the late Middle Ages.[2]

To be quite fair, we should not single out the tritone, since on the well-tempered scale, *all* the intervals correspond to irrational numbers, starting with the semi-tone: the twelfth root of two. The difference, presumably, is that the other intervals are quite close to simple ratios of whole numbers, whereas the augmented fourth is not. For example, a fourth consists of five semi-tones, and a fifth consists of seven. The corresponding numerical values are therefore

	"just" intonation	well-tempered
a fourth	4/3 ~ 1.3333	$(\sqrt[12]{2})^5$ ~ 1.3348
a fifth	3/2 = 1.5	$(\sqrt[12]{2})^7$ ~ 1.4983

both remarkably close to the exact ratios. Nevertheless, some musicians experience well-tempered intervals as jarring, and continue to press for just intonation. Some, like Harry Partch, build their own musical instruments, based on one or another modification of the Pythagorean ideal, and compose and perform music for those instruments. Incidentally, Partch's book, *Genesis of a Music*, is a wonderful source for history and commentary on the matters we have been discussing. His chapter on "The Language of Ratios" is particularly appropriate.

One last comment before turning to another aspect of our subject: one might ask, in view of the incommensurable lengths of the strings needed to produce, say, an augmented fourth on the well-tempered scale, how one could ever construct such a pair of strings. There are at least two answers. First, the distinction between rational and irrational is not physically meaningful. In practice, all lengths are approximate, and we can as easily approximate $\sqrt{2}$ as any other number. Second, there is a simple *geometric* construction of the desired ratio. Just stretch a string around a right isosceles triangle. Then by the Pythagorean theorem, the ratio of the hypotenuse to either of the sides is precisely $\sqrt{2}$, so that plucking the corresponding strings should produce a perfect tritone.

A striking feature of this whole story is the leading role played by mathematicians and scientists in uncovering and elaborating the deep bonds between mathematics and the musical notions of consonance and dissonance. There is another connection between mathematics and a different aspect of music that lies much nearer the surface, and that can be described much more briefly. That aspect is the use of rhythm, and more particularly of *polyrhythm*. By polyrhythm, we mean the practice of playing two lines of music in two different rhythms. For example, Brahms was especially fond of the combination "three against two", where one instrument plays three notes in the same time that another plays just two. Chopin frequently wrote pieces pitting four against three, and occasionally five against three.

Modern composers have ventured further and further in the same direction. One of them, Conlon Nancarrow, finally despaired of finding players who could handle the complexities of his rhythms and turned to composing for a pair of player pianos. Once freed from human limitations, he could make really daring leaps, culminating in his Study #33: "Canon—$\sqrt{2}/2$". It was his first one involving irrational tempo relations; each piano is moving along at a fixed speed, where the ratio of the two speeds is $\sqrt{2}/2$!

To understand better what that means, let us think of polyrhythms as being represented by a pair of metronomes, one of which we keep always at the same speed. The second one is then adjusted to give us the desired rhythm. For example, if we want a Brahmsian three against two, we simply set the second metronome to strike three times for each pair of beats of the first. Similarly, we can produce four against three, and so on. In each case, we start the two metronomes together, and we see that they repeatedly come back together after a certain number of beats. Note that the more complicated the ratios, the longer we have to wait before the two metronomes come back together. What happens when the tempos are related irrationally, as in Nancarrow's piece, is that, if they start together, then they will *never* come back together, no matter how long we wait.

The metronome analogy may be the most illuminating way there is to illustrate the difference between rational and irrational numbers. That the distinction should arise in both aspects of music—in harmony, comparing just intonation with equal temperament, and in rhythm, contrasting standard polyrhythms with Nancarrow's $\sqrt{2}/2$—may seem initially surprising. What is even more surprising is that both of these aspects turn out to be separate manifestations of one and the same

phenomenon. To explain how that comes about, we must return once more to the subject of consonance/dissonance.

For the ancient Pythagoreans, numbers possessed a semi-mystical significance. That numbers were revealed to underlie as distant-seeming a human activity as music greatly added to their sense of mystery and power. The Pythagoreans reveled in that sense of mystery, as did many scientists subsequently, most notably, perhaps, Johann Kepler. But other scientists, more in the Galilean tradition, wanted to find *physical* causes that would dispel the mystery, and reveal the reasons behind the role played by numbers. In fact, Galileo himself, in *Two New Sciences*, was one of the first to explain the physical origins of the mathematical ratios occurring in music. He pointed out that what really counts in determining what we perceive as the pitch of a note is the number of vibrations per second made by the string. If one divides a string in half, each half will vibrate twice as fast (assuming always that the tension remains the same) and we will hear the note we identify as an octave above the original. In general, all other factors being equal, the number of vibrations will be proportional to the reciprocal of the length. Thus a ratio of lengths 3:2 corresponds to the same ratio of frequencies of vibrations. Galileo uses this physical fact as the basis for a theory of consonance and dissonance. The musical interval of a fifth sounds consonant to our ear because the two series of vibrations in the air set up by the strings will strike our eardrums in a regular fashion, coming together on every third beat. The more complicated the ratios, the more irregular the effect on our eardrum, and the more dissonant our perception of the tones.

We now easily see the connection between musical intervals and polyrhythms. Two tuning forks tuned a fifth apart are nothing but a pair of metronomes beating three against two, speeded up by a factor of several hundred. Complicated rhythms representing ratios of large numbers, when speeded up produce more dissonant intervals, and Nancarrow's canon on $\sqrt{2}/2$ turns into a perfect augmented fourth.

As a final note, we should point out that the perception of consonance and dissonance changes from generation to generation and from culture to culture, and that we should not make the mistake of assuming that consonance is "good" or "beautiful" in music, and dissonance "bad" or undesirable. In fact, what we seem to respond to is a deliberate tension between the two. Perhaps it is appropriate to let Galileo have the last word. Writing in *Two New Sciences*, he points out that *too* much consonance strikes us as "too bland, and lacks fire." What we seek in our music is just the right amount of dissonance. "This produces a tickling and teasing of the cartilage of the eardrum, so that the sweetness is tempered by a sprinkling of sharpness, giving the impression of being simultaneously sweetly kissed, and bitten."

Acknowledgement: I would like to thank Nancy Hechinger and Tom Lehrer, who read an earlier version of this paper and made helpful suggestions that have been incorporated here, and also Brian Osserman for preparing the illustrations.

[1] I shall use the phrase "Pythagorean Theory" throughout, since that is the standard term in Western writing, although it cannot really be justified on historical grounds. Pythagoras himself was a shadowy and enigmatic figure who lived some two hundred years before Euclid. We have no clear evidence that he produced any written documents, and, in any case, none have come down to us. What he did have was a large following of "Pythagoreans" for many generations afterwards, and the general consensus of scholars today is that the "Pythagorean Theorem", as we know it from Euclid, is far more likely to have originated with them, perhaps a century after Pythagoras, than with Pythagoras himself. There is somewhat more reason to believe that Pythagoras did expound the "Pythagorean" theory of music, although it should be added that the first *written* record that we have occurs in Euclid: *Section of the Canon.* To be more precise, Euclid is the first written record in the *West*, but ancient Chinese historians document a much earlier development of the same theory, attributed to *Ling Lun*, perhaps two thousand years before Pythagoras.

[2] According to the Harvard Dictionary of Music, "It has always been considered a 'dangerous' interval, to be avoided or treated with great caution", while the Grove Dictionary states, "In 19th-century Romantic opera the tritone regularly portrays the ominous or evil."

PART III

Inner Life of Mathematics

Modernizing the Philosophy of Mathematics

Nicolas D. Goodman
State University of New York at Buffalo
Buffalo, New York

The traditional view of the nature of mathematics, which goes back at least to Plato and which is still current in mathematics departments today, is that mathematics is the purely rational study of immaterial forms. Mathematics, on this view, is concerned exclusively with objects like numbers, shapes, and functions which do not occur in the physical world, although they may have imperfect instances in the physical world. Thus the geometer studies straight lines and perfect circles, but the lines he draws on the blackboard are not straight and the circles he draws are not perfect. Since the objects the mathematician studies are not physical, there is no way for him to have empirical knowledge about those objects. Thus mathematics is an a priori discipline, independent of all experience. The Euclidean methodology, though rarely applied in practice, is still the ideal in principle. Mathematicians should deduce their theorems by logical inference from self-evident axioms. Any other source of mathematical knowledge may be heuristically useful, but is not strictly correct. As Plato says in the *Republic* (Book VI, 510), the objects that mathematicians study are "ideals which can be seen only by the mind" [11, p. 310].

On the traditional view, then, mathematical knowledge is certain but devoid of content about the physical world. Applied mathematics, in so far as it provides information about physical phenomena, is not genuinely mathematics. Pure mathematical research is a wonderful game and a high art. But it is not science.

Perhaps the most obvious interpretation of the traditional view of the nature of mathematics is that mathematics is about mental objects which are illuminated, so to speak, by the inner light. Straight lines and perfect circles, transcendental numbers and Hilbert spaces, are all ideas in the psychological sense. They have their existence in the mind of the mathematician. This view was quite popular in the nineteenth century, but was discredited at the end of the last century by the powerful arguments and biting polemic of the German logician Gottlob Frege. (See Frege [2].) If right triangles exist only in our minds, asks Frege, must we speak of my Pythagorean theorem about my right triangles and your Pythagorean theorem about your right triangles? The fact that we can unambiguously communicate about mathematics but not about purely mental phenomena like our emotions makes it implausible that mathematical objects are mental. The Dutch mathematician L. E. J. Brouwer held a mentalistic view as late as 1948. Nevertheless, the strong consensus among philosophers throughout this century has been that, if there are such things as mathematical objects, then they are not mental. Instead, they must be immaterial abstract objects existing somehow eternally outside of space and time. Mathematics consists of eternal truths about eternal objects. This was surely Plato's view. The mystery is how we could possibly come to know such truths.

An appeal to self-evidence or to mathematical intuition as the ground of our mathematical knowledge seems to beg the question. If we have no experience of mathematical objects, how is it that we have reliable intuitions about them? The standard solution to this problem in the twentieth century has been to claim that mathematics is somehow true by virtue of the meanings

> **If right triangles exist only in our minds, asks Frege, must we speak of my Pythagorean theorem about my right triangles and your Pythagorean theorem about your right triangles?**

of the words we use when we talk about mathematics. Mathematics is said to be true by convention, true by definition, or logically true. In the standard philosophical terminology, going back to Immanuel Kant, sentences which are true by virtue of the meaning of the terms involved are said to be *analytic*, whereas sentences which are true by virtue of the facts of the matter are said to be *synthetic*. In this terminology, then, the standard account of the foundations of mathematics throughout most of this century has been that mathematical theorems are analytic.

Since the Second World War, however, there has been a systematic attack, led by the philosopher W. V. Quine, on the distinction between analytic and synthetic truths. (For an eloquent account of the breakdown of the analytic-synthetic distinction and of the larger philosophical movement of which the attack on this distinction has been a central part, consult Rorty [10].) I will not rehearse the arguments which have led many philosophers to abandon the distinction. It should suffice to point out that there is no known account of the concept of meaning which is sufficiently clear to found a precise sense in which, say, Zermelo-Fraenkel set theory is analytic. Certainly set theory is not logically true and not, in any standard logical sense, true by definition.

So long as we remain within the empiricist tradition which has dominated Anglo-Saxon philosophy since the eighteenth century, however, it is not at all clear that the analytic-synthetic distinction can be given up while maintaining a distinction between a priori and a posteriori knowledge. That is to say, if we hold that all substantive knowledge about the world must be derived from experience, and if we cannot any longer make sense of the claim that mathematical knowledge is merely analytic, then we must conclude that mathematical knowledge is also, at least in part, empirical. (For this view, see Goodman [4] and Kitcher [7].)

Mathematicians now rely on extensive automatic computations, at least when the domain over which the computation takes place is discrete. An example is the Haken-Appel computer-assisted proof of the four-color theorem [1]. Even in continuous mathematics, where problems of round-off error become prominent, we find that mathematicians are increasingly concerned to justify reliance on computer-generated solutions, since often no other solutions are available. (For an interesting recent example, see [6].) We would normally hold that the facts revealed by automatic computation are part of mathematics. Yet, when we rely on the computer to tell us those facts, we are assuming that the computer works as advertised, and, therefore that the laws of physics work, in this instance,

---

Indeed, the increasing tendency of mathematicians to rely on computers... clearly undermines the a priori character of their discipline.

---

as the engineers who designed the machine expected them to. Indeed, the increasing tendency of mathematicians to rely on computers to do their computational dirty work clearly undermines the a priori character of their discipline. It is hard to believe that a

result which can be established only by hours of computer time is true by virtue of the meanings of the symbols used in its formulation. An analytic truth, after all, should be recognizable as such by merely thinking.

The traditional examples of analytic truths were statements like "Every bachelor is unmarried" in which, as Kant said, the meaning of the predicate is contained in the meaning of the subject. It is hard to see how mathematical truths could be taken to be analytic in that sense, since all non-trivial mathematical proofs involve the construction of something that is not contained in the meaning of the terms. Even a pencil-and-paper computation has that synthetic quality insofar as we do not foresee its outcome with confidence before actually carrying it out. We learn mathematics by doing computations and other, less routine, constructions and by being surprised by the results of those constructions. Surely this is naturally described as learning from experience.

Once we recognize that we have no clear account of meaning to offer which would underpin the doctrine that the results of computations are analytic, we are at a loss, it seems to me, to offer any difference in principle between computations and physical experiments. In both cases the results are determined ahead of time, not by us but by the nature of reality. In both cases it is possible to obtain erroneous results by blundering, and, therefore, if the results are important, we rely on them only after they have been checked by others.

There are even cases in which a change in point of view can turn a particular procedure from a computation to an experiment, or vice versa. If I add the number of coins I just counted in my left-hand pocket to the number of coins in my right-hand pocket, I may be determining how much money I have. In that case I am doing a computation. But if I know how much money I am supposed to have, I may be checking to see whether any of the coins have been lost. In that case I am doing an experiment. The assertion that five pennies plus seven pennies make twelve pennies may be an instance of the mathematical fact that five plus seven make twelve, or it may be an instance of the physical fact that pennies do not merge or multiply or evaporate.

Of course, examples based on counting change are not examples of pure mathematics but of applied mathematics. That distinction, however, must also go the way of the analytic-synthetic distinction. Most of the objects considered by applied mathematicians are not strictly speaking physical at all. There are no frictionless pulleys or ideal gases in nature. Theories of computability, of computational complexity, or of program verification characteristically ignore the fact that real computers have malfunctions and even sometimes break down completely. Nevertheless, theories of frictionless pulleys, ideal gases, or perfect computers are surely part of applied mathematics. It is difficult, indeed, to find any mathematical discipline short of the most nitty-gritty engineering which does not involve considerable idealization of its subject matter. That does not mean that it is not about its subject matter.

Let us ask ourselves why it is that geometers talk about straight lines and perfect circles. A completely precise description of any macroscopic physical situation would require more bits of information than we ever have at our disposal. Ordinary language copes with this problem by being vague. For many purposes this solution is adequate, but for scientific purposes it has the disadvantage that vague concepts do not support long chains of reasoning. (Recall the so-called "paradox of the heap." A bald man with one more hair is surely still bald. By induction, therefore, all men are bald. The problem is the vagueness of the predicate "bald.") It is characteristic of mathematical language that it attempts to be completely precise and thereby makes long chains of reasoning possible. When we describe a physical situation mathematically, then, where does the vagueness go? It cannot just disappear, since that would involve giving an exact description of a real situation, which is almost never possible. The answer is that the vagueness goes into the word "approximately." We say that the differential equation we write "approximately describes" or "approximately models" the physical process we are discussing. We do not say, because we cannot say, in exactly what way or exactly to what extent the differential equation fails to describe the physical process. When the Greek geometer drew a circle in the sand, he meant his statements about the circle approximately to describe the figure before him. Plato turned this around into the claim that the circle drawn in the sand was only an approximation to the ideal circle the geometer meant to discuss. Actually, no such ideal circle is involved.

Many of the objects mathematicians discuss, from Turing machines to Hilbert spaces, can in this way be interpreted as approximate descriptions of physical objects or processes. But there are other mathematical objects which, at least apparently, do not correspond to anything in the physical world. A striking example is provided by the large cardinals of set theory, which are not the cardinalities of any collections of physical objects, no matter how we broaden the concept of a physical object. Such abstract mathematical objects, however, are no mere figments. They are introduced to meet theoretical needs arising in the description of other, more basic, mathematical objects which do arise as approximate descriptions of parts of the physical world. The epistemological status of such large cardinals, for example, is similar to the status of quarks, say, which are also quite far removed from our ordinary physical experience. In both cases we are dealing with theoretical entities introduced for the sake of producing a smooth and workable theory of a domain previously introduced. We believe in them because we do not see any other reasonable way to make the theory work, but they are not directly accessible to us. (For more on this line of thought see Goodman [5].)

In most respects, therefore, mathematical theories resemble other scientific theories. They are constructed to solve particular problems and then develop a life of their own. They are not originally deductive structures based on axioms, but rather informal bodies of reasoning based on conjecture and bold extrapolation. One sees this very clearly when reading the papers of Georg Cantor, for example. He began with problems in analysis, specifically in the theory of Fourier series. He was led to introduce more and more daring conceptual innovations by the needs of his growing theory. It was

We learn mathematics by doing computations and other, less routine, constructions and by being surprised by the results of those constructions.

not until he had been working on set theory for thirty years that he tentatively offered some axioms. The first clear axiomatizations of set theory came another decade later. (One may follow this development best by reading Cantor himself in [3].)

As was pointed out by Imre Lakatos in [9], we miss this analogy between how mathematical and other scientific theories evolve because of the way that the history of mathematics is usually written. For example, Leonhard Euler had a theory of infinite series quite different from ours. He freely manipulated divergent series and came to conclusions which we find absurd. (For example, he formally summed the geometric series $1 + 2 + 4 + \ldots$ and gave it the value -1.) When this theory was replaced by the modern one at the beginning of the nineteenth century, essentially by the work of Cauchy, many of Euler's results were simply discarded. Almost all of his arguments, which, as we say, were "merely formal," had to be replaced by new arguments in the modern style. (For a lively account of these developments, see Kline [8].) But the histories do not describe this as a scientific revolution like the rejection of the phlogiston theory in favor of the theory of oxygen. Instead, we say that Euler made some mistakes which were later corrected by Cauchy. Or we ignore the point entirely.

Thus the traditional philosophy holds that mathematical theorems, once established, are true forever, whereas scientific laws are often overthrown. But Newton's laws of motion, though refuted by Einstein, still hold approximately. The classical theorem that the sum of the angles of a triangle is two right angles holds only approximately because also refuted by Einstein. The phlogiston theory is completely dead, as is Euler's theory of infinite series. The true parts of the phlogiston theory were incorporated into the new chemistry of Lavoisier, just as the true parts of Euler's theory were incorporated into the new function theory of

---

In most respects, therefore, mathematical theories resemble other scientific theories. They are constructed to solve particular problems and then develop a life of their own.

Gauss and Cauchy. The history of mathematics could be written in such a way as to make it look very similar to the history of physics. Then we might be less inclined to think of mathematical theorems as eternal.

The results of mathematics are no more certain or everlasting than the results of any other science, even though, for sociological reasons, our histories of mathematics tend to disguise that fact. Indeed, I suggest that, once we apply the insights of modern analytic philosophy to mathematics, we will see that mathematics is no more different from physics than physics is from biology. There is no such thing as non-trivial a priori truth, and mathematics is our richest and deepest science of nature.

REFERENCES

1. K. Appel, W. Haken, and J. Koch, "Every planar map is four-colorable," *Illinois Journal of Mathematics*, v. 21(1977), pp. 429-567.
2. G. Frege, *The Foundations of Arithmetic*, trans. J. L. Austin, second ed., Oxford U. P., Blackwell, 1959.
3. G. Cantor, *Abhandlungen mathematischen und philosophischen Inhalts*, ed. E. Zermelo, Olms, Hildesheim, 1966.
4. N. D. Goodman, "The experiential foundations of mathematical knowledge," *History and philosophy of Logic*, v.2(1981), pp. 55-65.
5. N. D. Goodman, "Mathematics as natural science," to appear in *The Journal of Symbolic Logic*.
6. S. M. Hammel, J. A. Yorke, and C. Grebogi, "Numerical orbits of chaotic processes represent true orbits," *Bulletin of the A. M. S. (New series)*, v. 19(1988), pp. 465-469.
7. P. Kitcher, *The Nature of Mathematical Knowledge*, Oxford U. P., Oxford, 1984.
8. M. Kline, *Mathematical Thought from Ancient to Modern Times*, Oxford U. P., New York, 1972.
9. I. Lakatos, *Proofs and Refutations*, Cambridge U.P., Cambridge, 1976.
10. R. Rorty, *Philosophy and the Mirror of Nature*, Princeton U. P., Princeton, 1979.
11. W. H. D. Rouse, trans, *Great Dialogues of Plato*, Mentor, New York, 1956.

Value Judgments in Mathematics: Can We Treat Mathematics as an Art?

Thomas Tymoczko
Smith College
Northampton, Massachusetts

To paraphrase Bishop Bulter, mathematics is what it is and not another thing. Nevertheless, it is natural to compare mathematics to other things, and one inviting comparison is to science. Mathematics is grouped with science in our curricula. Mathematics is enormously useful in science. We measure the "hardness" of a science by the extent to which it uses mathematics, and a significant portion of mathematics education is service to science (calculus for engineers). Like science, mathematics seems to yield knowledge, and, like science, it is cumulative. In this essay, I want to explore the possibility of comparing mathematics, or aspects of mathematics, to fine art. Under the arts I include literature, painting, music, and sculpture. This comparison is hardly original. The poet Edna St. Vincent Millay has said, "Euclid alone has looked upon Beauty bare," and many mathematicians have agreed. But often they make the comparison for reasons I am not interested in pursuing. I do not intend, here, to defend pure mathematics as worthwhile apart from immediate practical applications or to pay homage to the imaginative or nonmechanical aspects of mathematics, even though scientists do pure research without invoking art, and notwithstanding the fact that science, too, is creative. The interesting case for mathematics as an art is the possibility of regarding at least some of its products as objects of aesthetic enjoyment. My project is very tentative and at best will be incomplete. I offer no definitions of "mathematics", "science", or 'art', but rely on the good

> **The interesting case for mathematics as an art is the possibility of regarding at least some of its products as objects of aesthetic enjoyment.**

graces of my readers to follow me along in this initial foray. And I want to hold open the possibility that the project can fail without dire consequences befalling anyone. Mathematics is quite all right by itself; we don't need to pretend it is an art if it's not.

In the first part of the essay, I explore the rationale for an aesthetic approach to mathematics. I take my start from a provocative paper by Borel [Borel, 1983]. His key idea is that value judgments are necessary in mathematics. Borel proposes the idea of aesthetic criticism as a clear alternative to von Neumann's claim that only the relation of math to science can provide the source of value judgments in mathematics. Later in that essay, Borel suggests as a third alternative that the existence of mathematical reality, in and of itself, might provide mathematics with the kind of critical self assessment needed to keep it on the straight and narrow. I will argue the case for aesthetic criticism.

In Part II I ask how the terms of art can gain so much as a foothold in mathematics. Even the simple artist-product-audience relationship generates problems when applied to mathematics. Whatever we take the mathematical products to be, it is tempting to identify mathematicians as both the artists and the audience. But it is hard to take aesthetic criticism seriously if the artists are the only ones doing it ("Gee, we're all great!"). I attempt to resolve this problem by borrowing an idea from Livingston who characterizes proofs as paired objects, a kind of template and the actual, lived presentation of that template. [Livingston, 1986] I try to develop an analogy to music whereby proofs are divided into two parts, composition and performance and explain how this way of looking at proofs can resolve the problem of the audience and establish the notion of criticism in mathematics.

Part III concludes the essay with an attempt to look at one very familiar proof as a work of art, by trying to describe it in terms of art criticism. My attempt is tentative and perhaps points up the difficulties and risks of aesthetic criticism in math. But the difficulties must be met and the risks taken if we wish to maintain that any aspect of mathematics is, in any serious sense, a form of art.

Why am I even tempted to regard mathematics as an art? Because so many mathematicians and philosophers have made the comparison.

> Mathematics, rightly viewed, possesses not only truth, but supreme beauty—a beauty cold and austere, like that of sculpture, without

appeal to any part of our weaker nature, without the gorgeous trappings of painting or music, yet sublimely pure, and capable of a stern perfection such as only the greatest art can show. [Russell, p. 57]

If Bertrand Russell compared mathematics to sculpture, Henri Poincare compared it to painting, and A. Borel, to both painting and poetry (albeit, poetry in a language only mathematicians can understand). Many less eminent mathematicians feel compelled to use aesthetic

---

Mathematics, rightly viewed, possesses not only truth, but supreme beauty—a beauty cold and austere.

---

terms, at least in a general way, to describe mathematics. So I feel confident to take as part of my data the artistic concern of practicing mathematicians. Even as an amateur, I note that there are certain proofs that strike me as beautiful, or more accurately, they strike me as variously elegant, clean, simple, economical, powerful, subtle or surprising. I confess to liking to read them over more than once, obviously not to become more certain of the conclusion, but for the pure pleasure gained by following the flow of ideas.

Mathematicians' intuitions that what they are doing is art, or a kind of art, or that it can be regarded as an art are worth examining in their own right. But there is another factor that I want to bring in: the need to ground value judgments in mathematics. By value judgments in the broad sense, I mean those positive judgments (like importance, elegance, relevance, promise) and those negative judgments (like inconsequentiality, triviality, crudity, sterility) that are necessary to characterize a discipline and to shape its progress. As Borel puts it:

> Surely not all concepts and theorems are equal; as in G. Orwell's *Animal Farm*, some must be more so than others. Are there then internal criteria which can lead to a more or less objective hierarchy? You will notice that the same basic question can be asked about painting, music, or art in general: It thus becomes a question of aesthetics. [Borel, p. 11]

Borel's suggestion is two fold: some value judgments are necessary in mathematics and second, by regarding mathematics as an art, we can find that aesthetic criteria are able to support such judgments.

Since I share Borel's contention that value judgments are as necessary in mathematics as in painting, music or art, I would like to offer two arguments to support that assertion.

The first is a positive argument; the practical need for selection processes in mathematics. It has been estimated that 200,000 theorems are proved every year. [See Davis and Hersh for a discussion of Ulam's estimate.] If mathematics is a living discipline, and not simply an accumulation of records in the archives, or on computer disks, then obviously mathematicians must choose or select what to pass on—not only in undergraduate courses, but in seminars and supported research projects. Mathematicians cannot avoid being critical, that is, selective, if they wish to perpetuate their field. But such criticism must be based on value judgments in the broad sense: "this is more important than that; it deserves to be taught." Without such value judgments, there could be no selectivity and mathematics would sink under the weight of 200,000 theorems per year! [Tymoczko, 1986a gives a more extended version of this argument.] (Later in the paper I will argue that aesthetic judgments—value judgments in the narrow sense—can serve as selection criteria for mathematics. I suggest that mathematicians can be seen as using aesthetic criteria in creating the 'mathematical canon', that body of great mathematical proofs, theories and concepts which guides teaching and research. In literature and music, the idea of a canon of great works, a set of paradigms against which to measure new additions, is a key feature of the art.)

At this point, some readers might concede the need for value judgments, so broadly construed, in mathematics. But they might object that such value judgments could be supplied by mathematical reality itself. We have no need for aesthetic criteria, they might say, but can read off the value of mathematical products directly by seeing how they match up with mathematical reality. In other words, truth itself is a value and the only one needed in mathematics. Borel himself seems tempted by this idea and we can understand its appeal. [Borel, p. 13,14] If belief in an objective world of mathematical entities could obviate the appeal to subjective preferences, why not take the former route, no matter how mysterious mathematical nature is?

Unfortunately this suggestion is even less plausible than is the analogous suggestion in art (a good painting is one that depicts reality accurately). The basic objection to it is not that mathematical reality does not exist, but that, if it does, it is too rich and too mute. It is too rich in that, besides the basic objects that mathematicians like to study, it contains every bizarre permutation, combination and collection of them. Mathematical reality is too mute in that it comes without labels. There are no little tags reading "this connection is important", "this one is trivial". We don't perceive that some structures are important, some trivial, some constructions elegant, some crude, without bringing our values to our perceptions. These are not features of mathematical reality if the latter is construed as independent of us. It is only in relation to actual mathematicians with actual interests and values, that mathematical reality divides up into the basic and bizarre, the important and the trivial. Quine pointed out long ago that, if mathematics is just the creation and study of formal systems, then a legitimate piece of mathematics would be the formal theory of Greek mythology! [Quine, 1936] With equal force, we can say that, if any description of an abstract structure counts as mathematics, then there is a mathematics for the trivial structure isomorphic to the Greek gods. Conversely, if that is what mathematics is, then the important question for philosophy is: what constitutes good—that is valuable—mathematics.

My conclusion is that mathematical reality itself cannot ground value judgments. It cannot exert the kind of critical controlling functionof which Borel speaks, and this is true even if we believe in a platonic reality or a formalist reality for mathematics. We have to search elsewhere for this critical function. Borel's original suggestion was to view mathematics as an art. But there is a very popular alternative: see math as a science and look to practical applicability to provide the controlling function. Von Neumann defended this view (called "mathematical maoism" by Davis and Hersh!)

> As a mathematical discipline travels far from its empirical sources, or still more, if it is second and third generation only indirectly inspired by ideas coming from "reality", it is beset with very grave dangers. It becomes more and more purely aestheticizing, more and more purely 'l'art pour l'art'... there is a great danger that the subject will develop along the line of least resistance... will separate into a multitude of insignificant branches... In any event... the only remedy seems to me to be the rejuvenating return to the source: the reinjection of more or less directly empirical ideas. [in Borel, p. 12]

Quite clearly, von Neumann accepts Borel's demand for a controlling function in mathematics but disagrees on the advisability of aesthetics's providing it. Indeed, von Neumann goes so far as to rule out the possibility of aesthetics's providing it. If von Neumann were right, it would seem that art itself is not possible or that anything counts as art, since pure art itself "develops along the lines of least resistance." With such a low view of art, it is no wonder that von Neumann turned to science for comfort. We'll take up this negative charge in Part II, but now I want to focus on the idea that mathematicians should depend on their relation to science to guide their choices.

It is not an implausible idea; there are enormous affinities between mathematics and science as we earlier noted. Indeed, it might well be argued that mathematics is an essential part of science, or a very useful tool of science. There is a natural continuum—engineering, physics, mathematics—and some philosophers and mathematicians have felt that mathematics is just a bit more abstract than science but nonetheless connected to the space-time world. In this regard, we can mention the views of Quine, Kitcher and others (including myself) in the quasi-empiricist school. [See Tymoczko, 1986b for a general account of quasi-empiricism.] Still, it is one thing to note the connection between some aspects of mathematics and science and quite another to see this connection as essential to mathematics. Von Neumann takes the latter step.

As an ad hominum argument against von Neumann, we can point to those areas of modern science like cosmology or string theory where practical applicability seems out of the question. But then von Neumann might respond by noting that even if cosmology has no practical applications, still it is multiply connected to the rest of modern science. In the first place cosmology

Aesthetic judgments—value judgments in the narrow sense— can serve as selection criteria for mathematics.

depends on, and to some extent influences, such fields as relativity theory, quantum mechanics, and astronomy, fields which are in turn connected to practical applications. Moreover, in the other direction, cosmologists, in their reliance on astronomy, rely on the latest in technological advances, computer models and radio telescopes. If cosmology is pure science without any practical applications itself, still it is part of a "seamless web", to use Quine's phrase, with very prominent strands of the most modern engineering and technology. Cosmology is subject to many checks and balances that themselves are tied to practical concerns. It is just such checks and balances that von Neumann wishes to apply to mathematics. Without the constraints and the suggestions provided by science, mathematics could lose its moorings and become idle symbol play.

On the other hand, I'd like to suggest that what the previous response really points to is not practical applicability but connectedness to earlier work. This might be taken as a worthy criterion but why not call it an aesthetic one? To be sure, we would be suspicious of a mathematical field that developed sui generis without any ties to traditional mathematics. But I do not see why a connection must be made to science rather than to a branch of mathematics. It might even be that, say, string theory in physics is ultimately justified by providing a satisfactory mathematical account of certain matters, by relating physics to mathematics rather than vice-versa. Similarly, we might justify set theory, in part, because of its connections to what is historically regarded as mathematics. To make an analogy with music, I might justify something as good modern music because of its relation to a musical tradition that goes back to Bach's composing liturgical music for the practical purposes of the Church. However, this does not mean that I should insist that all modern music meet the test of practical applicability and be playable in church services.

If we took von Neumann seriously, there are many areas of mathematics we would find it difficult to preserve because their ties with science are so weak, such as transfinite set theory, pure number theory. But then this provides us with a nice test case for von Neumann's position against Borel's. For Borel wants to argue that transfinite set theory can be justified in its own terms, better, in aesthetic terms: because it is a beautiful theory with many elegant mathematical proofs and constructions. Cantor's diagonal argument is striking; Gödel's constructible universe has a purity and simplicity; large cardinals seem breathtaking in themselves yet their study bears a comforting if paradoxical relation to classical combinatorics and counting arguments. If I can enjoy reading Joyce again and again, if I can lose myself in Picasso's paintings without worry about their practical applicability, then why can not my awe before the transfinite universe (more prosaically, my enjoyment of set theory) count as sufficient justification? "From the Paradise created for us by Cantor, no one will drive us out." [Reid, p. 177] Hilbert's stirring words are surely an aesthetic defense of set theory; science examines the world as it is, only art can provide us with a hint of paradise.

In this regard, it is interesting to look at intuitionistic critiques of set theory as aesthetic criticisms—this is not how mathematics ought to be done. One is tempted to compare such criticisms to criticisms of atonal or electronic music by conservative music critics. They both have the same moral overtones, the sense of outrage, that sometimes characterizes aesthetic criticism. The analogy threatens to break down, however, because intuitionistic mathematics is itself so aesthetically unattractive. On the other hand, contemporary set theory seems to be going through a crisis of sorts. From the outside, it seems to be "separating into a multitude of insignificant branches" as Von Newmann warned. Can the application of aesthetic criteria reverse this trend?

But if we defend mathematics on aesthetic merits, then we should follow Borel in seeing that defense as committed to developing aesthetic criticism in mathematics. Mathematicians need to be able to criticize on aesthetic grounds for this alternative to von Neumann to work. Not just everything we like is beautiful. If it were, mathematics could develop along the lines of least resistance. "I can prove this, therefore I like it, therefore it is good mathematics." To make good on Borel's alternative, on the alternative that mathematics is like an art, we must be able to discuss it as art is discussed.

What keeps an art alive and growing is, I claim, the ancillary practice of criticism. In painting, literature, music, dance, it is criticism that helps to provide the constraints and encourages the directions for the art to develop. Now if mathematics were regarded as an art and we could develop a practice of aesthetic criticism for it—or we could recognize such critical practices that we already engage in—then we would satisfy von Neumann's deep concern without surrendering mathematics to science. In the rest of this paper I will explore the possibilities for an aesthetic of mathematics, techniques for criticizing mathematical works analogous to those developed for the arts.

II.

Philosophers of art distinguish between aesthetics and criticism. Criticism is concerned with particular works of art. Aesthetics studies the general terms and concepts used in artistic criticism. So I might write a critical study of Baudelaire's poems, and an aesthetician might study my use of terms like "meter", "theme" and "tone". Thus criticism is prior to aesthetics. If we never talked about works of art at all, it would be impossible to do aesthetics. Indeed, I am tempted to say that, if we never talked about works of art at all, if we never discussed them critically, then they would not be works of art. Criticism, on my view, is an essential feature of art. While it is tempting to speak of the aesthetics of mathematics, we can see that in terms of the preceding distinction this is, strictly speaking misleading. What interests us is mathematical criticism. If we are to treat mathematics as an art, then we must learn to talk about

> **Borel wants to argue that transfinite set theory can be justified in its own terms, better, in aesthetic terms: because it is a beautiful theory with many elegant mathematical proofs and constructions.**

it critically or to recognize that critical talk that already occurs. It is mathematical criticism (criticism of mathematics as an art form) that might serve as an alternative to science in von Neumann's scheme.

By criticism I do not mean a thumbs-up or thumbs-down response to some putative artistic product. That simple pro or con approach might better be characterized as the response of the marketplace to the product (one buys it or not). Von Neumann suggested that science is the ultimate marketplace for mathematics, and I do not intend to oppose him by suggesting a marketplace of detached likes and dislikes that are called 'aesthetic' for no other reason than that they are detached. That would be the route to naive subjectivity. By criticism, I mean a way of talking about works of art. A good critic provides us with a way of seeing or hearing a work of art, and good art is subject to multiple analyses. A critic draws our attention to certain features of the work

---

I will explore the possibilities for an aesthetic of mathematics, techniques for criticizing mathematical works analogous to those developed for the arts.

---

and in so doing enhances our appreciation or articulates for us the movement of our own feelings. So, when a critic juxtaposes the bedroom scene implicit at the beginning of Shakespeare's *Othello* and the horrifyingly explicit bedroom scene at the conclusion, when he draws our attention to the images of sheets, marriage sheets and winding sheets, and to the incriminating handkerchief stained with strawberries (red, like blood), we are forced to acknowledge the validity of the connections (the shock of recognition), much as we are forced to acknowledge the validity of a proof. Individuals can disagree in their likes and dislikes for even great works of art, but it is a misconception of critics to regard them as if they were defending their own preferences. A great work has a richness and complexity that a sensitive audience can be brought to acknowledge, and it is the function of criticism to bring them to this stage. [See Cavell for a discussion of the role of criticism in art.]

To have an art, we need artists, products and audience. In the case of mathematics, the artists, let's assume initially, are mathematicians. The artistic products on which I will focus are proofs. Other things might qualify as well: theories, methods of construction or computation, perhaps even certain concepts. Nevertheless, proofs are paradigm in mathematics: if anything in mathematics is beautiful, proofs are. We shall say more about what we take proofs to be, but first let us ask about audience. What is the audience of mathematics?

Listen to Borel on this:

I already mentioned the idea of mathematics as an art, a poetry of ideas. With that as a starting point, one would conclude that, in order for one to appreciate mathematics, to enjoy it, one needs a unique feeling for intellectual elegance and beauty of ideas in a very special world of thought. It is not surprising that this can hardly be shared with nonmathematicians: Our poems are written in a highly specialized language, the mathematical language; although it is expressed in many of the more familiar languages, it is nevertheless unique and translatable into no other language; unfortunately, these poems can only be understood in the original. The resemblance to an art is clear. One must also have a certain education for the appreciation of music or painting, which is to say one must learn a certain language. [Borel, p. 15]

I find something very disquieting in this remark: it is the disappearance of the audience. If mathematicians are poets they are poets who write only for other poets. The resemblance to art is not clear. I can't imagine an art where a necessary condition of enjoyment is the ability to create it. To be sure, Borel's point about education is well taken. Music needs an educated, even a gifted listener. [Copland, Ch. 1, discusses how serious a need this is] We are not to judge an art by the reaction of people brought in off the streets; that is the marketplace idea. But a style of music accessible only to composers of that very style would seem to be in danger of losing its status as an art. Here indeed is the kind of degenerate aesthetizing about which von Neumann was worried. If aesthetic criticism is to get off the ground in mathematics, there must be some independence between artists and critics.

The idea of a music accessible only to composers of that music is the idea of a dry, sterile music. But suppose we modify the image slightly. What of a music accessible only to performers—what of the idea of a music accessible to people with some experience in making music? This idea seems not so farfetched. And, if I weaken the requirement still further, if I count as performers even those who learned a little piano or who sing, I start to come close to the actual audience of music. If I might switch analogies, a poetry intelligible only to poets seems scholastic. But it seems quite reasonable to insist on an audience who can perform poetry, who can read it aloud. Surely this is not asking too much. It is not as if we are ruling anyone out of this audience, as if we were to prohibit anyone's enjoying poetry if she did not read some aloud to herself. But we might say that she would not be our primary audience, that we would not take her criticisms so seriously.

Two ideas are beginning to emerge, albeit slowly. One is that we might have taken a false step in identifying the artists of mathematics with mathematicians. It might well be that mathematicians are better identified as the audience of mathematics, that a mathematician is

Criticism is an essential feature of art.

one who enjoys mathematics, a connoisseur of mathematics. Again a comparison to literature might be illuminating. There are many departments of literature in colleges and universities. However these departments are not staffed by artists but by critics. The typical literature professor is not a writer, not a poet or novelist, but is a critic of literature, a sensitive and appreciative reader; similarly for departments of visual arts or music. I wonder whether it is possible to look at mathematics in this way. What if we saw the typical mathematics professor as a sensitive and appreciative reader or critic of mathematics? Then the inventive artists in mathematics might be only a subset of mathematicians. I do not want to say "creative mathematicians" since that term is so loaded. Many are happy to be literary critics instead of poets, but who would be content to be an "uncreative mathematician"?

Here the second idea might aid us. Recall in our discussion of music we distinguished between composers and performers. Music intelligible only to composers seemed almost dishonest, a usurpation of ordinary terms, but a music intelligible only to performers, or better, a music written to be enjoyed by performers seemed not so restrictive. Can we find an analogy to the composition/performance distinction inside mathematics? If so, we might then distinguish between composing mathematicians and performing mathematicians both of whom would be creative albeit in different ways. And in performing the proofs, whether presenting them for a wider audience or for oneself, how can a mathematician be engaging at the same time in the vital function of criticism.

To answer this we must return to the second term in the series artist-product-audience which in our case was a proof. Now what is a proof? How can we look at proofs in such a way as to make it plausible to view them as works of art?

Livingston has suggested that a proof "is an intrinsically paired object", the material proof and the practices of proving to which it is intrinsically tied. [Livingston, p. 171] In Livingston's words, a proof is a "pedagogic object—it teaches provers how to prove [a given] theorem, and it does this by providing, metaphorically, in the material proof, a template of that

course of action". [loc.cit.] Perhaps this idea of a pair of objects is too strong, perhaps we should say that a proof has two aspects, the composition and the performance. As a first approximation, let us say that the mathematician as composer creates a proof, the template or, in terms of our music analogy, a score, and that the mathematician as performer creates a version of that proof, a performance which can be enjoyed and appreciated. The composition side of a proof faces metaphysics and can even be seen as a mathematical structure in and of itself. The performance side of a proof faces epistemology and must be a convincing argument; it must be capable of persuading mathematicians to accept it just as aesthetic criticism can persuade us to accept a work of art. No structure, in and of itself, can accomplish the latter task; for that we need a human activity.

In foundationalist philosophy it was the metaphysical aspect of proofs that dominated. For a while it seemed that a proof could even be identified with a mathematical structure: "a finite sequence of wffs (well formed formulas) each of which is either an axiom of the theory or follows from preceding wffs by a rule of inference and the last wff in the sequence is the theorem proved". The foundationalist ideal was a completely self-interpreting proof, one that needed no performance. Yet insofar as that ideal is realized, "the proof" recedes into the domain of mathematical structures, becomes yet another mathematical object which we can only approach via proofs of the more ordinary sort. On Livingston's account, proofs become lived works, historical objects that can exist in time, are susceptible to variations, can be approached from different directions and can be left in different ways. (There is a noticeable similarity to Lakatos's ideas here. [Lakatos, 1976]) It is just this aspect of proofs that is susceptible to aesthetic criticism, and to aesthetic play and enjoyment ("Look at this neat proof!")

If we can follow out this flight of fancy a bit more, we can begin to see the mathematics lecture hall, be it a class room or a conference, as if it were a concert hall or small parlor where some have gathered to attend a performance. The (performing) mathematician presents a proof, recreates for the audience the lived work of a discovery composed by himself or another. The audience, whether it is faculty or students, we imagine listening for the pure enjoyment of the piece. In presenting the proof, the mathematician is also functioning as a critic, not only in his initial selection of the proof, but in his organization and emphasis. If we are very bold, or very foolish, we might even glimpse the possibility of a more humanistic mathematics (is it radically new or a return to forgotten prefoundational roots?) moving a little more slowly, being a little more patient with itself, a little more grateful for the pearls that have been obtained (as opposed to that marketplace attitude, "What next?").

Now I want to return to earth from this fanciful flight and reconsider the notion of proof just adumbrated. The analogy with music is not exact. It is possible to compose a musical piece without performing it; here is the score, there the performance of that score. But as we use the terms, it is not possible to compose a proof without also presenting it. We can't quite lift off what Livingston calls the template, the material proof, from all of its presentations. This fact is tied to the question, what makes two performances of a proof, performances of the same proof? To some extent, formal proofs might be considered as providing an answer to this, the formal proofs being the template that the actual presentations rely on. Indeed, I've heard some mathematicians say that what they do is present descriptions of proofs, the real proofs remaining behind the scenes in a kind of formalist-platonic heaven. But there is something wrong with this besides the gut feeling that what we are actually presented with is the actual proof. It is, as De Millo, Lipton and Perlis pointed out, that the formal proofs do not function as templates, they are rarely, if ever, written down or spoken aloud. ["Social Processes and Proofs of Theorems and Programs"] In fact, really formal proofs

---

A good critic provides us with a way of seeing or hearing a work of art, and good art is subject to multiple analyses.

---

are usually merely assumed to exist to validate the actually presented proof. It is not that the assumption is wrong, but that it is idle, as if all mathematicians were to begin their proofs by asserting that 'all ideas come from God who is the true Author of the following proof'. The belief in perfect formal proofs, like the belief in perfect Authors, can be very comforting, but it turns no philosophical wheels.

But if formal proofs aren't the 'templates' of ordinary proofs, what then becomes of our notion of proof as paired object? What binds the presentations, whether written in a text or spoken in a classroom into a unity? The best answer I have is that for most presentations, the template, the material proof is a previous presentation. It is as if composers did not write scores but only taught musicians how to play a piece and others learned from previous performances. The presentations of a proof form a causal, historical series, and this series we might take to constitute the proof (following Gadamer who identifies the 'eminent' literary text with the history of its (competent) readings). [Gadamer, p. xix] If this conception is vague, its

vagueness matches our actual experience of proofs. It highlights the achievement of an original proof, a proof presented without antecedent. It allows for the modest achievement, a new proof presentation whose template is a previous proof of some other theorem. It allows for variants in proofs, the simplifications and generalizations that so many proofs go through. (Who would want to learn Gödel's theorem by following the original proof?) And it leaves a place for insignificant proofs, presentations that are never followed but performed only once and consigned to the archives. Finally, it can give rise to what critics in other arts call 'the canon' the body of lived proofs, the presentations still going on, that we want to teach to our students so that they can become gifted listeners of mathematics, sensitive critics able to judge new works as they appear.

We have been exploring one way of looking at mathematics as an art. While we have been able to identify mathematicians as artists, we did this by distinguishing two types of mathematicians, or better, two types of artistic achievements in mathematics. One was the composition of proofs, and one was the presentation/performance of proofs. Identifying a proof with a historical object, a sequence of presentations of that proof, turned out to be reasonably faithful to the actual practice of mathematicians. The audience of mathematical art we identified also with mathematicians, but we can extend the community to include student apprentices and eventually anyone who can find aesthetic enjoyment in math. Finally we toyed with the possibility of teaching mathematics in more strict analogy with teaching art, such as literature, where the instructor-critics introduce students to great works from the canon for their own sake, to be appreciated in themselves.

In so doing, we have also connected the two themes of this paper, comparing mathematics to art by establishing a place for criticism in mathematics. The locus of criticism turns out to have been an essential part of mathematics all along, the presentation of proofs. Only now the presented proofs are regarded as aesthetic objects, their presentation as aesthetic performances. They are valuable in themselves and not merely as steps for obtaining results—the theorems. Perhaps we might say that from a scientific point of view, it is the mathematical results that matter. The results count as the advances in knowledge, but it is the proofs that can count as achievements in art. It is now time to turn to the serious work of mathematical criticism and to see whether we can talk about mathematics in an aesthetic way.

III.

If any proof counts as a work of art, the standard proof of the Fundamental Theorem of Arithmetic does. [See

Fisher, Rosen or Uspensky for this proof] In saying this, I do not mean to praise that particular proof but to test the claim that proofs can be works of art. Can we present this proof (any proof) as an artistic achievement? My goal here is not to do anything new in mathematics, but only to draw attention to certain ways that mathematicians actually proceed.

The proof of the FTA arises as a response to a natural conjecture. Given the definition of primes and our experience with the smaller natural numbers, it is immediately plausible that any natural number greater than 1 is either prime or a composite of prime powers with a unique canonical factorization. When we attempt to prove this conjecture, our first steps proceed with a certain inevitability. That every number greater than 1 is a product of primes is an easy consequence of definitions and induction (or the least number principle). Anyone who has trouble with this has trouble with appreciating mathematics, more specifically mathematical proofs, at all.

But in attempting to prove the uniqueness of the prime decomposition, we encounter a problem—what aesthetic criticism calls tension. We come up against the conjecture-lemma that if p is prime, p divides ab, and p does not divide a, then p divides b. Given this lemma the tension is resolved and the proof proceeds to a conclusion with the same inevitability with which it began. So our attempted proof has created a new need, one that we might not have recognized at first, of establishing this particular lemma, (unless, like my colleague, Jim Henle, you find this particular tension unnecessary and find another route to the Fundamental Theorem that avoids the lemma altogether. Alternatives are almost always available in mathematics.)

We can attack the new problem by a strategy that is familiar (or one that will become familiar, depending on the sophistication of the audience). We can replace our original problem by a more general problem, one that brings the essential features of the case more clearly into focus. The more general problem is that if p and a

---

A great work has a richness and complexity that a sensitive audience can be brought to acknowledge, and it is the function of criticism to bring them to this stage.

---

are relatively prime, have no factors in common, then if p divides ab, p divides b. Either we have or can introduce the idea of greatest common divisor (GCD) to

capture the idea of common factors. The tension in our original proof can be resolved if we can prove that if GCD(p,a)=1 and p divides ab, p divides b.

The problem has now been transposed to a new key. The original tension remains in the background but now we have the problem involving GCD and division and we can cast about for theorems relating, x,y and GCD(x,y). For a while, we can recapture that inevitability again since the most obvious relation between x,y and GCD(x,y) is the process of obtaining GCD(x,y) from x,y and Euclid's Algorithm provides a straight forward answer to that. Now just a little insight, a little innovation, allows us to reverse the system of equations in Euclid's Algorithm to establish that GCD(x,y) is a linear combination of x and y. There are integers m,n such that GCD(x,y)=mx+ny. Applying this to the case at hand, we conclude that there are m,n such that 1=mp+na. We are near the conclusion (is it possible to feel this?). If one stares at the linear combination and the assumption that p divides ab and the goal of showing that p divides b, one can be struck by the happy thought of multiplying through the equation by b to get b=mpb+nab. Now all is obvious. p divides both factors on the right, hence p divides b, the lemma is proved and the Fundamental Theorem of Arithmetic is established.

I have just described a proof that every mathematician knows. Is there anything about this description that inclines us to acknowledge the proof as artistic, as locatable in an aesthetic realm?

For one thing, the proof is described as a lived work (in Livingston's sense), something that one experiences in a way that one experiences a (short) piece of music or a poem. The proof existed in time with a tempo or density, speeding up at certain points and slowing down at others. I tried to be patient with the proof, letting it last, attending to the different rhythms in it, now quickly and easily moving along, now encountering a stumbling block, tension or subsidiary tension. This description would be totally lost on one who was in a hurry, who was interested only in establishing the conjecture so he could use the result for other work. To one who cared only for getting results, who had the attitude that all proofs were alike, the above exercise would seem pointless and melodramatic.

Also the description has a certain memorable quality to it, it is coherent, it hangs together. It is easier to remember because the steps have been colored differently, some highlighted, some background. (Isn't it natural to regard a formal proof as colorless?) At the least the proof is organized, it has lemmas and sublemmas. Since lemmas imply organization and value judgement, orderings of importance, the organization of a proof into lemmas is an aesthetic matter. Lemmas are a matter of presentation (and composition when the composer is writing with the performance in mind).

Furthermore the description attributes to the proof a certain obviousness or inevitability. This is the way it has to be. There are no false steps or unnecessary detours. Various tensions are created and resolved in different ways. The first imaginative strategy was to prove a generalization of the needed lemma. An apprentice might not have thought of that but, having seen it, will remember the technique for it is common in mathematics. One of the pleasures of mathematics is its ability to achieve economy, the strongest results with the least means. The last imaginative strategy was multiplying through the equation by b. Apprentices might not have thought of that either, but, once they have seen it, they must, if they are to enjoy mathematics, accept it as natural—"I could have done that" they must be able to say. If they can't say that— if they say "I could never have thought of that in a million years", we'd question their ability to appreciate mathematics. It doesn't get much easier than that, we want to tell them, there are some gaps you just must be able to jump. Notice the difference between these two groups of students, neither of which, after all, anticipated the step. The former is at least learning to be critical ("I could have done that" = "that is a natural step"), the latter remains outsiders. (Of course I am concerned here with one step. I am not suggesting that to appreciate a mathematical proof, one must at the end conclude "I could have done that"!)

In passing we might note that not only does the proof involve the concept of GCD in a natural way (in the generalization from primes, to relatively prime numbers with no common factors) but what gets so involved, the GCD, turns out to be met with in numerous other proofs of number theory. I am tempted to praise the power of the proof. Not only is it complete in itself, but in its own unwasted motion it lays the groundwork for further proofs. This points to a peculiar feature of mathematical aesthetics, one that Borel describes in the following paradoxical fashion:

> I would naturally find a method of proof more beautiful if it found new and unexpected applications, although the method itself hadn't changed. It may have become more important, but in and of itself not more beautiful. [Borel, p. 16]

Perhaps it is the word 'beauty' that is misleading here. Certainly connections to the rest of mathematics can enhance the aesthetic character of a proof, for the connections help color the proof in different shades, revealing what is more and what is less important. In the same way, if there were only one impressionistic painting, it might not strike us as marvelous, but as anomalous, outlandish. New Critics aside, the connection of one work to others in a corpus or genre can help us to see the work differently.

In summary, the proof meets Hardy's standards of "a very high degree of unexpectedness, combined with inevitability and economy". Hardy [1967, p. 113] continued:

> The arguments take so odd and surprising a form; the weapons used seem so childishly simple when compared with the far-reaching results; but there is no escape from the conclusions. There are no complications of detail—one line of attack is enough in each case;.... A mathematical proof should resemble a simple and clear-cut constellation, not a scattered cluster in the Milky Way.

Now what are we to make of all this? If readers find the above description appropriate to the classical proof of the FTA, then they will be well disposed to further explore the aesthetic dimension of mathematics. The description uses critical terminology borrowed from the arts, and this terminology describes aesthetic features. If readers find the description appropriate but wrong, then the result is even more interesting. Such readers would be challenging my aesthetic judgments: "Tymoczko's just mistaken about what's important in the proof, where the key steps are". I would be astonished if there would be such disagreement over so basic a proof. But surely the fact of two critics disputing the aesthetical details of a particular piece would count towards the relevance of aesthetic criticism.

A third possibility is to acknowledge the appropriateness of this account of the Fundamental Theorem's proof—which after all is beautiful—but to worry that it is a special case, an unusual mathematical

---

The formal proofs do not function as templates; they are rarely, if ever, written down or spoken aloud. In fact, really formal proofs are usually merely assumed to exist to validate the actually presented proof.

---

proof. There is little I can do in this paper to satisfy that worry. True, that proof is unusual, just as any proof is or any work of art is. I would not hold that proof up as a standard of comparison, any more than I would so hold up a Bach fugue. Many proofs proceed entirely differently, some pulling together very divergent strands into a unified whole, as in Gödel's First Theorem, rather than moving steadily towards a natural goal. Ultimately, the only answer to this concern is to try to see more of mathematics as artistic creation and performance; to try to present proofs critically and to assess the results.

A final possibility is that the reader finds the above description totally inappropriate, in bad taste, an embarrassment. If one adheres to a very rigid form for proofs, one might feel that the above description is an unwanted intrusion on that form. In the extreme case, the form would be a finite sequence of wffs, etc. Anything added to that might be classified as purely emotive content, something which math could do without quite well, thank you. But such an extreme case is not very plausible; who composes or presents one's proofs in first order logic or set theory? (The recalcitrant formalist—"True we don't present proofs formally, but we should. What we do is a weakness on our parts, a concession to human frailty, and a falling away from real mathematics." I admire the intellectual purity of this view, but I think it very wrongheaded.) Will any less extreme rejection do? I would not accept the criticism that such descriptions as I have given are too colorful for serious mathematics but might be perfectly palatable in the classroom. For a central thrust of the thesis that mathematics is art is that serious mathematics does go on in the classroom, and that the classroom encompasses the seminar room and international conference hall as well. I prefer the more extreme antagonist. Mathematics is no serious art if such aesthetic descriptions as I attempted are totally inappropriate.

If one is to make a successful comparison of mathematics to fine art, much more work needs to be done. I have only scratched the surface here. No doubt many analogies and disanalogies have occurred to readers as they read along. (What of the crisis in modern art— what counts as a work of art? Silent pieces of music? Large soup cans? But what counts as modern mathematics? Computer proofs? Rubik's cubes? Might there be an aesthetic crisis in modern math?) However, the goal of this essay is quite modest, to open an area of discussion, not to conclude one. In the end, no one can prove that a mathematical proof is an art, any more than one can prove that a particular painting is beautiful. All one can do is to describe it in the manner I've called aesthetic criticism and hope this leads the reader to see the object in a new way. On the other hand, if Wittgenstein is right, perhaps this is all that ever happens in mathematical proofs. As he said in a slightly different context [1953, section 144]:

> I wanted to put that picture before him, and his *acceptance* of the picture consists in his now being inclined to regard a given case differently: that is, to compare it with *this* rather than *that* set of pictures. I have changed

his *way of looking at things*. (Indian mathematicians: "Look at this.")

ACKNOWLEDGMENTS

I would like to thank my colleagues, John Connolly and Murray Kiteley of the Smith College Philosophy Department, and Jim Henle of the Mathematics Department for many helpful suggestions. This paper is much improved because of their critical readings.

REFERENCES

Borel, A., "Mathematics: Art and Science", *The Mathematical Intelligencer*, Vol. 5, No. 4, (1983) 9-17.

Cavell, Stanley, "Music Discomposed", reprinted in his *Must We Mean What We Say?*, Charles Scribner's Sons, New York, 1969 180-212.

Copland, Aaron, *Music and the Imagination!*, Harvard UP, Cambridge, 1953.

Davis, P. and R. Hersh, *The Mathematical Experience*, Birkhauser, Boston, 1981.

De Millo, R., R. Lipton and A. Perlis, "Social Processes and Proofs of Theorems and Programs", reprinted in Tymoczko, 1986b.

Fisher, Alec, *Formal Number Theory and Computability*, Oxford UP, Oxford, 1982.

Gadamer, Hans, *Truth and Method*, Continuum Crossroads, New York, 1975.

Hardy, G.H. *A Mathematician's Apology*, 1940, reprinted Cambridge UP, Cambridge, 1984.

Kuhn, Thomas, *The Structure of Scientific Revolutions*, 2d ed. U of Chicago Press, Chicago 1970.

Lakatos, Imre, *Proofs and Refutations*, Cambridge UP, Cambridge, 1976.

Livingston, Eric, *The Ethno-methodological Foundations of Mathematics*, Routledge & Kegan Paul, London, 1986.

MacIntyre, Alasdair, *After Virtue*, U of Notre Dame Press, Notre Dame, 1981.

Putnam, Hilary, *Reason, Truth and History*, Cambridge UP, Cambridge, 1981.

Quine, W. V., "Truth by Convention" (1936) reprinted in his *The Ways of Paradox*, Random House, New York, 1966.

Quine, W. V., "Two Dogmas of Empiricism" *Philosophical Review*, 60 (1951), 20-43.

Reid, Constance, *Hilbert*, Springer-Verlag, New York, 1970.

Rosen, Kenneth, *Elementary Number Theory*, Addison-Wesley, Reading, Ma., 1984.

Russell, Bertrand, "The Study of Mathematics", (1902) reprinted in his *Mysticism and Logic*, Doubleday, Garden City, NY, 55-69.

Tymoczko, Thomas, "Making Room for Mathematicians in the Philosophy of Mathematics", *The Mathematical Intelligencer*, Vol. 8, No. 3, 1986, 44-50.

Tymoczko, Thomas. *New Directions in the Philosophy of Mathematics*, Birkhauser, Boston, 1986.

Uspensky, J.V. and M.A. Heaslet, *Elementary Number Theory*, McGraw-Hill, New York, 1939.

Wittgenstein, Ludwig, *Philosophical Investigations*, Macmillan, New York, 1953.

Thirty Years After The Two Culture Controversy: A Mathematician's View

Philip J. Davis
Brown University
Providence, Rhode Island

The great cleavage between the provinces of the sciences and the humanities was, for the first time, made, or at least revealed, for better or for worse, by Giambattista Vico. Thereby he started a great debate of which the end is not in sight.

Sir Isaiah Berlin
The Sciences and the Humanities

The Two Nations: An Early Victorian Novel

In 1845, with the publication of the novel *Sybil, or the Two Nations,* by Benjamin Disraeli, a new phrase entered the consciousness of the British reading public. The novelist, who was an enigmatic personality, an epigrammatist of the highest order, a consummate politician and diplomatist, and who was destined to become twice Prime Minister of England, wrote that Queen Victoria reigned over two nations. Here is how Disraeli put it:

> ...two nations; between whom there is no intercourse and no sympathy; who are as ignorant of each other's habits, thoughts and feelings as if they were dwellers in different zones, or inhabitants of different planets; who are formed by a different breeding, are fed by a different food, are ordered by different manners, and are not governed by the same laws...THE RICH AND THE POOR.

In this novel, Disraeli the writer demonstrated his human concerns a generation before Disraeli the politician was able to translate this concern into political practice.

The Two Cultures

In 1959, shortly after the launching of the first Sputnik, a debate, known as the "Two Culture Controversy", broke out and was part of the British and American university scene for perhaps a decade. Briefly, this controversy related to a perceived "split" between the humanistic and the scientific-technological cultures. The controversy then quieted down and, as often happens to such things, seems to have been abandoned in favor of alternate rhetorical sensations. At the time, members of the mathematical profession, indeed of the scientific professions, whether inside the University or out, hardly responded to the debate; hardly uttered a peep.

The expression "two cultures" entered the language as a residue of the debate, but the debate itself, carried out primarily in an academic context that ignored the larger issues of economics and politics, was consigned to the ashbin of the irrelevant.

I think it of some importance to review the Two Cultures Controversy so as to speculate on how the two cultures—if indeed they can be separated into two—have fared vis-a-vis one another in the intervening years and also to offer one mathematician's view of where his subject fits with regard to the perceived split.

Enter: C. P. Snow

The opening blows were delivered in 1959 in a set of prestigious lectures known as the BBC Rede Lectures. The lecturer was Charles P. Snow (later knighted). Snow (1905-1980) was a man who had a variety of careers in his lifetime: scientist, science administrator, novelist, essayist, and towards the end of his life, a media personality and a generalized guru. He wrote a number of very successful novels *à clef* centering around Oxbridge politics. These novels were made into a television series which continues to be shown.

I shall give in summary form what I understand to be the gist of Snow's argument.
* There are two cultures currently in operation in the intellectual world: the scientific-technological and the humanistic-literary.

* Science has remade and continues to remake the world. The future lies with it.

* Science has profundity and intellectual beauty.

* Science is the greatest creation of the collective human mind.

* A humanist who knows nothing about, say, The Second Law of Thermodynamics must be regarded as uneducated as a scientist who knows nothing about, say, Shakespeare.

* The two cultures exist as discrete and separate groups. They do not communicate with each other.

* There is mutual incomprehension and hostility between the two groups.

* The literary culture (cult) is backward looking. It consists of exclusivists who view themselves as the New Aristocrats.

* The split between the two cultures is a very serious one for mankind.

It is clear to me, though there is no direct allusion, that in his choice of the expression "Two Cultures", Snow was echoing the words of Disraeli a century before. What lends additional force to this conjecture is that Snow's Fourth Rede Lecture, entitled "The Rich and The Poor", states that the world is now divided into the

---

The Two Culture Controversy appeared as the latest in a long series of intellectual conflicts that have shaken the Western World. There was the conflict around 50 B.C., with Cicero as hero, between The Word and The Deed.

---

rich nations and the poor nations and discusses the role that science might play in alleviating the plight of the poorer.

The claims of the "Two Culture" hypothesis were sweeping. They were offensive to a great number of thoughtful people, and they were probably meant to be. Other people, including many on the university scene, agreed totally with Snow. The gauntlet having been thrown down, who would respond to Snow?

Leavis

The first person whose answer received wide publicity was F. R. Leavis, (1895-1978), professor of English at

Cambridge University, literary critic and philosopher. Leavis had a reputation for brilliance and mordant wit. This much was commonly allowed. Other observers of the scene added that while this was the case, Leavis was also a snob, an autocrat, a crank, a paranoid, an anti-American, and quite generally an unpleasant fellow. Nonetheless, they admitted that he was an accomplished and forcible writer.

In his 1962 Richmond Lecture, Leavis attempted to deflate Snow's claims and to defend the humanistic claims. The Two Culture Controversy also became known as the Snow-Leavis controversy. I shall present a precis of Leavis' rebuttal.

* Snow's lectures and subsequent volume were accompanied by a great deal of publicity and media "hype" which inflated the claims beyond what they were worth.

* Snow compared Rutherford (Ernest Rutherford, 1871-1937, Cambridge theoretical physicist and Nobelist who investigated the atomic nucleus) with Shakespeare. Such a comparison is totally meaningless.

* There is a more basic creation of the human mind than science. It is language. Without language, the scientific edifice is impossible.

* Humanistic values are those which promote a consciousness of full human responsibility. This consciousness goes beyond psychology or social sciences or any of the "humanistic disciplines".

* The values embodied in the consciousness of full human responsibility are fostered by the interplay between living language and genius of a creative type: call this the literary sensibility. Literary sensibility deals with thought that is untranslatable into logic, or into mathematical symbols. (Some people added that what Leavis was asserting was simply that the only road to an education was through English literature, and that he, Leavis, was the final arbiter of what counted in that field. Thus: Dickens out; this in, that out, this in.)

* The mass university is in trouble and it would be hard to redeem it.

* The mass university is responsible for the cretinization of the public.

* The civilization it represents has, almost overnight, ceased to believe in its own assumptions. It

advocates ethical relativism, dehumanized computerization. It places all cultural values on a basis of equal validity.

* Nonetheless, we must attempt to redeem it. We must have an educated public and this can come about only through the University which must inculcate the values of the arts, language, ideas, criticism, etc.

Underlying both the attack and the rebuttal, there was a strong undercurrent of *ad hominem* argumentation which was entertaining at the time, but which I shall skip here.

An Old Conflict In New Garb?

To some students of the history of ideas (cf. O.A. Bird) the Two Culture Controversy appeared as the latest in a long series of intellectual conflicts that have shaken the Western World. There was the conflict around 50 B.C., with Cicero as hero, between The Word and The Deed. The word, embodied in rhetoric, is central. (Or so it was claimed.) The word is prized for its beauty, its embodiment of knowledge, its power over human affairs. In opposition, there are the deeds, the beliefs, the customs and the institutions.

In the 13th century, one perceives the seeds of the conflict between science and religion. The theological ideal asserts that religion is the goal of life towards which all things must tend. All forms of knowledge must conform to a religious purpose; the arts and sciences, therefore, are the handmaidens of theology.

The central personality in this conflict is Saint Thomas Aquinas (1225-1274) and it is interesting that in Aquinas and quite generally in the Mediaeval Scholastics, we see religion itself, bending slightly toward the ideals represented by classical Greek geometry, insofar as theology was bolstered by an appeal to logic. As a result, it has been said that Thomas represented a theology of the mind whereas Augustine (354-430) represented a theology of the heart.)

The Two Culture Controversy then is, in a sense, an extension of this medieval argument. Modern claims against the sciences charge them with wanton destructiveness, depersonalization, a mechanistic, materialistic view of nature and of people (note the saying of Marvin Minsky, one of the leading computer theoreticians, that people are just meat machines), the introduction of thoughtless mathematical criteria, and the intellectual imperialism of a positivist philosophy which asserts that the only possible knowledge is scientific knowledge.

Scientists were also charged with social indifference and a kind of intellectual hauteur as exemplified by the initial euphoria of the creators of the A-Bomb who asserted that the confirmation of a theory of physics of such dimensions was well worth putting mankind at risk. The fact that there was then formed a *special and separate* organization known as the Union of *Concerned* Scientists may be taken as a measure of the general indifference of scientists of a generation ago.

Claims against the humanities, presented for the most part by the humanists themselves, charge them with outworn values, and a production in art and literature which is of low stature and content. Literature is either of the mass market type with debased values, or has split into small, mutually non-communicating, incomprehensible coteries.

A generation ago, Sir Herbert Read, art and literary critic and philosopher, wrote that current art and literature were characterized by excessive egotism, by excessive excitation and sensationalism, by escapism. Literature did not, in Read's opinion, confront the human condition, and it denied the tragic sense of life.

Jacques Barzun, professor of literature at Columbia University, in his A.W. Mellon Lectures of 1972 entitled *The Use and Abuse of Art*, opined that the current state of art was terrible. It was done in a vacuum of belief. It was overly self-conscious and self-examining, and self-consciousness was held by many to

---

[Art] exhibited both redemptive and destructive features. Finally, it was tempted by science, in that it proposed to bring in scientific methodological modes and modes of interpretation. This was perceived to be fatal for the humanities.

---

be an indication of a loss of vitality. It exhibited both redemptive and destructive features. Finally, it was tempted by science, in that it proposed to bring in scientific methodological modes and modes of interpretation. This was perceived to be fatal for the humanities.[1]

Apart from the claims for and against, the two cultures were said to have generated two entirely different kinds of languages, and two different sets of adjectives were thought to describe these languages. Thus, humanistic language was said to be metaphorical, ambiguous, obscure, extra-logical, concrete, organic, personal, aesthetic, non-verifiable, expressive, subjective.

Scientific language, on the other hand, was said to be literal, unambiguous, clear, precise, abstract, mechanological, nonaesthetic, verifiable, descriptive and objective.[2]

Reconciliations

The gauntlet having been thrown down by Snow and picked up by Leavis, over the next decade journals and books, from popular to professional, were filled with thoughtful partisan responses and reconciliations. A complete bibliography would be vast. I shall allude briefly to two: the views of R.G. Collingwood (enunciated before the Snow-Leavis controversy) and of Aldous Huxley.

R.G. Collingwood (1889-1943) was a British historian and philosopher. To reduce his main points to a few words, Collingwood said

* There are not two cultures but five cultures. They are: art, religion, science, history, and philosophy.

* Each of these has claimed at some time to be *the sole, unequivocal,* possessor of knowledge.

* The fight between art and science was inevitable and it will continue. It is based on the misunderstanding that art and science are independent.

* One should cultivate the metaphor of "the country of the mind". Arts and sciences reside in the same country.

If Collingwood has discovered five cultures, what is to prevent us from discovering ten? And perhaps one should even update his argument by alluding to a possible relationship between these "cultures" and the idea put forward by Howard Gardner that human intelligence is not a single monolithic entity but that there are many possible dimensions of intelligence.

Aldous Huxley (1894-1963), who as a distinguished novelist and critic, and who as the grandson of Thomas Henry Huxley was quite immersed in science, had a foot in both camps. He proposed a rapprochement along lines appropriate to a writer. Consider, said Huxley, the birds singing in the garden. The older, romantic view, was that the birds sang to their lady loves. The newer view is that they sing to lay down territorial claims. The writers of the future, Huxley hoped, growing up with this knowledge, will find interesting ways of

having the birds sing—for territorial reasons.

These, then, are the opinions of a generation ago. Through the fray, as I have indicated, most scientists, like B'rer Fox, lay low, said nothing, were hardly aware of the controversy. Perhaps they had no need to, convinced as most of them were, that the age was a scientific age, that the major modalities of expression whether in act or in language were scientific, and that the future lay with it.

Where Do We Stand Today?

How would one characterize the world since the Rede Lectures of C. P. Snow? In the wider world, there has been terrorism, both of the body and of the mind, increased nationalisms, and a deepening split between the haves and have-nots, both individual and national. There is an international drug culture (illicit) which keeps millions of humans in thrall and has important economic implications.

Malaise with present values is overturning heartless and dogmatic political and social arrangements. The spiritual vacuum left in the wake of the death of idealisms is often filled by a return to mysticism, to religious fundamentalism and to an intolerance that has

---

In technology, there is no question but that the major event has been the widespread computerization which is entering every aspect of human life and changing it.

---

found little reverberation and support in current intellectual thought but which can be fierce enough in its violent and unyielding claims to tear apart humans and nations. On the side of sanity, there has been a quiet search for transcendental values and a search for new intellectual justifications for such values.

What have been the major events of the past generation in the arts and sciences and how have they affected the perception of the Two Culture Controversy? In technology, there is no question but that the major event has been the widespread computerization *which is entering every aspect of human life and changing it.* The potential for invasive, algorithmic social manipulation on the part of governments has, for better or for worse, been increased tremendously by computers.

Medical technology has often made great strides by

postulating the human body as the source and the recipient of spare parts, by regarding it as an object for procedures (i.e., medical algorithms) and for techno-biomedical manipulation. Dean of Brown Medical Program Stanley Aronson recently pointed out that the only part of the human body that cannot now be replaced is the brain.[3]

Other technological advances have been equally spectacular, if slightly less visible on the surface: think of the laser, the semi-conductor, and superconductivity. Think of the miraculous drugs (licit) that modern pharmacology has created. Think of the jet plane, nuclear reactors and bio-engineering.

Each of these seems to have revolutionized our lives even as each presents its problematic aspects. I say "seems" because I do not think that the full force of these developments will be felt for many years. It is clear, for example, that raising the average life span by, say, ten years, is bound to have profound economic, cultural and moral aspects that are only now beginning to surface.

The technology of space proceeds shot by shot, and is now followed with enthusiasm only by a small fraction of thinking people, including scientists, and is agreed to by politicians mainly in the context of defense. It requires a great leap of the imagination to believe that a great leap into space will solve any human problems, except possibly as a result of spin-offs.

In theoretical physics, many ideas have been put forward often with great eclat; brilliant, but occasionally of ambiguous stature and certitude. One critic[4] has observed that in striking a balance between the empirical and the theoretical, contemporary physics has

---

One critic has observed that in striking a balance between the empirical and the theoretical, contemporary physics has moved substantially toward the latter, being inspired (and tempted) more and more by the formalisms of mathematics.

---

moved substantially toward the latter, being inspired (and tempted) more and more by the formalisms of mathematics. Theoretical physics deals increasingly at the edges of the universe, the very small and the very large, and as it does, real world experience becomes more and more attenuated, and physical intuition is tempted and pulled by mathematical theory. In many instances it is hard to discern what is the borderline between what is physics and what is mathematics.

The literary imagination, when projected into the arena of space, appears to create more dystopias than utopias, and wallows in a mixture of advanced techno-cosmology and the evocation of primitive social organizations and ethics. It would not be inappropriate to call it the vision of *techno-barbarism* and to judge that in a world that possesses ultimate weapons it is almost an artistic necessity.[5]

The graphic arts/film have remained on a spiritual plateau even as the potentialities of image creation have multiplied wondrously. Technological advances, including those which derive from the mathematization of images has enabled us to create visual realities that exist only as bits in a computer memory but which seduce the eye into belief in their real-world existence. Despite this, graphics, which suffered a very serious jolt with the invention of photography in the 1830's, has wandered bereft of a valid iconography and has settled at an imbalance of the destructive/redemptive ratio greatly in favor of the former.

Literature has split into the banal, the esoteric, and the sci-fi, with sci-fi commanding a wider and wider audience and developing its own world of publishing, distribution, and its own critical apparatus. It itself has split into hard sci-fi, sci-fantasy, "new wave", and other subfields. Little by little, in the same way that some classical composers in the 1920's adapted and incorporated the jazz idiom, there has been a cozying up to sci-fi material by authors and by critics who began their careers by dealing with "conventional" material.

Playwrights and directors have substituted motion and sensation for words; have substituted audience shock and guilt and participation for audience elation. There is very little for the ear in the average new play. One hundred years of movies and fifty of the computer have killed the static image. I was appalled the other day when visiting a small and nicely set out local museum when I realized that I was totally bored by a cabinet full of stuffed birds. The dynamic, video surrogate was far more engaging than the real thing.

The very act of reading has itself been threatened by the development of high-tech modalities of information transfer, paradoxical in view of the fact that the gross amount of paper consumed has gone up remarkably. As verbal material has become easier and easier to produce mechanically, the act of reading confines itself more and more to the trivial (but often critical: read and fill out such and such a form in many, many copies).[6]

In a minor way, the increased frequency of abstract hieroglyphs displayed publicly (men, women, telephone, food, no right turn), all employed in the name of universal comprehensibility, seems to augur a new age in which, by a perverse operation of Occam's razor, the number of concepts widely available and in the common vocabulary will have been substantially reduced.

At the same time, the Apple Computer Company in a widely circulated brochure teases us with the assertion "there are just some things you can't get from books", as a come on to buy "courseware" which offer a "round trip ticket to 17th Century France", a "half million dollars worth of lab instruments" (for surrogate lab experiments), "a closeup look at a moving eyeball", and "a private performance of your latest minuet", with music derived from a score. This kind of thing is not produced without considerable moral support from the professional humanists.

Speaking of music, in the past thirty years, instrumentalists of the traditional variety have increased in numbers and in quality. There are now more performers of concert calibre than in the whole history of music. Busking is back, often of fine quality the spirit finds expression in music but not much bread and butter. On the other hand, musical composition has retreated into the primitive or the impossible. And those who produce the impossible often do it to the accompaniment of breast beatings that take the form: Why am I producing this; for whose benefit; is such and such a musical cadence moral (!). Even Broadway, which was once the source of much that was original, accessible, pleasurable and middlebrow has failed to come up with a tune that one can retain in one's memory for more than a half second.

Personal singing, recitation of poetry died of embarrassment in the West.[7] Passive listening is in; active performing in the living room is out.

The vast auditory potentialities presented by electronic and digital techniques result mostly in live performances that are so loud that they are ear-destroying, or has resulted in reissues of Bach or Beatles according to the latest Fourier digital purification, the latest redefinition of the orchestral "A" sound, or the latest frisson of personal interpretation.

Mathematical analysis, resulting in music as a probabalistic process, has yielded very little to attract attention. Musical training and composition has been tempted by mathematics and physics and despite the fact that Mozart wrote an article on the mathematization of composition, it has failed to deliver a single composition of an electronic variety that anyone would care to give a K. Number to.

The word "experimental" which for a while seemed to be the property of the scientists has been so overworked in music, in literature, in the arts, in education, even, that the term now evokes indifference or derision even as it may attract foundation money in substantial quantities. To be funded, what is "experimental" must often be backed up by "scientific" evaluations that display standard deviations of this and that computed routinely and most often thoughtlessly and pointlessly. The ritualistic display of such numbers that gives the process a cachet of the scientific and objective.

On top of all this, there has evolved a criticism of literature and art and music, that is itself tempted by science, but has not yet discovered the productive lines along which it should yield to such temptation. Contemporary criticism is logico-analytic in spirit. It derives from or is allied to the structuralist schools of the Bourbaki mathematicians and the Piagetian structuralist school of cognition and learning, and looks at the manner in which texts are constructed and not what they are saying in human terms.

Current criticism, whose conceptual novelty and brilliance is often great, constitutes a vast apparatus of self-justification. It is also a criticism which refuses to admit that historical ages differ in what they are able to bring forth, that forms of expression may suffer from exhaustion and dry up. An individual is permitted to suffer from burnout, and to experience angst as a consequence, but our age is not. It is only human that those who live by ideas would like to assert they are living in a great age of ideas or that those who live by action should assert that they live in days of heroic actions.

Ours, then, is an age which wants desperately to create in every known modality, which often gives lavish support to its creations and to the cult of creation itself, and which, as a consequence, suffers terribly from grade inflation in judging what our children and what our professionals have created. Often the issue of censorship is raised and is a smokescreen for the fact that what is being argued for has little artistic merit.

On the university scene, (and here I follow the opinions of an undergraduate dean at Brown University, herself a humanist, as well as my own observations), one sees the humanities in disarray and in some decay. Less money is available for humanist salaries and grants than in the sciences. There are a declining number of majors in these areas and, recently, particularly among native born American students, in the sciences. The predominating desire of students is to be programmed for success, and parents, themselves, do not enjoy the thought of sponsoring a child who displays monastic tendencies or who is willing to endure the low status and low pay of a life of lonesome scholarship. The goal is the MBA degree with its mechanical thought (or lack of thought), and with its ultimately mathematically based concerns at a high metalevel of structural manipulation far above the mere production and distribution of food, shelter and clothing.

To be sure, some university courses have been offered that blend the sciences with the humanities in an attractive way. But they exist on the fringes; they are largely disregarded by the individual departments and disregarded by the students who themselves sense the disdain for such in an age when only the specialist can make "real" money.

A devoted teacher has written me "Individual thought and initiative are replaced by the idea of *procedures,* and this is bolstered in both the arts and sciences by the increased exposure to canned computer procedures."

At the undergraduate level, England, the birthplace of the Two Culture Controversy, seems to be more split into two cultures than the USA. An undergraduate in England does courses ONLY relevant to his major. While it is true that undergraduates enter college having

---

Literature has split into the banal, the esoteric, and the sci-fi.

---

done their 'A levels' which is roughly about sophomore college level in the USA, after you do your A levels, you are pretty much limited to those areas which will become your college speciality.

Where, then, does general education in England come from? From the O levels, completed at the age of fifteen? Generally speaking, an English college graduate will know more in his specialty and know less of a general nature. This was the background from which Snow was arguing.

This has changed a bit in recent years, but an English don of my acquaintance described the change as an attempt on the part of the government to buy college education on the cheap. After all, he added, the government consists almost exclusively of Oxbridge people who did the humanities (Snow was an exception), and "those fellows look after themselves".

Surveying the total scene, I think one is tempted to cry out, paralleling the Emperor Julian's comment on the triumph of Christianity: "*Vincisti, O Scientia*". Thou hast conquered, O Science. The battle is over, victory has been pronounced by Snow, and the shouting has long died down.

But the victory, if it is a victory, is hollow, possibly pyrrhic, not because as Collingwood suggested there is

but one country of the mind but more aptly, as Howard Gardner and Clifford Geertz both suggest, there may be many countries of the mind and one must live in close proximity to them all. What one wants is not to cry "victory", but "synthesis", and to assert that synthesis is possible.

Where Mathematics Fits In

It is now time to look specifically at the field of mathematics. In an age that is increasingly mathematized, with a subject that has been traditionally called "The Handmaiden of the Sciences", can it be asserted firmly that mathematics must be reckoned as one of the sciences? That would seem to be the case. I am familiar with many universities that have a Department of Mathematical Sciences, meaning that pure mathematics and applied mathematics, logic, statistics and computer science are all gathered under this one designation, but I am not aware of a single university that has a Department of Mathematical Arts.

And yet, is mathematics so bereft of humane values that it merits this isolation? Was not, in the period of classical education, mathematics part of the traditional quadrivium? Was there not, even within memory of living upper-class British, the psychological association of mathematics with the classics? While it would never have been appropriate to do "the stinks" (the pejorative word in the U.K. for chemistry, etc.) mathematics, although possessing a certain relationship to science and technology, remained pure and unsullied, with humane values attributed to it.

And the people in the field, what do they think about the split? Do mathematicians, contemplating the humanists in the next building feel the way Snow did, paralleling Disraeli, that they are

> moving among two groups, comparable in intelligence, identical in race, not grossly different in social origin, earning about the same incomes, who have almost ceased to communicate at all, who, in intellectual, moral and psychological climate, have so little in common, that...one might have crossed an ocean.

I think the answer to this question is, by and large, "yes". I shall elaborate with a number of true stories.

1. Several years ago, I submitted an article (rather like this one) to a mathematics journal whose readership was quite general (among mathematicians, that is). My article contained a number of literary allusions. The editor of the journal returned the article to me with the notation: "Get rid of all your references to literature. I am not familiar with them and I find their inclusion offensive." Although in my experience, I have found that while scientists are much more willing to embrace

the humanities than vice versa, I am appalled at the extent to which my fellow mathematicians, Continental as well as American, who are brilliant craftsmen in their fields, often exhibit little interest in general intellectual ideas.

2. I submitted to a publisher the manuscript of a book which I hoped and thought would have some possibilities as a trade book. My manuscript contained a few pages with some mathematical equations. The publisher begged me to get rid of those equations. "Equations in a trade book are poison. No one will buy the book."

3. An acquaintance who teaches philosophy and the philosophy of science at one of the English "red brick" universities told me that he was worried about his job in view of the government's hard nosed view of what university education should be all about. Another acquaintance at the same university, but in the mathematics department told me that she had no such worries. We're all right, Phil, she told me.

4. A colleague reports that when he spent a few months at the Institute for Advanced Studies in Berlin in 1986, an institute that embraces both the sciences and humanities, the split between the two was so severe that it was as though the institute membership had been cut through with a knife.

At this institute, a series of "general" lectures was announced. The scientists turned up at the humanists' lectures and were bored stiff. The humanists did not turn up at the scientific lectures. No way.

A Synthesis: Some Possibilities

There is temptation which may be counterproductive and there is genuine fusion and synthesis. There is reconciliation. There is rapprochement. Some poets and novelists—Kurt Vonnegut, for example, have gone

---

The very act of reading has itself been threatened by the development of high-tech modalities of information transfer

---

the route suggested by Huxley. Some writers, much more plugged into computers and going with what they sense is the wave of the future, are now producing "interactive novels". These computer programs exhibit the principal feature of postmodernism: the simultaneous operation on several conceptual levels; the reader/programmer who becomes an integral part of the

story he reads.

On the scientific side, there are writers like Oliver Sachs, Stephen J. Gould, Lauren Eiseley, who have tried to infuse human values into science—but very likely at the cost of loss of status within their respective research communities.

Turning again to mathematics. In the severe judgements just cited pronounced on art by Jacques Barzun, let us replace the word "art" by the word "mathematics" and try to formulate parallel and hopefully true statements. The aptness of some of the statements which follow is not limited to mathematics, but may also be valid for physics, etc.

* Research mathematics is not in terrible shape. It is in fine shape. There are lots of people working at it with energy, enthusiasm and brilliance. It is quite well supported by the US Government and other governments of the advanced nations. (To a considerable extent out of defense money. What will happen in a post-cold war age is anyone's guess.) Old problems are solved. Mathematics is full of new and interesting problems, new ideas, new methodologies. New applications emerge daily. Whether this century is, as is sometimes claimed, as great an age for mathematics as were the previous three centuries, is a matter for the future historians of the subject to say.

* Mathematics is done in a vacuum of belief—of sorts—if one interprets this vacuum to refer to a formalist philosophy which most (pure) mathematicians adopt when queried about the essential nature of the raw materials they work with.

* Mathematics is self-conscious and self-examining, but this is the hallmark of its methodology rather than an indication of an enfeebled art.

* Mathematics possesses an aesthetic component which is strong and which is immediately apparent to the practitioner. Perhaps the aesthetic of mathematics is more palpable at the higher levels of the subject and to those individuals of "mathematical receptivity". However, this type of sensibility can be nurtured from an early age.

Unfortunately, the aesthetic sense is rarely conveyed in course work, and judgement is often pronounced that while mathematics is necessary, like Metamucil perhaps, for the healthy operation of modern civilization, it has about as much aesthetic appeal as the contents of the New York Telephone Book.

As in the graphic arts, there is useful tension between the representational and the making of beautiful patterns, those in mathematics being for the most part logical rather than pictorial. In recent years, of course, computer graphics has contributed greatly to the latter.

Where mathematics is sometimes enfeebled is in the excessive stress that mathematical research often places on the aesthetic component at the expense of its other components. In stressing the aesthetic, mathematics is tempted by Art with a capital A, although one can say with considerable truth that the productive conjunction of mathematics and art (with a small a), architecture and design since antiquity is well attested.

* Mathematics exhibits both redemptive and destructive features. If there has been a shortfall of self-examination, it is not in the examination of its own inner material by its own methodology, but it is in a steady refusal to examine how the characteristic features of mathematical thought operate and affect us.[8]

This refusal is often accompanied by a head-in-the-sand ostrich stance that says: What I am doing could not conceivably do harm to a fly. It forgets the fact that every mathematician is a teacher and every student walks out of the classroom or laboratory as an independent spirit who will do what he will with the materials he has gained.

* Mathematics has a history. Except for portions that derive from deep antiquity, one knows fairly exactly who created what, where, when and how. The total context of the development is often not perceived, and it is to the credit of modern historians of the subject that they are trying to fill in the gaps.

I emphasize this point, not because history is something unique to mathematics—presumably human ideas of whatever sort all have histories—but because it is often asserted that the truths of mathematics are atemporal, and hence, stand outside history. If this were the case then mathematics, intellectually speaking, would stand outside both the two cultures.

* Like literature, mathematics can be an avenue of mental escape from this world. The unworldly mathematician, living on Cloud 9 among his theories, is a stock figure.

* Mathematics, like poetry, has both precision and ambiguity. (Will the true geometry please stand up. Or: consider the predictive ambiguity implied by the recently explored and developed "deterministic chaos". There is also deductive ambiguity in that in alternate presentations of one and the same mathematical area, what is definition and what is theorem may occupy interchanged roles.)

William Empson has written of seven types of ambiguity in poetry. The perceptive critic of

mathematics could easily double that number.

* Mathematics can be rhetorical. This seems strange, for if rhetoric is the art of persuasion, then mathematics would seem to be its antithesis. Not that mathematics does not persuade, but it is commonly thought that it needs no art to perform its persuasion. This is a misconception. (See Davis and Hersh)

* Mathematics, like literature, has metaphor. (E.g.: mathematical models of real world phenomena. Or: even terms such as 'open', 'closed', 'amicable', 'lifting', 'surgery', 'space' [as in a function space] which are used metaphorically. There are hundreds of such terms and they lend added force and picturesqueness to abstract conceptualizations and arguments.)

* Mathematics has a sense of outcome; a feeling of rightness; a sense of catharsis as the discovery or development of a process works its way to a conclusion. In mathematical physics, for example, the sense of "rightness" gained from experience is often a major principle of validation.

In a remarkable passage, Edmund Wilson (1895-1972) who, in his day, was the leading American literary critic, was able to discover the kinship of the sense of pattern and outcome in drama with that of mathematics:

> ... all our intellectual activity... is an attempt to give a meaning to our experience...; for by understanding things we make it easier to survive and get around among them. The mathematician Euclid, working in a convention of abstractions, shows us relations between the distances of our unwieldy and cluttered-up environment upon which we are able to count. A drama of Sophocles also indicates relations between the various human impulses which appear so confused and dangerous... (and) of the way in which the interaction of these impulses is seen in the long run to work out... (and) upon which we can also depend. The kinship, from this point of view, of the purposes of science and art appears very clearly in the case of the Greeks, because not only do both Euclid and Sophocles satisfy us by making patterns, but they make much the same kind of patterns. Euclid's *Elements* takes simple theorems and by a series of logical operations builds them up to a climax in the square on the hypothenuse. A typical drama of Sophocles develops in the same way.
> —The *Historical Interpretation of Literature.*

*   Mathematics has paradox.

*   Mathematics has mystery and can convey awe.

And the grand mysteries that spread over the whole enterprise are philosophical: why mathematics is true and why it is useful.

* Mathematics is a language—a perception that goes back as far as Aristotle. Mathematicians have perceived other similarities to language and its uses. The creator of semiotics was primarily a mathematician and the son of a mathematician: Charles Sanders Peirce.

Brian Rotman, who has recently advanced the studies of the semiotics of mathematics, has written "Are mathematical proofs akin to stories, dialogues, arguments constructible in natural languages? In what sense could there be substantive commonalities between

---

To be funded, what is "experimental" must often be backed up by "scientific" evaluations that display standard deviations of this and that computed routinely and most often thoughtlessly and pointlessly. The ritualistic display of such numbers that gives the process a cachet of the scientific and objective.

---

mathematical modes of discourse and those of other systems?" Rotman goes on to argue that semiotics provides a framework "specific enough to discuss natural language and sufficiently loose to include all sorts of other modes of sign use." Here we would have a reconciliation of Leavis' assertion of the primacy of natural language with the existence of an alternate mathematical symbolism.

But the claims of mathematics as the *Universal Language*, the *characteristica universalis* of Leibniz and others, falls to the ground when one realizes that the practice of mathematics is itself split into dozens of mutually incomprehensible specialties, each speaking its own dialect.[9]

* Mathematics is allied to and has contributed mightily to philosophy. (This is so well known that it hardly needs elaboration.) G-C. Rota, who is both a mathematician and philosopher, laments that in the past century philosophy has been greatly tempted by mathematics, much to its loss.[10]

I do not mean to imply here that morality is a

monopoly of the humanists. One may be reminded of Berthold Brecht's "Erst kommt das Fressen, dann kommt die Moral" and ask crudely how do the humanists and the technologists compare in delivering the bread. Or consider the enormous moral damage and physical suffering caused by some intellectuals, e.g., Joseph de Maistre. See, I. Berlin, l.c., or *The Crooked Timber of Humanity*, Knopf, New York, 1990.

* Mathematics has contributed to theology. It is seen as an instantiation of various theological ideals (On this point see Funkenstein.) Theology has contributed to mathematics. For example, deeply felt interpretations of man's fate have led to two major branches of applied mathematics: the deterministic and the probabilistic.

Mathematics grasps for the transcendental in many ways and in so doing, can be a surrogate for traditional forms of religious expression.

* Like anthropology and literature, mathematics embodies mythologies. I use the term "myth" not to mean that which is false, but that which is accepted as normative within the culture. My friend and occasional coauthor, Reuben Hersh, has written about the four myths of mathematics: its unity, objectivity, universality, and certainty.

* As in the arts, (goodbye rock; hello rap!), new forms and styles in mathematics emerge so fast that it is sometimes hard to distinguish what is fad and what is of lasting value. And this ambiguity is exacerbated by the myth that says that insofar as all mathematics is equally true, it is all potentially applicable in the physical or social worlds, and deserves an intellectual status in proportion to a perceived inner depth or complexity.

I have used the words "metaphor", "paradox", "ambiguity", "catharsis", etc., deliberately. These words are found in the humanistic vocabulary but not normally in the scientific; and the degree to which I am not understood across humanistic/scientific lines (over and above my own limitations as an expositor) may be taken as a measure of the extent to which the two culture split persists. It would be a rewarding enterprise to fill in and to carry forward in much greater detail the analogies that I have been able only to sketch lightly here.

Toward The Future

There is a growing, often inchoate, realization that a satisfactory life cannot be built exclusively on scientific values. The time of rebellion has run out on the values of such scientists as I.I. Rabi (1898-1988), Nobelist in Physics), who was reared in a religious atmosphere, rejected it outright, and who found in scientific positivism a surrogate religion. We require a synthesis of many elements: the five cultures of Collingwood, the modes of Geertz and of Gardner as starters. Part of the construction (or reconstruction) will come about from the realization that each culture contains elements of the other.

Despite the increasing mathematization of every aspect of our lives, from credit cards to medical diagnosis, from word processors to black holes (which is largely a mathematical construct and one that has a strong grip on the imaginative, fantasy-creating faculties) the humane values that mathematics exhibits, the points of contact between mathematics and the humanities are by and large ignored by mathematical educators and hence are totally unappreciated by the wider public. The tough minded might now ask: so what? To which I answer that we must believe that to change someone's perception makes a difference in this world.

It is vital that mathematicians come to realize once again and to teach that mathematics carries and always has carried humanistic seeds. These seeds must be nurtured much more than they have been in the past. If one can reposition mathematics as a human institution, and not an institution of Platonic remoteness, nor as an institution which looks exclusively to sell its products "off the rack", then the shade of a worried Giambattista Vico will look down and say "All is now well."

NOTES

1 Here is a strong formulation of this, taken from the recent writings of Sir Isaiah Berlin: "In ironical language... Maistre, no less than the German romantics and after them the French anti-positivists Ravaisson and Bergson, declared that the method of the natural sciences is fatal to true understanding. To classify, abstract, generalize, reduce to uniformities, deduce, calculate and summarize in rigid timeless formulas is to mistake appearances for reality, describe the surface and leave the depths untouched, break up the living whole by artificial analysis, and misunderstand the processes both of history and of the human soul by applying to them categories which at best can be useful only in dealing with chemistry or mathematics."—I. Berlin: "Joseph de Maistre and the Origins of Fascism", New York Review of Books, Sept. 27, 1990.

2 O.A. Bird. These parallel columns of adjectives, set in opposition to one another, may put us in mind of the famous parallel columns of William James when in 1906 he introduced the terms "tough minded" and "tender minded" into philosophy. The "tender minded" were characterized as being "rationalistic, intellectualistic, idealistic, optimistic, religious, free-willist, monistic, and dogmatical" while the "tough minded" were the opposite.

James goes on to say: "...you know what each example thinks of the example on the other side of the line. They have a low opinion of each other. Their antagonism, whenever as individuals their temperaments have been intense, has formed in all ages a part of the philosophic atmosphere of the time. It forms a part of the philosophic atmosphere today." William James, *Pragmatism*, Lecture I.

The Jamesian split seems to cut across the Snow-Leavis split. I would judge that the mathematicians who are my contemporaries are tough-minded in three of the Jamesian categories and tender-minded in five.

And how would James have dealt with this: that both mathematics and mathematical physics seem to create realities out of their own substance. Or that this author has characterized mathematics as "true facts about imaginary objects" ? (Davis and Hersh)

3 Just you wait! One of the dreams of Computer Science is to replace the brain by a mathematical surrogate.

4 David Berlinski. Personal communication.

5 Blossom Kirschenbaum. Personal communication.

6 No joke this: scientific proposals to the Army must be submitted in twenty copies. In some instances, there are not that many people in the world who are really qualified to pass judgement on the scientific merit of the proposal.

7 The American Poet Edwin Honig tells of a get-together in Moscow of Soviet and American poets not so long ago. The Soviet poets recited from memory seemingly infinite quantities of poetry. The American poets could hardly recite their own works.

It would seem that the demands on memory in the literary field have changed from dramatic recitation to knowledge of the hundreds of little dinky instructions and algorithms that must kept in mind in order to use a word processor.

The wide variety of computer "languages" has extended the meaning of the word "language" and has compromised the study of natural languages, e.g., the replacement of a natural language by a computer language in requirements for a degree.

8 See, e.g., P.J. Davis and R. Hersh: *Descartes' Dream*, Section VI, where the excesses of rationalism have been called into question. This questioning occurs also in Z. Bauman's Modernity and the Holocaust where Holocaust-style phenomena are seen not as a failure of "human, purposive, reason-guided activity" but as

"legitimate outcomes of the civilizing tendency and its constant potential".

[9] See, P.J. Davis: *The Tower of Mathematical Babel*, SIAM NEWS November 6, 1987. I recently heard of one mathematician who speaks of a neighboring subspecialty as being written in a "foreign terminology".

[10] "Philosophers of all times... have suffered from recurring suspicions about the soundness of their work and have responded to them as best they could. The latest reaction against the criticism began around the turn of the century and is still very much with us. Today's philosophers (not all of them, fortunately) have become great believers in mathematization. They have rewritten Galileo's famous sentence to read 'The great book of philosophy is written in the language of mathematics'." —G-C. Rota.

BIBLIOGRAPHY

Barzun, Jacques, *The Use and Abuse of Art*, A.W. Mellon Lectures in the Fine Arts, Bollingen Series LXXV, Princeton University Press, 1972.

Bauman, Zygmunt, *Modernity and the Holocaust*, Ithaca: Cornell University Press, 1989.

Bird, Otto A., *Cultures in Conflict*, University of Notre Dame Press, 1976.

Collingwood, R.G., *Speculum Mentis or the Map of Knowledge*, Oxford: Clarendon Press, 1924.

Davis, Philip J. and Reuben Hersh, *Descartes' Dream*, Harcourt Brace Jovanovich, San Diego, 1986.

Davis, Philip J. and Reuben Hersh, *Rhetoric and Mathematics*, in: The *Rhetoric of Human Sciences*, Nelson, McGill, and McCloskey, eds., University of Wisconsin Press.

Funkenstein, Amos, *Theology and the Scientific Imagination* Princeton: University Press, 1986.

Gardner, Howard, *Frames of Mind: the theory of multiple intelligences*, New York, Basic Books, 1983.

Geertz, Clifford, *Local knowledge: further essays in interpretive anthropology*, New York, Basic Books, 1983.

Hersh, Reuben, *Mathematics has a front and a back*, Department of Mathematics and Statistics, University of New Mexico, Albuquerque, New Mexico, January 1988.

Huxley, Aldous, *Literature and Science* London: Chatto and Windus, 1963

James, William, *Writings: 1902-1910*, New York, Literary Classics of the United States, Inc., 1987.

Leavis, F.R., Two Cultures? : *The Significance of C.P. Snow*. With an essay on Sir Charles Snow's Rede Lecture by Michael Yudkin. New York, Pantheon, 1962.

Ozik, Cynthia, *Science and Letters—God's Work—and Ours*, The New York Times Book Review, September 27,1987, p. 3, 51.

Read, Sir Herbert, *Art and Alienation: the Role of the Artist in Society*, London, Thames and Hudson, 1967.

Root-Bernstein, Robert Scott, *On Paradigms and Revolutions in Science and Art: The Challenge of Interpretation* , Art Journal, Summer 1984, pp.109-118.

Rota, Gian-Carlo, *Mathematics and Philosophy: The Story of a Misunderstanding*. Department of Mathematics, MIT, Cambridge. 1990.

Rotman, Brian, Personal communication, 1990.

Snow, C.P., *The Two Cultures and the Scientific Revolution*, Cambridge University Press, 1959.

Steiner, George, *A False Quarrel?*, In the Symposium on The Two Cultures Revisited, The Cambridge (England) Review, March 1987. See also the articles by Noel Annan, Michael Black, Steven Rose, and Philip Snow in this symposium.

Wilson, Edmund, *The Triple Thinkers* , New York: Farrar, Straus and Giroux, 1976.

ACKNOWLEDGEMENTS

I am grateful to the Alfred P. Sloan Foundation for supporting my studies relevant to this article.

I wish to acknowledge the valuable comments of Candace Kent, Blossom Kirschenbaum, Peter Lax, and Anneli Lax.

The Concept of Mathematical Truth

Gian-Carlo Rota
Massachusetts Institute of Technology
Cambridge, Massachusetts

dedicated to Jack Schwartz
on his sixtieth birthday

1. Introduction.

Like artists who fail to give an accurate description of
how they work, like scientists who believe in
unrealistic philosophies of science, mathematicians
subscribe to a concept of mathematical truth that runs
contrary to the truth.

The arbitrariness of the accounts professionals give of
the practice of their professions is too universal a
phenomenon to be brushed off as a sociological oddity.
We shall try to unravel the deep-seated reasons for such
a contrast between honest practice and trumped-up
theory. We shall not focus on the psychological
aspects of the problem. We shall look instead for the
philosophical source of this state of affairs.

We shall firmly take a view of mathematical activity
drawn from observed fact, in opposition to the
normative assertions of some recent philosophy of
mathematics. An authentic concept of mathematical
truth must emerge from a dispassionate examination of
what mathematicians do, rather than from what
mathematicians say they do, or from what philosophers
think mathematicians *ought to do*.

2. The ordinary concept of mathematical truth.

The description of mathematical truth that is nowadays
widely accepted runs roughly as follows. A
mathematical system consists of axioms, primitive
notions, a notation, and rules of inference. A
mathematical statement is held to be true if it is
correctly derived from the axioms by applications of the
rules of inference. A mathematical system consists of
all possible true statements that can be derived from the
axioms. The truth of the theorems of mathematics is
seldom seen by merely staring at the axioms, as anyone
who has ever worked in mathematics will tell you.
Nevertheless we go on believing that the truth of all
theorems can "in principle" be "found" in the axioms.
Terms like "in principle" and "found" are frequently used

to denote the "relationship" of the theorems of
mathematics with the axioms from which they are
derived, and the meaning of such terms as "found",
"relationship" and "in principle" is glibly taken for
granted. Discussions of conditions of possibility of
such "relationships" are downplayed; what matters is to
arrive by the shortest route to the expected conclusion,
decided upon in advance, which will be the peremptory
assertion that all mathematical truths are "ultimately"
tautological.

No one, fortunately, will go as far as confusing
tautology with triviality. The theorems of mathematics
may be "ultimately" or "in principle" tautological, but
such tautologies more often than not require strenuous
effort to be proved. The definitive proof of any major
mathematical theorem calls for the collective effort of
years of work of successive generations of
mathematicians. Thus, although the theorems of
mathematics may be tautological consequences of the
axioms, at least "in principle", nonetheless such
tautologies are not immediate, nor evident, nor easy to
come by.

But then, what is the point of the word "tautology" in
this discussion? Of what help to the understanding of
mathematics is it to assert that mathematical theorems
are "in principle" tautological? Such an assertion, far
from clearing the air, is a way of discharging the burden

The truth of the theorems of mathematics is seldom seen by merely staring at the axioms.

---

of understanding what mathematical truth is onto the
catchall expression "in principle".

What happens is that the expression "in principle" is
carefully cooked up to conceal a normative term,
namely, the word "should". We say that mathematical

theorems are "ultimately", "in principle", "basically" tautological, when we mean to say that such theorems *should* be evident from the axioms, that the intricate succession of sillogistic inferences by which we prove a theorem, or by which we understand someone else's theorem, is only a temporary prop that *should* sooner or later, *ultimately* let us see the conclusion as an inevitable consequence of the axioms.

The role of the implicit normative term "should" in what appears to be no more than a description has seldom been brought out into the open. Perhaps it will turn out to be related to the sense in which the word "truth" is used in the practice of mathematics. Let us see.

3. Mathematical truth vs. formal truth.

We shall establish a distinction between the concept of truth as it is used by mathematicians and another, superficially similar concept, also unfortunately denoted by the word "truth", the one that is used in mathematical logic and that has been widely adopted in analytic philosophy.

The narrow concept of truth coming to us from logic may more properly be called verification. The logician denotes by the word "truth" either the carried-out verification of the correctness of derivation of a statement from axioms, or else its semantic counterpart, the carried-out verification that said statement holds in all models.

This concept of truth is a derived one, that is, it already presupposes another concept of truth, the one that mathematicians tacitly abide by. To realize that the concept of formal truth is hopelessly wide off the mark, perform the following eidetic variation.

Imagine a mathematics lesson. Can you envisage a good mathematics teacher training pupils *exclusively* in their ability to produce logically impeccable proofs from axioms? Or even (to take a deliberately absurd exaggeration) a teacher whose main concern is the students' efficiency in spotting the consequences of a "given" set of axioms?

In the formal concept of truth, axioms are not questioned, and the proving of theorems is viewed as a game in which all pretense of telling the truth are suspended. But no self-respecting teacher of mathematics can afford to pawn off on a class the axioms of a theory without giving some motivation, nor expect the class to accept the results of the theory, the theorems that is, without some sort of justification other than formal proof. Far from taking axioms for granted and beginning at once to spin off their consequences, any teacher who wants to be understood

will instead engage in a back-and-forth game—what is known in philosophy as a *hermeneutic circle*—where the axioms are "justified" by the strength of the theorems they are able to prove, and, once the stage is set, the theorems themselves, whose truth had previously been made intuitively evident, are finally made inevitable by formal proofs which come almost as an afterthought, as the last bit of crowning evidence of a theory that has already been made plausible by non-formal, non-deductive, and at times even non-rational rhetoric. A good teacher who is asked to teach the Euler-Schläfli-Poincaré formula giving the invariance of

> **The theorems themselves, whose truth had previously been made intuitively evident, are finally made inevitable by formal proofs which come almost as an afterthought.**

the alternating sum of the number of sides of a polyhedron, say, will acknowledge to the class that such a formula was believed to be true long before any correct definition of a regular polyhedron was known, and will not hide the fact that the verification of the formula in a formal topological setting is an afterthought. The teacher will insist that the class shall not confuse such formal verification with the factual/worldly truth of the assertion. It is such a factual/worldly truth that motivates all formal presentations, not vice versa, even though formalist philosophies of mathematics dishonestly pretend it to be the case.

Thus, what matters to any teacher of mathematics is the teaching of what mathematicians in their shop talk informally refer to as the "truth" of a theory, a truth that has to do with the concordance of a statement with facts of the world, like the truth of any physical law. In the teaching of mathematics, the truth that is demanded by the students and provided by the teacher is such a factual/worldly truth, not the formal truth that one associates with the game of theorem-proving. A good teacher of mathematics is one who knows how to disclose the full light of such factual/worldly truth before students, while at the same time training them in the skills of carefully *recording* such truth. In our age, such skills happen to coincide with giving an accurate formal-deductive presentation.

Despite such crushing evidence from the practice of mathematics, there are influential thinkers who will claim that discussions of factual/worldly truth should be permanently confined to the slums of psychology. It is more comfortable to deal with notions of truth that are prescribed in a *pensée de survol,* and which save us from direct commerce with mathematicians.

The preference for the tidy notion of verification, in place of the unkempt notion of truth that is found in the real world of mathematicians, has an emotional source. What has happened is that the methods that have so brilliantly succeeded in the definition and analysis of formal systems, have failed instead in the task of giving an account of other equally relevant features of the mathematical enterprise. Philosophies of mathematics display an irrepressible desire to tell us as quickly as possible what mathematical truth *ought to be,* while bypassing the descriptive legwork to the truth that mathematicians live by.

All formalist theories of truth are reductionistic. They derive from an unwarranted identification of mathematics with the axiomatic method of presentation of mathematics. The fact that there are only five regular solids in three-dimensional Euclidean space may be presented in widely differing axiomatic settings; no one

---

All formalist theories of truth are reductionistic. They derive from an unwarranted identification of mathematics with the axiomatic method of presentation of mathematics.

---

doubts its truth, irrespective of the axiomatics which is chosen for its justification. Such a glaring example should be proof enough that the relation between the truth of mathematics and the formal axiomatic truth which is indispensable in the presentation of mathematics is a relation of *Fundierung* .

Nevertheless, in our time the temptation of psychologism is again rearing its silly head, and we find ourselves forced to revive old and elementary anti-psychologistic precautions .

No state of affairs can be "purely psychological": the psychological aspects of the teaching of mathematics must of necessity hark back to a worldly truth of mathematics. To view the truth that the mathematics teacher appeals to as a mere psychological device is tantamount to presuppose a worldly mathematical truth while refusing to thematize it. In the words of Edmund Husserl, there is no *Real* without a *reell* counterpart. *Nihil est in intellectu quod prius non fierit in mundo,* we could say, murdering an old slogan.

To conclude: mathematical truth is no different from the truth of physics or chemistry. Mathematical truth results from the formulation of facts that are out there in the world, facts that are unpredictable, independent of our whim or of the whim of axiomatic systems.

4. An embarrassing example.

We shall now argue *against* the thesis we have just stated, again by taking some events in the history of mathematics as our guide.

We shall give an example that displays how observation of the practice of mathematics leads to a more sophisticated concept of truth than the one we have stated .

Our example is the history of the prime number theorem. This result was conjectured by Gauss after extensive numerical experimentation, guided by a genial intuition. No one seriously doubted the truth of the theorem after Gauss had conjectured it and verified it as far as his calculations could carry. However, mathematicians cannot afford to behave like physicists, who take experimental verification as confirmation of the truth. In mathematics, only a formal proof may be taken as confirmation of the truth. Thanks to high-speed computers we know today what Gauss could only guess, that conjectures in number theory may fail for integers so large as to lie beyond the reach of even the best of today's computers, and thus, that formal proof is today more indispensable than ever.

The proof of the prime number theorem was obtained simultaneously and independently around the turn of the century by the mathematicians Hadamard and de la Vallée Poussin. Both proofs, which were very similar, relied upon the latest techniques in the theory of functions of a complex variable. They were justly hailed as a great event in the history of mathematics. However, to the best of my knowledge no one verbalized at the time the innermost reason for the mathematicians' glee. An abstruse theory, at the time the cutting edge of mathematics, namely, the theory of functions of a complex variable, which had been developed in response to geometric and analytic problems, turned out to be the key to settle a conjecture in number theory, an entirely different field.

The mystery, as well as the glory of mathematics, lies not so much in the fact that abstract theories do turn out to be useful in solving problems, but, wonder of wonders, in the fact that a theory meant for one type of problem is often the only way of solving problems of entirely different kinds, problems for which the theory was not intended. These coincidences occur so frequently, that they must belong to the essence of mathematics. No philosophy of mathematics can be excused from explaining such occurrences.

One might think that, once the prime number theorem was proved, other attempts at proving the prime number theorem by altogether different techniques would be abandoned as fruitless. But this is not what happened after Hadamard's and de la Vallée Poussin's discovery. Instead, for about fifty years afterwards paper after paper began to appear in the best mathematics journals that provided nuances, simplifications, alternative routes, slight generalizations, and eventually alternative proofs of the prime number theorem. For example, in the thirties, the American mathematician Norbert Wiener developed an extensive theory of Tauberian theorems, that unified a great number of disparate results in classical mathematical analysis. The outstanding application of Wiener's theory, widely acclaimed throughout the mathematical world, was precisely a new proof of the prime number theorem.

Confronted with this episode in mathematical history, an outsider might be led to ask: What? A theory lays claim to be a novel contribution to mathematics by proudly displaying as its main application a result that by that time had been cooked in several sauces? Isn't mathematics supposed to solve *new* problems?

To allay your suspicions, let me hasten to add that Wiener's main Tauberian theorem was and is still viewed as a great achievement. Wiener was the first to succeed in injecting an inkling of purely conceptual insight into a proof that had heretofore appeared mysterious. The original proof of the prime number theorem mysteriously related the asymptotic distribution of primes to the behavior of the zeros of a meromorphic function, the Riemann zeta function. Although this connection had been established some time ago by Riemann, and although the logic of such improbable connection had been soundly tried out by several mathematicians who had paved the way for the first proof, nevertheless the proof could not be said to be based upon obvious and intuitive concepts. Wiener succeeded in showing, by an entirely different route than the one followed by his predecessors, and one that was just as unexpected, that there might be a conceptual underpinning to the distribution of primes.

Wiener's proof had a galvanizing effect. From that time on, it was believed that the proof of the prime number theorem could be made elementary.

What does it mean to say that a proof is "elementary"? In the case of the prime number theorem, it means that an argument is given that shows the "analytic inevitability" of the prime number theorem (in the Kantian sense of the expression) on the basis of an analysis of the concept of prime, without appeal to extraneous techniques.

It took another ten years and a few hundred research papers to remove a farrago of irrelevancies from Wiener's proof, and the first elementary proof of the prime number theorem, one that "in principle" used only elementary estimates of the relative magnitudes of primes, was finally obtained in collaboration by the mathematicians Erdös and Selberg. Again, their proof was hailed as a new milestone in number theory. Unfortunately, elementary proofs are seldom simple. Erdös and Selberg's proof added up to a good fifty pages of elementary but thick reasoning, and was longer and harder to follow than any of the preceding proofs. It did, however, have the merit of relying only upon notions that were "intrinsic" to the definition of prime number, as well as on a few other elementary facts going back to Euclid and Erathostenes. In principle, their proof showed explicitly how the prime number theorem could be boiled down to a fairly trivial argument, once the basic notions had been properly grasped—but only in principle. It took another few hundred research papers, whittling down Erdös and Selberg's argument to its barest core, until, in the middle sixties, the American mathematician Norman Levinson (who, incidentally, was Norbert Wiener's only

---

Thanks to high-speed computers we know today what Gauss could only guess, that conjectures in number theory may fail for integers so large as to lie beyond the reach of even the best of today's computers, and thus, that formal proof is today more indispensable than ever.

---

student) published a note a few pages long bearing the title "A motivated introduction to the prime number theorem". This note was published in the American Mathematical Monthly, a journal whose readership consists largely of high school and college teachers in the United States. Despite its modest title, Levinson's note gave a full purely elementary proof of the prime number theorem, one that can be followed by careful reading by anyone with no more knowledge of mathematics than that provided by undergraduates at an average American college.

After Levinson's paper, research on proofs of the prime number theorem dwindled. Levinson's proof of the prime number theorem, or one of several variants that have been discovered since, is now part of the curriculum of an undergraduate course in number theory.

What philosophical conclusions can we draw from this fragment of mathematical history?

When one leafs through any of the 300-odd journals that publish original mathematical research, one soon discovers that few published research papers present solutions of as yet unsolved problems; fewer still are formulations of new theories. The overwhelming majority of research papers in mathematics are concerned not with proving, but with reproving, not with axiomatizing, but with re-axiomatizing, not with inventing, but with unifying, with streamlining, with adding marginal results to known theories; in short, with what Thomas Kuhn calls "tidying up".

In the face of this evidence, we are forced to choose between two conclusions. The first is that the quality of mathematical research in our time is lower than we were led to expect. But what kind of evidence could have led us to such expectations? Certainly it was not the history of mathematics in the eighteenth and nineteenth centuries. Publication in mathematics in these centuries followed the same pattern we have described, as a dispassionate examination of the past will show.

Only one other conclusion is possible. Our preconceived ideas of what mathematical research *should be* do not correspond to the reality of mathematical research. The mathematician is not a person who, by staring in rapt attention at a piece of blank paper, grinds out solutions to problems like a complex machine delivering a finished product. Nor is the mathematician the possessor of the secrets for inventing imaginative theories that will unravel the mysteries of nature.

We hasten to add that the opposite opinion of the mathematician's activity is equally wrong: it is not true that mathematicians do *not* solve problems or do *not* invent new theories. They do; in fact, they make their living by doing so. However, most research papers in mathematics are not easily classified in regards to their originality. A stern judge would class all mathematics papers as unoriginal, except perhaps two or three in a century; a more liberal one would find a redeeming spark of originality in most papers that appear in print. Even papers that purport to give solutions of heretofore unsolved problems can be severely branded as either exercises of varying degrees of difficulty, or more kindly as pathways opening up brave new worlds.

The value of a mathematics research paper is not deterministically given, and we err in forcing our judgment of the worth of a paper to conform to objectivistic standards. Papers which only twenty years ago were taken as fundamental are nowadays viewed as misguided.

Newer theories do not subsume and expand on the preceding theories merely by quantitative increases in the amount of information. The invention of theories and the solution of difficult problems are not processes evolving linearly in time. Never more than today has the oversimplification of linearity been off the mark. We witness today a return to the concrete mathematics of the nineteenth century after a long period of abstraction; algorithms and techniques which were once made fun of (as Hilbert did of Gordan's invariant theory) are now revalued after a century of interruption. Contemporary mathematics, with its lack of a unifying trend, its historical discontinuities and its lapses into the past, is a further step on the way to the end of two embarrassing Victorian heritages: the idea of progress and the myth of definitiveness.

5. The concept of mathematical truth.

We have sketched two seemingly clashing concepts of mathematical truth. Both concepts force themselves upon us when we observe the development of modern mathematics.

The first concept is similar to the classical concept of truth of a law of natural science. According to this first view, mathematical theorems are statements of fact; like all facts of science, they are discovered by observation and experimentation. The philosophical theory of mathematical facts is therefore not essentially distinct from the theory of any other scientific facts, except in phenomenological details. For example, mathematical facts exhibit greater precision when compared to the facts of certain other sciences, such as botany. It matters little that the facts of mathematics might be "ideal", while the laws of nature are "real", as philosophers used to say some fifty years ago. Real or ideal, the facts of mathematics are found out there in the world, and are not artificial creations of someone's mind. Both mathematics and natural science have set themselves the same task of discovering the regularities

> **Contemporary mathematics, with its lack of a unifying trend... is a further step on the way to the end of two embarrassing Victorian heritages: the idea of progress and the myth of definitiveness.**

in the big wide world. That some portions of such world may be real and others ideal is a remark of little relevance.

The second view seems to lead to the opposite conclusion. Proofs of mathematical theorems, such as the proof of the prime number theorem, are achieved at the cost of great intellectual effort. They are then gradually whittled down to trivialities. Doesn't this temporal process of simplification that transforms a fifty-page proof into a half-page of argument betoken the fact that the theorems of mathematics are nothing but creations of our intellect? Doesn't it follow from these observations that the original difficulty of a mathematical theorem, the difficulty with which we struggle when we delude ourselves to have "discovered"

Every mathematical theorem is eventually proved trivial. Every mathematical proof is a form of debunking.

---

a new theorem, is really due to human frailty alone, a frailty that some stronger mind will at some later date dissipate, by displaying the triviality that we, sinners that we are, had failed to acknowledge? Every mathematical theorem is eventually proved trivial. Every mathematical proof is a form of debunking, wrote the mathematician G.H. Hardy. The mathematician's ideal of truth is triviality, and the community of mathematicians will not cease their beaver-like work on a newly discovered result until they have shown to everyone's satisfaction that all difficulties in the early proofs were spurious, and only an analytic triviality is left at the end of the road. Isn't the progress of mathematics—if we can speak of progress at all—just a gradual awakening from *"el sueño de la razón"*?

The way I should like to propose out of the paradox of these two seemingly irreconcilable views of mathematics will use an argument originally due to Edmund Husserl. The same argument is useful in a large number of other philosophical puzzles, and justly deserves a name of its own. I should like to baptize it by the name *"the ex universali argument"*.

Let us summarize the problem at hand. On the one hand, mathematics is undoubtedly the recording of phenomena which are not arbitrarily determined by the human mind. Mathematical facts follow the unpredictable a posteriori behavior of a nature which is, in Einstein's words, *raffiniert* but not *boshaft*. On the other hand, the power of reason sooner or later reduces every such fact to an analytic statement amounting to a triviality. How can both these statements be true?

But observe that this duplicitous behavior is not the preserve of mathematical nature alone. The facts of other sciences, of physics and chemistry, and someday (we firmly believe) even of biology and botany exhibit the very same duplicitous behavior.

Any law of physics, when finally ensconced in a proper mathematical setting, turns into a mathematical triviality. The search for a universal law of matter, a search in which physics has been engaged throughout this century, is actually the search for a trivializing principle, for a universal "nothing but".

The unification of chemistry that has been wrought by quantum mechanics is not differently motivated. And the current fashion of molecular biology can be attributed to the glimmer of hope that this glamorous new field is offering biology, for the first time in the history of the life sciences, to finally escape from the whimsicality of natural randomness into the coziness of Kantian analyticity.

What is true is that *the ideal of all science, not only of mathematics,* is to do away with any kind of synthetic a posteriori statement, and to leave only analytic trivialities in its wake. Science may be defined as the transformation of synthetic facts of nature into analytic statements of reason.

Thus, the *ex universali* argument shows that the paradox that we believed to have discovered in the truth of mathematics is shared by the truth of all the sciences. Although this admission does not offer immediate solace to our toils, it is nevertheless a relief to know that we are not alone in our misery. More to the point, the realization of the universality of our paradox exempts us from even attempting to reason ourselves out of it within the narrow confines of the philosophy of mathematics. At this point, all we can do is turn the whole problem over to the epistemologist, or to the metaphysician, and with a reminder of the old injunction: *hic Rhodus, hic salta.*

Mathematics and Philosophy:
The Story of a Misunderstanding

Gian-Carlo Rota
Massachusetts Institute of Technology
Cambridge, Massachusetts

We shall argue that the attempt carried out by certain philosophers in this century to parrot the language, the method, and the results of mathematics has done harm to philosophy.   Such an attempt results from a misunderstanding of both mathematics and philosophy, and has done harm to both subjects.

1. The Double Life of Mathematics

Are mathematical ideas invented or discovered? This question has been repeatedly posed by philosophers through the ages, and will probably be with us forever. We shall not be concerned with the answer. What matters is that by asking the question, we acknowledge the fact that mathematics has been leading a double life.

In the first of its lives, mathematics deals with facts like any other science. It is a fact that the altitudes of a triangle meet at a point; it is a fact that there are only seventeen kinds of symmetry in the plane; it is a fact that there are only five non-linear differential equations with fixed singularities; it is a fact that every finite group of odd order is solvable. The work of a mathematician consists in dealing with these facts in various ways. When mathematicians talk to each other, they tell the facts of mathematics. In their research work, mathematicians study the facts of mathematics with a taxonomic zeal similar to that of the botanist who studies the properties of some rare plant.

The facts of mathematics are as useful as the facts of any other science. No matter how abstruse they may appear at first, sooner or later they find their way back to practical applications. The facts of group theory, for example, may appear abstract and remote, but the practical applications of group theory have been numerous, and they have occurred in ways that no one might have anticipated.   The facts of today's mathematics are the springboard for the science of tomorrow.

In its second life, mathematics deals with proofs. A mathematical theory begins with definitions, and derives its results from clearly agreed upon rules of inference. Every fact of mathematics must be ensconced in an axiomatic theory and formally proved if it is to be accepted as true. Axiomatic exposition is indispensable in mathematics, because the fact of mathematics, unlike the facts of physics, are not amenable to experimental verification.

The axiomatic method of mathematics is one of the great achievements of our culture. However, it is only a method. Whereas the facts of mathematics, once discovered, will never change, the method by which these facts are verified has changed many times in the past, and it would be foolhardy not to expect that it will not change again at some future date.

2. The Double Life of Philosophy

The success of mathematics in leading a double life has long been the envy of philosophy, another field which also is blessed—or maybe we should say cursed—to live in two worlds, but which has not been quite as comfortable with its double life.

In the first of its lives, philosophy sets to itself the task of telling us how to look at the world. Philosophy is effective at correcting and redirecting our thinking. It helps us do away with glaring prejudices and unwarranted assumptions.   Philosophy lays bare contradictions that we would rather avoid facing up to. Philosophical descriptions make us aware of phenomena

> Whereas the facts of mathematics, once discovered, will never change, the method by which these facts are verified has changed many times.

that lie at the other end of the spectrum of rationality, phenomena with which science will not and cannot deal.

The assertions of philosophy are less reliable than the assertions of mathematics, but they run deeper into the

97

roots of our existence.
The philosophical assertions of today will be part of the common sense of tomorrow.

In its second life, philosophy, like mathematics, relies on a method of argumentation that seems to follow the rules of some logic or other. But the method of philosophical reasoning, unlike the method of mathematical reasoning, has never been clearly agreed upon by philosophers, and much philosophical discussion since the beginnings in Greece has been spent on discussions of method. Philosophy's relationship with Goddess Reason is closer to a forced cohabitation than to the romantic liaison that has always existed between Goddess Reason and mathematics.

The assertions of philosophy are tentative and partial. It is not even clear what it is that philosophy deals with. It used to be said that philosophy was "purely speculative," and this used to be an expression of praise. But lately the word "speculative" has become a Bad Word.

Philosophical arguments are emotion-laden to a greater degree than mathematical arguments. Philosophy is often written in a style which is more reminiscent of a shameful admission than of a dispassionate description. Behind every question of philosophy there lurks a gnarl of unacknowledged emotional cravings which act as powerful motivation for conclusions in which reason plays at best a supporting role. To bring such hidden emotional cravings out into the open, as philosophers have felt it their duty to do, is to call for trouble. Philosophical disclosures are frequently met with the anger that we reserve for the betrayal of our family secrets.

This confused state of affairs makes philosophical reasoning more difficult, but far more rewarding. Although philosophical arguments are blended with emotion, although philosophy seldom reaches a firm conclusion, although the method of philosophy has never been clearly agreed upon, nonetheless, the assertions of philosophy, tentative and partial as they are, come much closer to the truth of our existence than the proofs of mathematics.

3. The Loss of Autonomy

Philosophers of all times, beginning with Thales and Socrates, have suffered from the recurring suspicions about the soundness of their work, and have responded to them as best they could.

The latest reaction against the criticism of philosophy began around the turn of the century and is still very much with us.

Today's philosophers (not all of them, fortunately) have become great believers in mathematization. They have rewritten Galileo's famous sentence to read: "The great book of philosophy is written in the language of mathematics."

"Mathematics calls attention to itself," wrote Jack Schwartz in a famous paper on another kind of misunderstanding. Philosophers in this century have suffered more than ever from the dictatorship of definitiveness. The illusion of the final answer—what two thousand years of Western philosophy had failed to accomplish—was thought in this century to have come at last within reach by the slavish imitation of mathematics.

Mathematizing philosophers have claimed that philosophy should be made factual and precise. They have given guidelines to philosophical argument which are based upon mathematical logic. They have contended that the eternal riddles of philosophy can be definitively solved by pure reasoning, unencumbered by the weight of history. Confident in their faith in the power of pure thought, they have cut all ties to the past, on the claim that the messages of past philosophers are now "obsolete."

Mathematizing philosophers will agree that traditional philosophical reasoning is radically different from mathematical reasoning. But this difference, rather than being viewed as strong evidence for the heterogeneity of

---

> **Today's philosophers...have rewritten Galileo's famous sentence to read: "The great book of philosophy is written in the language of mathematics."**

---

philosophy and mathematics, is taken instead as a reason for doing away with non-mathematical philosophy altogether.

In one area of philosophy the program of mathematization has succeeded. Logic is nowadays no longer a part of philosophy. Under the name of mathematical logic, it is now a successful and respected branch of mathematics, one that has found substantial practical applications in computer science, more so than any other branch of mathematics.

But logic has become mathematical at a price. Mathematical logic has given up all claims to give a foundation to mathematics. Very few logicians of our day believe any longer that mathematical logic has anything to do with the way we think.

Mathematicians are therefore mystified by the spectacle of philosophers pretending to re-inject philosophical sense into the language of mathematical logic. A hygienic cleansing of every trace of philosophical reference had been the price of admission of logic into the mathematical fold. Mathematical logic is now just another branch of mathematics, like topology and probability. The philosophical aspects of mathematical

---

Mathematical logic has given up all claims to give a foundation to mathematics.

---

logic are qualitatively no different from the philosophical aspects of topology or the theory of functions, aside from a curious terminology which, by an accident of chance going back to Leibniz's reading of Suárez, goes back to the Middle Ages.

The fake-philosophical terminology of mathematical logic has misled philosophers into believing that mathematical logic deals with the truth in the philosophical sense. But this is a mistake. Mathematical logic does not deal with the truth, but only with the game of truth. The snobbish symbol-dropping one finds nowadays in philosophical papers raises eyebrows among mathematicians. It is as if you were at the grocery store and you watched someone trying to pay his bill with Monopoly money.

4. Mathematics and Philosophy: Success and Failure

By all accounts, mathematics is the most successful intellectual undertaking of mankind. Every problem of mathematics gets solved, sooner or later. Once it is solved, a mathematical problem is forever finished: no later event will disprove a correct solution. As mathematics progresses, problems that were once difficult become easy enough to be assigned to schoolboys. Thus, Euclidean geometry is now taught in the second year of high school. Similarly, the mathematics that mathematicians of my generation have learned in graduate school has now descended to the undergraduate level, and the time is not far when it may be taught in the high schools.

Not only is every mathematical problem solved, but eventually, every mathematical problem is proved trivial. The quest for ultimate triviality is characteristic of the mathematical enterprise.

When we look at the problems of philosophy, another

picture emerges. Philosophy can be described as the study of a few problems whose statements have changed little since the Greeks: the mind-body problem, or the problem or reality, to recall only two. A dispassionate look at the history of philosophy discloses two contradictory features: first, these problems have in no way been solved, nor are they likely to be solved as long as philosophy survives; second, every philosopher who has ever worked on any of these problems has proposed *the* "definitive solution," which has been invariably rejected as false by later philosophers.

Such crushing historical evidence forces us to the conclusion that these two paradoxical features must be an inescapable concomitant of the philosophical enterprise. Failure to conclude has been an outstanding characteristic of philosophy throughout its history.

Philosophers of the past have repeatedly stressed the essential role of failure in philosophy. José Ortega y Gasset, for example, used to describe philosophy as "a constant shipwreck." However, the fear of failure did not stop him or any other philosopher from doing philosophy.

Philosophers' failure to reach any kind of agreement does not make their writings any less relevant to the problems of our day. We reread with interest the mutually contradictory theories of mind that Plato, Aristotle, Kant, and Comte have bequeathed to us, and we find their opinions timely and enlightening, even in problems of artificial intelligence.

Unfortunately, the latter-day mathematizers of philosophy are unable to face up to the inevitability of failure. Borrowing from the world of business, they have embraced the ideal of success. Philosophy had better be successful, or else it should be given up, like any business.

5. The Myth of Precision

Since mathematical concepts are precise, and since mathematics has been successful, mathemitizing philosophers mistakenly infer that philosophy would be better off if it dealt with precise concepts and unequivocal statements. Philosophy will have a better chance at being successful, if it becomes precise.

The prejudice that a concept must be precisely defined in order to be meaningful, or that an argument must be precisely stated in order to make sense, is one of the most insidious of the Twentieth Century. The best known expression of this prejudice appears at the end of Ludwig Wittgenstein's *Tractatus*, and the author's later work, in particular the *Philosophical Investigations*, is a loud and repeated retraction of his earlier *gaffe*.

Looked at from the vantage point of ordinary experience, the ideal of precision appears preposterous. Our everyday reasoning is not precise, yet it is effective. Nature itself, from the cosmos to the gene, is approximate and inaccurate.

The concepts of philosophy are among the least precise. *The mind, perception, memory, cognition,* are words that do not have any fixed or clear meaning. Yet, they do have meaning. We misunderstand these concepts when we force them to be precise. To use an image due to Wittgenstein, philosophical concepts are like the winding streets of an old city, which we must accept as they are, and which we must familiarize ourselves with by strolling through them, while admiring their historical heritage. Like a Carpathian dictator, the advocates of precision would raze the city to the ground and replace it with a straight and wide Avenue of Precision.

The ideal of precision in philosophy has its roots in a misunderstanding of the notion of rigor. It has not occurred to our mathematizing philosophers that philosophy might be endowed with its own kind of rigor, a rigor that philosophers should dispassionately describe and codify, as mathematicians did with their own kind of rigor a long time ago. Bewitched as they are by the success of mathematics, they remain enslaved by the prejudice that the only possible rigor is that of mathematics, and that philosophy has no choice but to imitate it.

6. The Misunderstanding of the Axiomatic Method

 The facts of mathematics are verified and presented by the axiomatic method. One must guard, however, against confusing the *presentation* of mathematics with the *content* of mathematics. An axiomatic presentation of a mathematical fact differs from the fact that is being presented just as medicine differs from food. It is true that this particular medicine is necessary to keep the mathematician at a safe distance from the self-delusions

---

...eventually, every mathematical problem is proved trivial.

---

of the mind. Nonetheless, understanding mathematics means being able to *forget* the medicine, and to enjoy the food. Confusing mathematics with the axiomatic method for its presentation is as preposterous as confusing the music of John Sebastian Bach with the the techniques for counterpoint in the Baroque age.

This is not, however, the opinion held by our

mathematizing philosophers. They are convinced that the axiomatic method is a basic instrument for discovery. They mistakenly believe that mathematicians *use* the axiomatic method in solving problems and proving theorems. To the misunderstanding of the role of the method, they have added the absurd pretension that this presumed method should be adopted in philosophy. Systematically confusing food with medicine, they have pretended to replace the food of philosophical thought with the medicine of axiomatics.

This mistake betrays the philosophers' pessimistic view of their own field. Unable or afraid as they are of singling out, describing and analyzing the structure of philosophical reasoning, they seek help from the proven technique of another field, a field that is the object of their envy and veneration. Secretly disbelieving in the power of autonomous philosophical reasoning to arrive at the truth, they have surrendered to a slavish and superficial imitation of the truth of mathematics.

The negative opinion that many philosophers hold of their own field has caused damage to philosophy. The mathematician's contempt at the philosopher's exaggerated estimation of a method of mathematical exposition feeds back onto philosophers' inferiority complex, and further decreases the philosophers' confidence.

7. "Define your terms!"

This old injunction has become a platitude in everyday discussions. What could be healthier than a clear statement, right at the beginning, of what it is that we are talking about? Doesn't mathematics start with definitions and then develop the properties of the objects that have been defined, by an admirable and inexorable logic?

Salutary as this injunction may be in mathematics, it has had disastrous consequences when carried over to philosophy. Whereas mathematics *starts* with a definition, philosophy *ends* with a definition. A clear statement of what it is we are talking about is not only missing in philosophy; such a statement would be the end of all philosophy. If we could define our terms, then we would dispense with philosophical argument.

Actually, the "define your terms" imperative is deeply flawed in more than one way. While reading a formal mathematical argument, we are given to believe that the "undefined terms," or the "basic definitions" have been whimsically chosen out of a variety of possibilities. Mathematicians take mischievous pleasure in faking the arbitrariness of definition. In actual fact, no mathematical definition is arbitrary. The theorems of mathematics motivate the definitions as much as the definitions motivate the theorems. A good definition is "justified" by the theorems one can prove with it, just as the proof of a theorem is "justified" by appealing to a previously given definition.

There is thus a hidden circularity in formal mathematical exposition. The theorems are proved starting with definitions, but the definitions themselves are motivated by the theorems that we have previously decided ought to be right.

Instead of focusing on this strange circularity, philosophers have pretended it does not exist, as if the axiomatic method, proceeding linearly from definition to theorem, were endowed with a definitiveness which is instead, as every mathematician knows, a subtle fakery to be debunked.

Perform the following thought experiment. Suppose that you are given two formal presentations of the same mathematical theory. The definitions of the first presentation are the theorems of the second, and vice-versa. This situation frequently occurs in mathematics.

Our everyday reasoning is not precise, yet it is effective. Nature itself, from the cosmos to the gene, is approximate and inaccurate.

Which of the two presentations makes the theory "true?" The answer is: neither, evidently. What we have is two presentations of the *same* theory.

This thought experiment shows that mathematical truth is not brought into being by a formal presentation. Rather, formal presentation is only a technique for displaying mathematical truth. The truth of a mathematical theory is distinct from the correctness of any axiomatic method that may be chosen for the presentation of the theory.

Mathematizing philosophers have missed this distinction.

8. The Appeal to Psychology

What will happen to the philosopher who insists on precise statements and clear definitions? Realizing after futile trials that philosophy resists such a treatment, said philosopher will proclaim that most problems previously thought to belong to philosophy are heretofore to be excluded from consideration. He will claim that they are "meaningless" or, at best, that they

can be settled by an analysis of their statements that will eventually show them to be vacuous.

This is not an exaggeration. The classical problems of philosophy have become forbidden topics in many philosophy departments. The mere mention of one such problem by a graduate student or by a junior colleague will result in raised eyebrows, followed by severe penalties. In this dictatorial regime, we have witnessed the shrinking of philosophical activity to an impoverished *problématique*, mainly dealing with language.

In order to justify their neglect of most of the old and substantial questions of philosophy, our mathematizing philosophers have resorted to the ruse of claiming that many questions formerly thought to be philosophical are instead "purely psychological," and that they should be dealt with in the psychology department.

If the psychology department of any university were to consider only one tenth of the problems that philosophers are pawning off on them, then psychology would without question be the most fascinating of all subjects. Maybe it is. But the fact is that psychologists have no intention of dealing with problems abandoned by philosophers who have been derelict in their duties.

One cannot do away with problems by decree. The classical problems of philosophy are now coming back with a vengeance in the forefront of science. For example, the Kantian problem of the conditions of possibility of vision, after years of neglect, is now again rearing its old head in brain science.

Experimental psychology, neuro-physiology and computer science may turn out to be the best friends of traditional philosophy. The awesome complexities of the phenomena that are being studied in these sciences have convinced scientists (well in advance of the philosophical establishment) that progress in science will crucially depend on philosophical research of the most classical vein.

9. The Reductionist Concept of the Mind

What does a mathematician do when trying to work on a mathematical problem? An adequate description of this event might take a thick volume. We shall be content with recalling an old saying, probably going back to the mathematician George Pólya: "Few mathematical problems are ever solved directly."

Every mathematician will agree that an important step in solving a mathematical problem, perhaps the most important step, consists in analyzing other attempts, either attempts that have been previously carried out or else attempts that one imagines might have been carried out, with a view to discovering how such "previous" attempts were misled. In short, no mathematician will ever dream of attacking a substantial mathematical problem without first becoming acquainted with the *history* of the problem, whether the real history or an ideal history that a gifted mathematician might reconstruct. The solution of a mathematical problem goes hand-in-hand with the discovery of the inadequacy of previous attempts, with the enthusiasm that sees through and does away with layers of irrelevancies inherited from the past that cloud the real nature of the problem. In philosophical terms, a mathematician who

---

A good definition is "justified" by the theorems one can prove with it.

---

solves a problem cannot avoid facing up to the *historicity* of the problem. Mathematics is nothing if not an historical subject *par excellence.*

Every philosopher since Heraclitus has stressed with striking uniformity the lesson that all thought is constitutively historical—until, that is, our mathematizing philosophers came along, claiming that the mind is nothing but a complex thinking machine, not to be polluted by the inconclusive ramblings of bygone ages. Historical thought has been dealt a *coup de grace* by those who today occupy some of the chairs of our philosophy departments. Graduate school requirements in the history of philosophy have been dropped, together with language requirements, and in their place we find required courses in mathematical logic.

It is important to single out the myth that underlies such drastic revision of the concept of mind. It is the myth that believes the mind to be a mechanical device. This myth that has been repeatedly and successfully attacked by the best philosophers of our time—Husserl, John Dewey, Wittgenstein, Austin, Ryle, to name only a few.

According to this myth, the process of reasoning is viewed as the functioning of a vending machine which, by setting into motion a complex mechanism reminiscent of those we saw in Charlie Chaplin's film *Modern Times,* grinds out solutions to problems, like so many Hershey bars. Believers in the theory of the mind as a vending machine will rate human beings according by "degrees" of intelligence, the more intelligent ones being those endowed with bigger and better gears in their brains, as can of course be verified by administering I.Q. tests.

Philosophers believing in the mechanistic myth believe that the solution of a problem is obtained in just one way: by thinking hard about it. They will go as far as asserting that acquaintance with previous contributions to a problem may bias the well-geared mind. A blank mind, they believe, is better geared up to initiate the solution process than an informed mind.

This outrageous proposition originates from a misconception of how mathematicians work. Our mathematizing philosophers behave like failed mathematicians. They gape at working mathematicians in wide-eyed admiration, like movie fans gaping at posters of Joan Crawford and Bette Davis. Mathematicians are superminds who turn out solutions of one problem after another by dint of pure brain power, simply by staring at a blank piece of paper in intense concentration.

The myth of the vending machine that grinds solutions out of nothing may perhaps appropriately describe the way to solve the linguistic puzzles of today's impoverished philosophy, but this myth is far off the mark in describing the work of mathematicians, or any other serious work.

The fundamental error is one of reductionism. The *process* of the working of the mind, which may be of interest to physicians but is of no interest to mathematicians, is confused with the *progress* of thought that is required in the solution of any problem.

This catastrophic misunderstanding of the nature of knowledge is the heritage of one hundred-odd years of pseudo-mathematization of philosophy.

10. The Illusion of Definitiveness

The results of mathematics are definitive. No one will ever improve on a sorting algorithm which has been proved best possible. No one will ever discover a new finite simple group, now that the list has been drawn, after a century of research. Mathematics is forever.

We could classify the sciences by how close their results come to being definitive. At the top of the list we would find the sciences of lesser philosophical interest, such as mechanics, organic chemistry, botany. At the bottom of the list we would find the more philosophically inclined sciences, such as cosmology and evolutionary biology.

The old problems of philosophy, such as mind and matter, reality, perception, are least likely to have "solutions." In fact, we would be hard put to spell out what might be acceptable as a "solution". The term "solution" is borrowed from mathematics, and tacitly presupposes an analogy between problems of philosophy and problems of mathematics that is seriously misleading. Perhaps the use of the word "problem" in philosophy raised expectations that philosophy could not fulfill.

Philosophers of our day go one step farther in their mis-analogies between philosophy and mathematics. Driven by a misplaced belief in definitiveness measured in terms of problems solved, and realizing the futility of any attempt to produce definitive solutions to any of the classical problems, they have had to change the problems. And where do they think to have found problems worthy of them? Why, in the world of facts!

Science deals with facts. Whatever it is that traditional philosophy deals with, it is not facts in the scientific sense. Therefore, traditional philosophy is worthless.

This syllogism, wrong on several counts, is predicated on the assumption that no statement is of any value, unless it is a statement of fact. Instead of realizing the absurdity of this assumption, philosophers have swallowed it, hook, line and sinker, and have busied themselves in making their living on facts.

But previous philosophers had never been equipped to deal directly with facts, nor had they ever considered facts to be any of their business. Nobody turns to philosophy to learn facts. Facts are the domain of science, not of philosophy. And so, a new slogan had to be coined: philosophy *should* be dealing with facts.

This "*should*" comes at the end of a long line of other "*shoulds*." Philosophy *should* be precise, it *should* follow the rules of mathematical logic, it *should* define its terms carefully, it *should* ignore the lessons of the past, it *should* be successful at solving its problems, it *should* produce definitive solutions.

"Pigs *should* fly," as the old saying goes.

But what is the standing of such "*shoulds*," flatly negated as they are by two thousand years of philosophy? Are we to believe the not so subtle insinuation that the royal road to right reasoning will at last be found if we follow these imperatives?

There is a more plausible explanation of this barrage of *shoulds*. The reality in which we live is constituted by myriad contradictions, which traditional philosophy has taken pains to describe with courageous realism. But contradiction cannot be confronted by minds who have put their salvation in axiomatics. The real world is filled with absences, with absurdities, with abnormalities, with aberrances, with abominations, with abuses, with *Abgrund*. But our latter-day philosophers are not concerned with facing up to these unpleasant features of the world, nor, to be sure, to any real features whatsoever. They would rather tell us what the world *should* be like. They find it safer to escape from distasteful description of what is into pointless prescription of what isn't. Like ostriches with their heads in the ground, they will meet the fate of those who refuse to acknowledge the lessons of the past and to meet the challenge of our difficult present: increasing irrelevance followed by eventual extinction.

PART IV

Teaching and Learning Experiences

Towards a Pedagogy of Confusion

Stephen I. Brown
University at Buffalo
Buffalo, New York

In a telephone interview several years ago with New Your Times correspondent Kim Heron (1987), Walker Percy, reflecting on his new novel *The Thanatos Syndrome*, attempts to locate what he considers to be the cause of much anxiety and depression in our society. He comments,

> My own private theory is that a great deal of the anxiety and depression people experience comes from watching television, where there's a certain aesthetic unity—everything works out. My life and yours are much more fragmentary, haphazard and incomplete (p. 22).

Percy is correct, I believe, in attributing a less than desirable psychological state to a mismatch between implicit popular messages regarding the nature of life and the lived experience itself. I believe that the source of the problem may be considerably more encompassing than he lets on in his admittedly brief interview. Unity, coherence and clarity are illusions that appear to be conveyed not only through television, but in most bureaucratized settings from media communication through the supermarket milieu to the organization of schools. It is an illusion which is prized by both the hosts and the recipients of these services.

In this paper I would like to explore the ways that our curricula stress the values of clarity and coherence, and unity in ways that disconnect learning from "real lives which are fragmentary, haphazard and incomplete." I intend to focus upon two recent educational documents that reflect this emphasis. One document is a brief research report in mathematics education dealing with the differences between expert and novice teachers at the elementary school level. The other is a report proposing major reconstruction of American teacher education in all disciplines and at all levels of instruction.

The juxtaposition of these two pieces will enable us to understand not only the pervasiveness of these values and the intensity with which they are held, but will reveal the fragility of an alternative emerging epistemology and associated pedagogy—one that suggests a more liberated view of mathematics, learning, self and human potential more generally. Though both reports will serve as a wedge into alternative epistemologies, I have considerably more sympathy with the teacher education report and my major criticism will be reserved for the havoc wrought by the generally non-reflective manner in which clarity is promoted in the report on experts and novices. In particular, I will suggest a major logical fallacy that such research perpetuates and will also argue for the incompatibility of clarity with what appears to me to be important goals of the latest bandwagon in mathematics education and in curriculum writ large—problem-solving in mathematics and critical thinking in general. Let us turn to the teacher education report.

Foster Holmes?

Tomorrow's Teachers: A Report of the Holmes Group (1986) (referred to as *The Holmes Report*), representing the opinions of the deans of education of fourteen major research institutions, was motivated originally by a desire to improve the state of teacher education within the context of the university setting. Before long, however, the group assumed a more ambitious stance. They realized that improving the initial training of teachers would have little impact on American education

> **Unity, coherence and clarity are illusions that appear to be conveyed not only through television, but in most bureaucratized settings.**

if one is not at the same time aware of and desirous of changing a host of other variables. Thus the report raises important issues not only about the recruitment and training of first-rate teachers but about redefining the role and occupation of teaching. The report stresses the need for creating greater autonomy on the job and for modifying what is perceived by many to be a dead-end career path as well.

It is in general a sensitively written report, one in which there is appropriate appreciation for the interrelationship between democracy and schooling; for the role of parents in the education of their children; for the value of a genuinely liberal but necessarily reconstructed undergraduate education as a pre-condition for professional training of teachers; for a model of professional training that is critical of naive one-way views of teaching; and for the contributions to research on teaching that may be made via intuitions of competent master teachers that may not be captured in the form of propositional knowledge alone.[1]

Now for the contamination. Let us examine assorted excerpts dealing with issues of clarity. Though they appear innocuous at first, these quotes provide an undertow that has the potential to submerge much of what is valuable in the rest of the document.

> Competent teachers are careful not to bore, confuse, or demean students, pushing them instead to interact with important knowledge and skill (p. 29).

> For many children, partial understandings of school subjects turn into hopeless confusions and obscure abstractions (p. 30).

> It takes training to increase the higher order questions a teacher asks; to decrease the preponderance of teacher talk; to provide advanced organizers, plans and clear directions (p. 52)

Notice how confusion is linked with boredom and acts of demeaning character. Partial understandings not only lead to confusion for some students, but to *hopeless confusions*. Plans and directions must not only be provided by the teacher, but they must be *clear*.

We would perhaps expect injunctions for teacher clarity to be invoked by those who conceive of the mind as a tabula rasa and of teaching as filling vessels. It appears to be an anomaly, however, to encounter such language

---

The naturalistic fallacy is one of the most often committed philosophical errors in everyday thinking. It occurs when-ever one draws an "ought" conclusion from an "is" argument.

---

in *The Holmes Report*—a document which lauds "powerful teaching strategies like maieutic methods, role playing and social interaction" (p. 52). It is surprising to find such conceptions of teaching embedded in what appears elsewhere to be a plea for a

personal and constructivist view of knowledge. We are told, for example, that

> Students do not approach teaching as empty vessels; they more likely present the teacher with initial conceptions that are incomplete, flawed, or otherwise in need of transformation (p. 53).

This report is surely not alone in simultaneously attempting to relinquish a mode of teaching, learning and mind that perpetuates a "received" view of knowledge and at the same time holds fast to outmoded metaphors from that very tradition. As has happened numerous times in the past, we appear once more to be in a state of transition in education, one in which we are attracted to a more humanistic mode of interaction in general, but we are not quite capable of creating the language to express that new vision with force and consistency. We seem not to know how to talk sensibly about the roles of question-asking, explaining, testing, and empathizing in our reconstructed visions of teaching. We seem to be incapable of appreciating what it means to be confused, what virtues may follow from it and what kind of pedagogy may make use of it. Perhaps few of us have honestly reflected upon the way in which confusion has contributed to our own experiences both in and out of formal educational settings. We will have more to say about these matters, especially as they bear on research on teaching effectiveness in the sections following the next. At this point however, we turn to a specific research program in mathematics education which advocates precisely those virtues that Walker Percy locates as the source of twentieth century angst.

Experts and Novices

The educational document I have in mind, authored by Kay McKinney (1986), is an official publication of the U. S. Department of Education. As a mathematics educator, I received it unsolicited. It bears the imprimatur of William J. Bennett and Chester E. Finn, Jr., Secretary and Assistant Secretary respectively of the U. S. Department of Education at the time of publication. The document is only one and a half pages long, and makes no pretense of presenting a complete argument. Aware of its brevity, the document encourages the reader to write to the principal investigator of the research for a copy of the complete set of studies.

Though my description and criticism will refer to the brief report rather than the original documents, I shall be generous in assuming that the studies are impeccable in terms of definition of key concepts, research design, sample size, and so forth.[2]

What does the report tell us? From an intensive study (over a period of six years) of seven expert mathematics teachers in "very poor neighborhoods with 'difficult' students...[they] discovered that the expert math teachers and novices proceed quite differently" (p. 1). The report focuses on differences in use of time, orchestration of lessons, and content competence of teachers in both groups. In particular we are told that experts make wiser use of time, organize lessons better, and have better knowledge of content than the novices. Furthermore, we are told that it is not only these categories in isolation that differentiate the two groups, but the manner in which they are integrated as well.

I would like to highlight briefly those findings that appear to me to raise the kind of specter that I have discussed in the previous subsection and to which Walker Percy has alluded in the brief remark at the beginning of this piece. In order to do so, I will cluster some relevant verbatim comments under categories that are convenient for the purpose of analyzing and criticizing the report (comments that were not explicitly so clustered by the author):

On Explaining:
(1)   Kids who love math—and are very good at it— have great teachers who can explain not only how to add but also the underlying concepts.
(2)   Good teachers explain to students the problem, the solution, and when to use a specific process.
(3)   Novices often are just one step ahead of the students and don't really know why certain things are done. Their explanations of new material often are fragmented and incomplete.

On Efficiency:
(1)   Expert math teachers use almost all of the class time for math rather than for settling down and getting organized.
(2)   Novices can take as long as 15 minutes to correct the prior day's homework; experts take two or three minutes.
(3)   Most expert teachers cover at least 40 problems a day through games, drills, or written work. Novice teachers on the other hand, may cover only six or seven.

On Clarity:
(1)   Good teachers...[have] lessons that are clear, accurate, and contain lots of examples and demonstrations.
(2)   Students with expert teachers are rarely "lost" during a math lesson, they know what is happening, what is going to happen, and what they are supposed to do.

Now what follows from these research findings? To the extent that the research is valid and reliable, there is a clear implication that what we ought to do is hire, train and reward teachers according to their actual ability or perhaps their potential to explain well, operate efficiently, and minimize confusion. In the next section, I would like to point out an essentially logical fallacy that is exhibited by such a reaction, while in the section that follows it, I would like to raise some substantive issues that relate these claims to the profession's concern with problem-solving.

On Committing the Naturalistic Fallacy

The naturalistic fallacy is one of the most often committed philosophical errors in everyday thinking. It occurs whenever one draws an "ought" conclusion from an "is" argument. The fallacy is exhibited in the following arguments:
(1)   It is the case that people respond more readily to praise than to condemnation. Therefore, as educators, we ought to praise rather than condemn.
(2)   Kohlberg's research on moral development shows that it is the case that people mature according to progressive stages of development. They begin with a rudimentary conception of "right" as that which is rewarded and progress to the highest stage in which issues of social justice are in the forefront of moral decision-making. Therefore, in designing curriculum to accelerate moral reasoning, educators ought to create dilemmas that would enable students to progress (perhaps at a faster pace) through these stages.

What goes awry in the above cases is not that the "ought" conclusions are undesirable, but rather that they do not follow from the assumed premises of the argument. It is surely not only that people respond more readily to praise than to punishment that drives us to recommend what educators ought to do. If, for example, the empirical research indicated the opposite (that punishment was more effective than praise), we would most likely not unflinchingly mount a policy consistent with the data.

With regard to the moral education argument, Peters (1976) has made it clear that, though Kohlberg's program derives from an interesting empirical perspective, it lacks philosophical grounding. That is, Kohlberg neglects to appreciate that there are many conceptions of morality that compete with the Kantian one that is consistent with Kohlberg's empirical finds. No amount of further experimentation alone will establish that an *a priori* Kantian point of view is to be supported over a utilitarian one, for example. Peters' essential argument is that Kohlberg has incorrectly identified morality with the ability to argue rationally, and has underplayed the importance of action regardless of the nature of the justification of the act. In this regard Peters comments,

> The policeman cannot always be present, and if I am lying in the gutter after being robbed it is somewhat otiose to speculate at what stage the mugger is. My regret must surely be that he had not at least got a conventional morality well instilled in him (p. 289).

As educators, we surely are not blind to the naturalistic fallacy when we have good reason to question the flow from "is" to "ought." For example, the rather recent realization that there exists a lack of equity in the treatment of the sexes in mathematics education has led us as a profession to try to rectify the situation rather than to find ways of perpetuating it. Moral principles such as the following lead us to re-examine and try to reconstruct what we see as a sorry state of affairs: People of comparable potential ought to be encouraged to have equal access to the goods of our society, and irrelevant differences ought not get in the way of such access. Now we need to muster good arguments (ones to which we lacked sensitivity as recently as a quarter of a century ago) to support the contention that gender is an irrelevant difference while such variables as intelligence may not be.

Now how does any of this bear on the issues we have isolated in the previous section? How does this affect our inclination to view good teaching as that which involves good explanations, clear expectations, and efficient operation?

What emerges is a realization that we are attracted to a program of teacher preparation, selection, and reward that derives not from the facts, conclusions, or

Table 1
Expert/Novice Research Report

An "Ought" Commitment of Clarity

Kind of Teacher	Teaching Style	Student Performance	Research Findings and Principles are
Expert	clear	superior	consistent
Novice	confused	inferior	consistent

Table 2
Reversal of Empirical Findings in Expert/Novice Research Report

An "Ought" Commitment of Clarity

Kind of Teacher	Teaching Style	Student Performance	Research Findings and Principles are
Expert	clear	inferior	inconsistent
Novice	confused	superior	inconsistent

Table 3
Reversal of Value Judgement in Expert/Novice Research Report

An "Ought" Commitment of Confusion

Kind of Teacher	Teaching Style	Student Performance	Research Findings and Principles are
Expert	clear	superior	inconsistent
Novice	confused	inferior	inconsistent

observation of the kind of research summarized in the McKinney report, but rather from other sets of deeply entrenched beliefs and principles. While one's life as a policy implementor may be made easier by virtue of the fact that expert/novice research data are consistent with those principles, it is important to appreciate the driving force of the principles themselves rather than the supposedly neutral research findings in leading one to the conclusions. It is a mere convenience (and not an argument for what we ought to do in teacher education) that what one believes to be important and how the world operates are compatible.

The state of affairs suggested by the government report is summarized in Table 1.

It would be revealing to imagine how one would react if the novice/expert findings were reversed. Let us maintain the same selection procedure (such as excellent student performance on standardized tests and superior assessment of supervisors by those deemed experts), but suppose experts in fact exhibited all the explanation, clarity and efficiency qualities of the novice teachers and vice versa. The situation is summarized in Table 2.

Such a conclusion would place "clarity addicts" in quite an uncomfortable position. Though it would surely be difficult to predict the outcome of a debate among policy makers with such an orientation, one thing is sure: there would be considerable effort to justify why clarity is better than confusion (despite the unexpected empirical finding); and much of that discussion would generate arguments involving a wide collection of other

---

A problem is, to some extent, a project for the future we commit our-selves to by an act of the will.

"oughts." We most likely would discover in the process that the "ought" of commitment to clarity does not derive from the "is" of student performance.

The other side of the coin, however, is equally revealing. How would we operate as professional educators if we came across a report's conclusions about expert/novice teachers but believed that efficiency, clarity and good explanations were not virtues but vices in educating? Do we have options, or would we admit that our world view had been crushed? The state of affairs is suggested in Table 3.

Once more, we most likely would be placed in the position of defending our commitment—this time to confusion and its like—in a way that goes beyond the

data. While it is not the case that the data are irrelevant, they do not in and of themselves destroy our commitment to confusion. What might we as teacher educators or researchers do? One possibility, of course, would be to call into question the selection procedure. We might argue that standardized tests do not measure excellence, or that the opinion of the supervisors is open to question. But suppose we accept the independent measure of excellence. Then what?

An option that has implications for both praxis and research would be to attempt to uncover strategies for re-educating the students. That is, let us admit that the goal of high performance on standardized tests is a laudable one, and that students taught by experts whose approach is clear, efficient, and accompanied by good explanations do in fact perform better on these tests than those students taught by novices whose approach is confused, disorganized, and accompanied by poor explanations (or perhaps none at all). One thing we could do is to mount a training/research program that would enable the students to learn to handle to their advantage the confusion of the novices. If such a program were successful, then we might reverse our training, hiring, and rewarding teachers.

I grant you that this model for operationalizing a program that puts a premium on confusion is a bit bizarre. It is bizarre, however, not only because the "incompetence" of novices may not be reducible only to the observation that they in fact operate in some ways that are antithetical to what the expert does, but also because we do have an ingrained commitment to clarity that is hard to shake. Once we appreciate, however, that the research on experts/novices alone, does not automatically commit us to the adoption of a "clarity" policy, and once we realize that there are ways of confronting the findings (both from the perspective of challenging the selection criteria and from that of invoking training and research that would support an alternative value system with regard to clarity), we are at least in a position to entertain the possibility of an alternative value system: one which prizes confustion and its correlates in manyh different aspects of teaching and learning. We turn to that issue now in the context of some substantive issues related to problem-solving in the curriculum.

Problem-Solving and Clarity

The 1980s were ushered in as the decade of problem-solving in school curriculum. While problem-solving has become *the* explicit focus among educators in mathematics and science (as attested by yearbooks, special conferences, special issues of journals, policy statements by national organizations), its near relatives such as inquiry and critical thinking have achieved central billing in most of the "softer" fields as well.

One cannot graduate from a state college in California, for example, without taking at least one critical thinking course. The Board of Regents of New York State has included critical thinking as a central theme in curriculum at all levels.

The problem-solving focus in mathematics education in particular was given a substantial boost by the now often quoted policy statement of the *National Council of Teachers of Mathematics* under the title, *Agenda for Action* (1980). The first of eight curriculum recommendations is:

> Problem-solving must be the focus of school mathematics in the 1980's. The development of problem-solving ability should direct the efforts of mathematics education through the next decade. Performance in problem-solving will measure the effectiveness of our personal and national possession of mathematical competence (p. 1).

One reason that the call for problem-solving has become so convenient a rallying cry is that the concept appears to be as all-encompassing as motherhood and apple pie. For some people, the concept represents the salvation of the back-to-basics movement. Students will finally learn to do their fundamentals of arithmetic. Others, however, see problem-solving as a kind of activity that represents the antithesis of algorithmic (in the non-computer science connotation of the word) performance. Worth (1982), for example, expressing a legacy of George Polya (1954, 1962) captures the spirit

---

[Problems] require attention and courage, and they involve a significant act of self-surrender which can seriously jeopardize the individual's sense of himself. Lastly, a problem is a hopeful enterprise which involves an act of faith.

---

of a number of mathematicians and educators concerned with such a conception of problem-solving when she comments:

> Students must accept that a problem is a situation for which they do not know how to get an immediate answer (p. 16).

Many advocates who so conceive of "problem" and "problem-solving" are motivated by the observation, that in numerous national and international tests, the average performance of students at tasks requiring thought is abysmal. Carpenter et al. (1980), reporting on the results of a national assessment examination in which 70,000 United States students of ages 9, 13 and 17 were tested, comments:

> If it were necessary to single out one area that demands urgent attention, it would be problem-solving. At all ages and in virtually every content area, performance was extremely low on exercises requiring problem-solving. ... *In general, respondents demonstrated a lack of the most basic problem-solving skills* [emphasis theirs]. Rather than attempting to think through a problem and figure out what needed to be done to solve the problem, most respondents simply tried to apply a single arithmetic operation to the numbers in the problem (p. 338).

Most educators committed to problem-solving as *thinking* have a conception of problem-solving which involves explicit use of strategies. Students are taught to make use of heuristics such as asking themselves, "What is an example of a simpler case than the one I am having trouble with, and how might I use that knowledge to reduce my present state of ignorance?"

Some of those committed to such a view of problem-solving go one step beyond and invoke the concept of metacognition—providing a managerial perspective in which students are to step back to reflect upon and to evaluate the sequencing of heuristics they in fact might use in relation to what they have achieved (or might achieve further) in the solution of a problem.

It is an appreciation for both the heuristics of the problem-solving and metacognition that represents a major thread in the curriculum revisions proposed by the National Council of Teachers of Mathematics (1989) as well as the National Research Council (1990). The focus is set in *Curriculum and Evaluation Standards for School Mathematics,* a visionary document of the NCTM, in the discussion of what ought to be emphasized in mathematics for grades K- 4. They comment:

> Problem-solving should be the central focus of the school curriculum. As such, it is a primary goal of all mathematics instruction. Problem-solving is not a distinct topic but a process that should permeate the entire program.... Ideally, students should share their thinking with other students and with teachers, and they should learn several ways of representing problems and strategies for solving them. In addition, they should learn several ways of representing problems and strategies for solving them. In addition, they should learn to value the process of solving problems as much as they value the solutions (p. 23).

Much bolder claims are made for the centrality of problem-solving, so conceived, in the above document in the context of the curriculum for grades 9 through 12. Mathematics is essentially equated with such problem-solving behavior in the following remarks:

> Mathematical problem-solving, in its broadest sense, is nearly synonymous with doing mathematics. Thus whereas it is useful to differentiate among conceptual, procedural and problem-solving goals for students in the early stages of mathematical learning, these distinctions should begin to blur as students mature mathematically. ... From this perspective, problem-solving is much more than applying specific techniques to the solution of classes of word problems. It is a process by which the fabric of mathematics as identified in later standards is both constructed and reinforced (p. 137).

This theme is repeated in the recent publication of the National Research Council (1990):

> Problem-solving is a central focus of the mathematics curriculum. There is by now an extensive body of literature...indicating that strategies for problem-solving can be taught effectively. ...An effective approach to solving problems is provided by metacognition, the self-conscious ability to know when and why to use a procedure. (p. 31).

As enticing as such a perspective on problem-solving may be, the concepts of "exercise", "problem", "solution", "heuristic" and "metacognition" are all ones that are in need of more careful analysis from a strictly epistemological point of view than they have received by most advocates of curriculum reform. Contrary to Worth's contention, for example, a problem is not merely a situation for which one does not know how to get an immediate answer. There are many situations for which one may not know how to obtain an immediate answer because, for example, no question or implied question has been asked.

In fact, there is considerable debate within the field of philosophy of science regarding (1) what it means for something to be a problem [Lauden (1977), Nickles (1981), Popper (1972), Siitonen (1984), Sintonen (1985)], (2) the extent to which science (and inquiry more generally) is essentially defined by problem-solving, and (3) what it means for something to be a solution as well.

Valuable as such research may be, however, an appropriate analysis of teaching and learning in the context of problem-solving requires much more than a clarification of what a problem is and what problem-solving and thinking are all about from the perspective of philosophy of science. There are important educational issues that we must begin to invoke once we appreciate the kind of observation that Walker Percy makes with regard to our plastering over the reality of "fragmented, haphazard, and incomplete lives." Consider, for example, the following conception of problem suggested by Richard Wertime (1979), a professor of the humanities:

> A problem is, to some extent, a project for the future we commit ourselves to by an act of the will. This means by implication that a problem entails risk, since all "future projects"—to use Hannah Arendt's term—involve uncertainty. There is an essentially binding or promissory dimension to the act of facing a problem. A problem is not an entity which has its existence independent of a person regardless of the illusion fostered by textbooks full of problems; it is an intensely personal and passionate affair, one which is deceptively hard to break off at will. ... [Problems] require attention and courage, and they involve a significant act of self-surrender which can seriously jeopardize the individual's sense of himself. Lastly, a problem is a hopeful enterprise which involves an act of faith (p. 29).

To think of taking on a problem as entailing risks, as being intensely personal and passionate, as having the potential to jeopardize one's sense of self, all suggest educational dimensions that are a quantum leap removed from the essentially cognitive and metacognitive conceptions of problem and problem-solving that are proposed even by most radical mathematics educators. They are highly personal dimensions, and, even if philosophers of science can persuade us that they are dimensions that do not reside in the concept of problem itself, they surely belong to the discourse of education in relationship to problem-solving. What kinds of mathematically derived problems would we have to make available to our students and how would we talk about their connections with these problems (no less their efforts to solve them) if we were sensitive to how it is that problems have the potential to jeopardize one's sense of self? The dominant heuristic view of problem and problem-solving among educators surely suffers from a false sense of aesthetic unity. Problem-solving within which heuristics are emphasized tends to have the following components:

- Problems are generally clearly stated.
- They are assigned with little context and rationale.
- There is little choice offered regarding the student's inclination to receive the problem.

- They are expected to be resolved within a relatively short period of time with machinery that has already been created.
- They are not designed with the intention of raising important human questions of a personal or interpersonal nature.[3]

Perhaps one might argue that, while the above may be true, the implied criticism is irrelevant since it is the nature of mathematics to depersonalize one's relationship to it.

Other essays in this volume argue for the inadequacy of such a point of view, and, while we shall not repeat such arguments at length, we shall, in the next two sections, suggest the outline of a pedagogy which more reasonably reflects a reconstructed view of mathematics.

Before turning to such considerations, however, we should stress that even a less radically personal view of problem and problem-solving than that suggested by Wertime would lead us to reconsider the manner and the

---

We run a greater risk of misrepresenting the lived experience of inquiry when we discourage students from honoring their own doubts, ambivalences, disharmonies.

---

terrain within which problem-solving functions in the curriculum. In an important sense, while disciples of Polya have explored the computer science implications of the kinds of problem-solving heuristics he has advocated, they have given the more human dimension of his scheme short shrift. Consider, for example, Dewey's conception of *reflective thinking*— a conception that was proposed half a century ago, and one that is viewed by many as a general scheme that is compatible with Polya's more specialized application to the field of mathematics.

Dewey (1933) comments,
> Reflective thinking... involves (1) a state of doubt, hesitation, perplexity, mental difficulty, in which thinking originates, and (2) an act of searching, hunting, inquiring, to find material that will resolve the doubt, settle and dispose of the perplexity (p. 12).

> Thinking begins in what may fairly enough be called a forked-road situation, a situation that is ambiguous, that presents a dilemma, that proposes alternatives. As long as our activity glides smoothly along from one thing to another... there is no call for reflection (p. 14).

Now doubt, hesitation, perplexity, ambiguity, mental difficulty are states of mind that are associated with a degree of unpleasantness. One has to be ready to be placed in a state of confusion if problem-solving is to be handled in a non-mechanical way. In addition, one has to acquire and learn to cope with a variety of associated emotions. Unless one is to run rough-shod over the less radical Deweyan conception of problem depicted above, one has to be prepared to be disappointed, frustrated, angry, as one comes to appreciate not only that schemes and plans lead to dead ends, but that a supposed problem being investigated may make little sense.

None of this is to imply that problem-solving need be all doom and gloom. There is of course the joy associated with discovery, and even the process of working in and through confusion can be both intellectually and psychologically uplifting. These states of mind, however, require cultivation, a cultivation in which the cognitive states both inform and are influenced by the emotional ones. Scheffler (1977) has argued for the sense in which some emotions are in fact cognitive emotions—emotions that are compatible with one's cognitive state—and it would seem that such emotions are particularly significant in the context of problem-solving. In particular, the joy of verification and the functioning of surprise bear a strong relationship to how it is that one functions as a problem solver. Summarizing how it is that surprise and theory (of which problem-solving is a component) relate, he comments:
> The constructive conquest of surprise is registered in the achievement of new explanatory structures, while cognitive applications of these structures provoke surprise once more. Surprise is vanquished by theory, and the theory in turn, overcome by surprise. Cognition is thus two-sided and has its own rhythm; it stabilizes and coordinates; it also unsettles and divides. It is responsible for shaping our patterned orientation to the future, but it also must be responsive to the insistent need to learn from the future (p. 186).

Though one may not be in a position to provide a blueprint for a teaching regimen that orchestrates the cultivation of emotions and cognitions associated with problem-solving, it is hard to imagine that cognitive competence can be acquired in a training program that essentially ignores psychological components and that creates the illusion that the trip is one of clear sailing and that navigation is the primary responsibility of the teacher.

Perhaps there are numerous advantages to being exposed to "expert" teachers. What is hard to imagine, however, is that a diet of clear explanations of "the problem, the solution, and when to use a specific process" (as depicted in the government report) will result in the ability to cope with problems mathematical or otherwise. It is also difficult to imagine how one who is exposed to forty problems a day learns to cope with doubt and uncertainty.

What are we to make of the fact that "students with expert teachers are rarely 'lost' during a math lesson?" I suppose this means both that they find the mathematics relatively clear and that they know what to expect in terms of classroom organization. What this means then is that their teachers are providing not one but two Percy-type angst-producing messages simultaneously— that there is aesthetic unity and that there is essentially nothing problematic either with the subject matter or with the pedagogical act.

A New Pedagogy of Mathematics

As indicated in the discussion of *The Holmes Report*, we appear to be in a state of transition—one in which the old metaphors on teaching and learning (in all areas, but especially in mathematics education) have begun to crumble—but we appear to be incapable of creating ones that join an emerging epistemology with an accompanying pedagogy. It may be helpful, in an effort to find significant ways of relating the two, to take seriously the concept of confusion—not as something to be *avoided*, but rather as a force to be embraced. Issues about the functioning of confusion in educational settings run deep and have for the most part been ignored even by those who would abide by a less constricted view of mind than appears in the McKinney report. As educators, we need to give serious thought to how it is that confusion operates not only in problem-solving, but in a variety of cognitive and emotional states that are part of the educational terrain.

While we as teachers may create a state of angst by hiding our legitimate mathematical and pedagogical confusions, we run a greater risk of misrepresenting the lived experience of inquiry when we discourage students from honoring their own doubts, ambivalences, disharmonies. No field of inquiry grows in the absence of perceived anomalies; neither does the individual develop in a non-problematic environment.

Brown (1985), Davis and Hersh (1981), Hanna (1989), Kitcher (1983), Kline (1980), Lakatos (1977), Tymoczko (1986) and others have argued for the sense of mathematics as growing through a social and dialectic process in which confusion, uncertainty, and competing conceptions undergo a kind of metamorphosis that is characteristic of all fields of

human inquiry. Essentially all the fundamental mathematical concepts studied by students in grade school have such a history.

Even the names of number systems presently in use— *negative* numbers, *rational* and *irrational* numbers, *complex* or *imaginary* numbers—hint at their shaky inception. The original Latin name for negative numbers was "numeri ficti" or fictitious numbers. In fact, every number system had a long incubation period before its epistemological status was considered to be on firm footing.

One need not argue as Kline (1966) did at the inception of the "modern math" movement that the curriculum ought to unfold according to a "genetic principle" of pedagogy—in which the curriculum follows its evolution—in order to include the major confusions.

Some of these confusions are in fact experienced (though not acknowledged as "problems" in the curriculum) by students today. Take, for example, the early established and well entrenched intuition that a smaller number divided by a large number cannot have the same value as a larger number divided by a smaller one. Now in an effort to "preserve" another well-accepted property—that of "cross multiplication"—we find ways of persuading students that $(-1) / (1)$ and $(1) / (-1)$ are the same!

What is the logic that establishes the equality, and what is the price we must pay for reifying it? What is interesting about the above example (and all such examples requiring that we relinquish prior intuition as we expand what was previously not permissible) is that there is nothing God-given about the way in which one decides to adjudicate conflicting intuitions. In fact, we could create a respectable system of negative numbers which maintained the prior conception of the ratio of larger to smaller numbers, but once again some other price would have to be paid.

There is a critical issue that lurks beneath the surface here that re-appears every time we try to extend number systems (in the above case to fractions and or negative numbers) so as to enable us to solve heretofore unsolvable problems. Consider, for example, at a slightly more advanced level, the question of how we go about assigning meaning to numbers such as $\sqrt{-1}$. Early on in student's education we find ways of persuading them that $\sqrt{-1}$ has no meaning since the equation $x^2 = -1$ has no solution. If x is either positive or negative, its square is positive, and therefore cannot equal -1.

As students mature, however, we find ways of persuading them that in fact what had no prior meaning acquires meaning if we merely "extend" our number system to imaginary numbers. Well, how do we justify that extension? The issue is frequently clouded by the creation of some scheme (e.g., ordered pairs in which the first element is real and the second an imaginary number) that establishes the system, but by-passes the most fundamental logical issue: It appears that the new system (imaginary numbers) has all that the old scheme had and *more*—the more being that what was previously meaningless now acquires meaning through what appears to be an act of naming.

In fact, we *do* create a scheme that has logical force but at a price—the price being that we must relinquish something that we previously held to be dear—the property of *order*. That is, we can give $\sqrt{-1}$ and its fellow travelers meaning but only if we agree that we can no longer relate these numbers to zero nor to each other with regard to the concept of "greater than" of "less than."

What is at stake here is an issue that is delicate but has more potential to humanize the act of problem-solving than heuristic and matacognitive schools have ever imagined. That is, extension of number systems is *not* merely a logical act by any means. In any act of extension, we are forced to confront an issue that resides at the intersection of logic and aesthetics. In realizing limitations of the machinery that we have created up to the extension point, we must do more than forge ahead and "create" something that was nonexistent or meaningless beforehand. In fact we must *relinquish* some property(ies) from the old system in such an act of creation. One reason I suspect that the curriculum has not acknowledged such a reality is that there is something "un-American" about it all. That is, it runs against the grain of the American spirit to believe that in extending our desires we must relinquish part of what we have so far achieved.

While we must give up *some* cherished properties in extending number systems, there is nothing "God-given" about what in particular must be relinquished. In fact, from an historical point of view, there was always a matter of considerable debate, risk-taking, and confusion when the mathematics community decided what was so precious that relinquishing it would no longer enable us to consider the new system to be a number system at all.

There are surely ways of sharing some of the anguish, uncertainty, and confusion that derive historically from efforts to extend systems. Anyone, for example, who perceives the history of non-Euclidean geometry as one of merely modifying some of the postulates of Euclidean geometry ought to examine the correspondence between one of the founders of the field and his father, also a mathematician.

In a poignant letter from Papa Bolyai to his son, in the middle of the nineteenth century, we find the following advice (Boyer, 1968) regarding the latter's desire to pursue his exploration of non-Euclidean geometry:

> For God's sake, I beseech you, give it up. Fear it no less than the sensual passions because it, too, may take all your time, and deprive you of your health, peace of mind, and happiness in life (p. 589).

This communication, like the names used to refer to number systems themselves, is a wonderful clue that mathematics, far from being a "merely logical" domain of inquiry, is one that is driven by social construction of knowledge accompanied by all of the human traits—courage, fear, debate, confusion—that drive all other fields of inquiry.

There is much in the history of Euclidean geometry itself that it equally poignant. In most of the present-day curriculum on the secondary school level and beyond, Euclidean geometry and arithmetic are well integrated. Line segments have lengths, and most closed regions have areas. It came as a considerable shock to me to discover several years ago in looking through Euclid's *Elements* that, though he addresses questions of area with considerable sophistication he does so with no machinery more sophisticated than a grade school student would appreciate. Consider, for example, the following theorem:

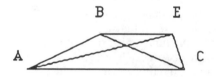

Theorem:
If line BE is parallel to line AC, then triangle ABC has the same area as triangle AEC.

Given our twentieth century mentality the theorem is easily established. BE is parallel to AC, the altitudes BF and EH are equal, and, since the two triangles have the same base (AC) and the same altitude, and since the area of a triangle is one half the product of base and altitude, they have the same area.

But guess what!!! Though the statement of the theorem appears in Euclid's *Elements*, the proof is *entirely* different. In fact Euclid does (as I intimated above) nothing with numbers and formulas for areas. Instead he finds a way to "cut and paste" pieces from the two triangles so that they occupy the same region!

Why did Euclid do that? Why did he not find a way of assigning lengths to segments so that he could get areas by multiplying altitudes and bases as we do today? In some very important sense, Euclid did have considerable knowledge of the set of real numbers.[4] The answer in part at least is that Euclid had a strongly held intuition that for any two line segments, like AB and CD below,

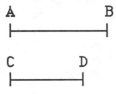

it should be possible to "replicate" each of them so that they would eventually form equal train tracks as below:

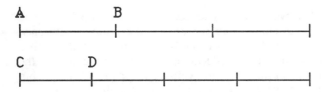

The realization that certain line segments (like the diagonal and side of a square) do not behave that way was enough to cause great suspicion about one's ability to associate lengths with line segments and consequently areas with regions! As with extensions of the number systems, there were aesthetic properties that ancient Greeks so revered that they were unwilling to acknowledge systems that did not exhibit those properties.

While a pedagogy which *informs* students of the human frailties that historically accompanied any extension of old systems to new terrain might be a step in linking pedagogy to a more honest view of the nature of mathematics, there is much more that we must imagine if problem-solving is to find ways of acknowledging our "fragmentary, haphazard, and incomplete lives."

That is, there is a need for students to debate honestly many of the issues that derive from the observation that logic, aesthetics, courage, compassion, blindness, and confusion are all part of the fabric of human existence even in the domain of mathematical thought. Where do we go, for example, with the "un-American" observation made earlier regarding extension of number systems? We could remain on purely mathematical terrain, and encourage students to explore various "mixes" of their own that enable them to adjudicate logic and aesthetics. That is, despite the fact that mathematicians have agreed to relinquish the concept of order in extending number systems to the set of imaginary numbers, is that a property that students themselves feel comfortable relinquishing? If not, what are the consequences? One possibility, of course, would

be for them to deny the validity of the system of imaginary numbers.

But the discussion could be moved in other directions. That is, they might be encouraged to explore what there is about the concept of order that is so appealing, and in turn they might wish to consider the relative "holiness" of other properties.

Such conversation cannot help but invoke what are essentially contestable and aesthetic dimensions of mathematical thought. One could push further, however, in defining the kinds of problems that are generated by efforts to integrate mathematics not only with science and the physical world, but with personal dimensions as well.

One could, for example, pursue questions like the following: How do you react to the proposal that extension of number systems in mathematics is deeply "un-American"—that you cannot merely "add on" desired properties without losing others? Is mathematics alone in its ungenerous conception of progress? What other areas of your personal experiences are like this conception of extending mathematical systems? For which ones is there greater hope of "having it all"?

Though deliberation of problematic conceptions of mathematics and exploration of personhood may be part of what is needed in an extended pedagogy that encourages the incorporation of a reconstructed view of mathematics, they are surely only a beginning. In addition, we must find ways of expanding and clarifying the criticism and of confronting the epistemology we described in the section entitled *Problem-Solving and Clarity*. As part of the process of personalizing mathematical concepts, we need to find ways to encourage students to cope with problems that may not be clearly stated at all. There are may ways of achieving that goal, but one pedagogical avenue for such realization would be to encourage and to teach students not only to solve but to pose problems of their own [Brown (1987) and Brown and Walter (1989)]. Such a pedagogy would be supported by an organization of the environment that provides greater opportunity for students to locate what may be personally and epistemologically interesting and important (as well as dull and unilluminating) about problems that they or their colleagues find worth investigating; and that encourages collaboration in all aspects of mathematical thinking and experiencing.

Though we are surely not in a position to provide a blueprint for bringing about such a reconstructed pedagogy, we can offer something of a prolegomenon. One key ingredient would be the reconceptualization of what counts for a text. The explicit format and implicit message of most texts at all levels convey the wrong set of emotional and epistemological messages. A collection of theorems, definitions, and proofs or a boxed summary of the "key ideas" isolated from the controversial context from which they were derived conveys a depersonalized sense of truth and a false sense of certainty that the new epistemology is attempting to challenge.

Students need to be provided not only with biographical sketches of the accomplishments of mathematicians, but with examples of deliberation and suffering that accompanied the development of ideas of professionals as well as students like themselves.

Essays, novels, and historical reflections will enable students to see that mathematical thinking is not expressed exclusively in depersonalized, linearly organized logical formats. (See Borasi and Brown (1985).) Anyone who has read Abbott's (1952) novel, *Flatland*, has come to understand issues of dimensionality in a more personal and breathtaking way than could be achieved by any "n-tuple" exposition.

Students not only need the opportunity to *read about* mathematical ideas expressed in less conventional formats, but we must also create new outlets for their own oral and written expression. Diaries are one such format within which students may have the opportunity

---

The first concern of the education of teachers is less with input of information about teaching and more with the creation of a climate for unearthing what they already believe about teaching.

---

to react to the emotional climate of the classroom and to the confusion, insights, desires they have encountered in the context of learning mathematics.

Yet another possible format is suggested by Brown and Walter (1990), who advocate the organization of classrooms around the concept of student as both *author* and *critic*. In their scheme, the class is divided into several editorial boards, and students' writing is criticized (and they are encouraged to revise their early drafts) by those who belong to other editorial boards. Though there is nothing particularly sacred about the editorial board scheme, the student as both creator of ideas and critic of the ideas of others is consistent with a pedagogy for a revised view of knowledge—one that is

both social and constructed rather than depersonalized and received.

As part of an emerging epistemology of mathematical knowledge as social and created, we also need to re-examine how it is we think of the concept of *error* in the context of mathematics education. Earlier pedagogical concerns with errors focused primarily upon the development of schemes to *diagnose* student errors, and the creation of teaching regimens to discourage their appearance. In short, errors were things to be avoided if at all possible. Borasi (1986, 1987) has begun to explore a reconstructed view of the place of errors in mathematics education—a view which conceives of error creation as an invitation to devise new questions and to challenge accepted truths.

How we view knowledge, what it is we read, what we expect students to write, how we interact with each other in the classroom are all issues that require education and re-education of teachers. It is beyond the scope of this essay to explore such matters in depth, but we should perhaps stress that new paradigms of teaching and learning mathematics cannot be realized through old patterns of teacher education.

A view of mathematics teacher education that reflects the new pedagogy would not only have to provide a vision that is consistent with some of these reconstructions, but would have to treat the teacher as someone with a full agenda rather than as someone who represents a *tabula rasa*. Borasi and Brown (1989a, 1989b) and Brown, Cooney and Jones (1990) discuss some of the dimensions of teacher education that would appear to be necessary components of a reconstructed view of pedagogy. They argue that the first concern of the education of teachers is less with input of information about teaching and more with the creation of a climate for unearthing what they already believe about teaching, learning, mathematics and education and how those beliefs are embedded in a world view that incorporates but reaches beyond both formal and informal views of education.

Confusion Conclusion

As soon as we ask students (and ourselves as well) to explore the sense in which their approaches, confusions, surprises, delights, and anguish associated with mathematical ideas informs them of how their minds operate and who they are as people, we open up new and educationally important curriculum terrain.

What is needed is a full-blown analysis of the place of confusion in educational settings—*a pedagogy of the confused*. We need a philosophically based analysis which lays out the terrain of educational confusion within which ideas from the previous section could be further developed and criticized. For example, what are the kinds of objects to which confusion may refer? Obviously, we can talk of confused statements or confused questions. They may create confusion for various reasons. For example, they could be ambiguous, or vague or false or meaningless.

To what objects other than statements might the concept of confusion apply? Is there a sense in which we might classify art objects such as diagrams or sketches or artwork with a scheme that is analogous to one we create for statements? What about humor? Is there perhaps a sense in which humor by definition must have an element of confusion? There certainly are aspects of it that do involve confusion of categories as Paulos (1980) argues in a ground-breaking work in which he relates humor to mathematical thinking.

How does the concept of confusion apply to each of the different pedagogical activities we consider valuable? In addition to the posing and solving of problems, we need to reconsider the meaning and place of such concepts as understanding, empathizing, proving, intuiting, explaining, and so forth. What is needed is not only an analysis of the sense in which confusion applies to each of these categories, but a kind of pedagogical imagination as well. For example, imagine a scene in which a teacher has "clearly" presented a concept such as that of a fraction, or a function, or congruence. Imagine further that the students find the concept understandable and "clear" as if taught by an "expert" teacher. A particularly enlightening activity might be to encourage the students either to find confusion in the category that appeared at first unproblematic, or to modify the concept so that it in fact acquires a healthy degree of confusion. Think, for example, of our discussion of the alleged equality of (-1) / (1) and (1) / (-1).

In addition to philosophical analysis, basic research of a psychological nature that explore the potentially healthy uses of confusion, and reviews of biographies that explores the role played by confusion in coming to important understandings about the mathematical world, we might profit from cross-cultural studies of the relationship of clarity to confusion in different organizational contexts.

Consider the possibility of devising a curriculum in various fields in which the central focus is the concept of confusion. Such a course might inquire into the political, psychological, epistemological causes and consequences of portraying disciplines and problems in school settings with considerably greater clarity than is experienced by most practitioners in the field.

Students might explore not only what it takes to create a feeling of confusion and to inquire into the advantages of such a state, but programs might be created to stretch the bounds of their tendency to "give up" when they perceive chaos in their search for understanding.

In "confusion" (pun intended), it is worth noting that establishing a vision of curriculum and of teacher-training upon an illusion of clarity and harmony is not unique to the expert/novice domain of research. Essentially all of the curriculum improvement programs in the past forty years have had clarity as their bedrock. It was implied in the rallying cries of the "structure of the disciplines" movement, the "student discovery" program, the "teach problem-solving" era, and the antithesis of these cries as well. What we are calling for is a full blown analysis of the concept of confusion and its associates, an analysis which not only makes use of research of a psychological, anthropological, philosophical nature, but which creates new kinds of questions and new pedagogical forms for the study and the experiencing of those qualities of human existence that enable us to better understand and cope with where it is that most of us live most of the time.

Footnotes

1. A recent publication by the same group entitled, *Tomorrow's Schools: Principles for the Design of Professional Development Schools* (1990) works through the parameters of an environment—the professional development school—within which these principles might be realized. The professional development school combines the attributes of a laboratory school for university research, a demonstration school and a clinical setting for the improvement of teaching for interns as well as experienced teachers.

2. This brief report does in fact accurately capture one stream of research in the area of teacher effectiveness. My invited participation at a *Research Agenda Conference on Effective Mathematics Teaching* at Columbia, Missouri, in March 1987 (co-sponsored by The National Science Foundation and the National Council of Teachers of Mathematics Foundation) which focused upon the kind of research summarized by the McKinney report, supports this contention. Though there was considerable criticism offered by fellow researchers, all of them drew the implication that, if technical flaws in the research were rectified, then the major conclusions of the research would be justified. In fact most of the discussion centered around the search for parameters that would enable researchers to predict with a degree of confidence those inexperienced teachers who would eventually look like "the experts." But we are getting ahead of our

story. I merely wish to claim here that, even if the McKinney report were an inaccurate rendering of a stream of teacher effectiveness, it would serve my purposes here with integrity since it represents a point of view that is well entrenched in educational discourse—a point of view that I believe to be fundamentally mistaken and in need of criticism and reconstruction.

3. The recent documents of the National Council of Teacher of Mathematics (1989) and the National Research Council (1990) come closer than any defined by the profession to challenging some of these criteria. Nevertheless, they do shy away from perceiving mathematical inquiry as a vehicle for coming to some of the personal dimensions implied by the list.

4. Though his intuition about commensurability of two line segments precluded the possibility of Euclid conceiving of the ratio of two line segments in general, he did have the machinery to talk intelligently of the concept of "equal ratio" for any four line segments—even if the two separate "ratios" were of pairs of incommensurable segments. In a disguised way then, Euclid had all the machinery he needed to impose the concept of length on that of segment, though it took over 2000 years for mathematicians to feel comfortable to modify their intuition in order to make use of it. One cannot but be awestruck at the similarity between the ancient "equal ratio" construction of Eudoxus and Richard Dedekind's treatment in his 1872 publication entitled *Continuity and the Irrational Numbers*.

References

1. Abbot, E.A., *Flatland*, New York, Dover Publications, 1952.

2. Borasi, Raffaella, "Exploring Mathematics Through the Analysis of Errors," *For the Learning of Mathematics*, Vol. 7, No. 3, 1987, p. 1-8.

3. Borasi, Raffaella, "Using Errors as Springboards for Learning Mathematics," Special Issue of *Focus* (*On Learning Problematic Mathematics*), Vol. 7, Nos. 3 and 4, 1986.

4. Borasi, Raffaella and Brown, Stephen I., "A 'Novel' Approach to Texts," *For the learning of Mathematics*, Vol. 5, No. 1, 1985, pp. 21-23.

5. Borasi, Raffaella and Brown, Stephen I., "Reflections On the Continuing Saga of Mathematics Teacher Education," *New York State Mathematics Teachers Journal*, Spring 1989a.

6. Borasi, Raffaella and Brown, Stephen I., "Soundoff: Mathematics Teacher Preparation: A Challenge," *Mathematics Teacher*, Vol. 82, No. 2, 1986b, p. 88-89.

7. Boyer, Carl. *History of Mathematics*, Boston, Wiley, 1968.

8. Brown, Stephen, "Liberal Education and Problem-Solving: Some Curriculum Fallacies," *Proceedings of the Philosophy of Education Society*, (David Nyberg, editor), Normal, Illinois, 1985.

9. Brown, Stephen I., *Student Generation*, Arlington, Massachusetts, *Committee On Mathematics and Its Application*, 1987.

10. Brown, Stephen I., Cooney, Thomas J., Jones, Douglas, "Research in Mathematics Teacher Education," in *Handbook of Research on Teacher Education*, (W. Robert Houston, Martin Haberman, John P. Sikula, editors), Macmillan Publishing Co., Reston, VA, 1990.

11. Brown, Stephen I., and Walter, Marion I., *The Art of Problem Posing Second Edition*, Hillsdale, NJ., Lawrence Erlbaum and Associates, 1990.

12. Burton, Leone (editor), *Girls Into Maths Go*, London; Holt, Saunders Publishing Co., 1986.

13. Carpenter, Thomas P., Corbitt, Mary K., Kepner, Henry, S. Jr., Lindquist, Mary M., Reys, Robert, "Results of the Second NAEP Mathematics Assessment: Secondary School," *Mathematics Teacher*, May 1980, pp. 329-338.

14. Davis, Philip J., Hersh, Reuben, *The Mathematical Experience*, Cambridge, Massachusetts, Birkhauser, 1981.

15. Dewey, John, *How We Think*, Boston, D. C. Heath and Company, 1933.

16. Hanna, Gila, "More Than Formal Proof," *For the Learning of Mathematics*, Vol. 9, No. 1, 1989, pp. 20-23.

17. Heron, Kim, "Technological Hubris," *The New York Times Book Review*, April 5, 1987, p. 22.

18. Kitcher, Philip, *The Nature of Mathematical Knowledge*, Oxford, Oxford University Press, 1980.

19. Kline, Morris, "A Proposal for the High School Mathematics Curriculum, *Mathematics Teacher*, Vol. 59, No. 4, 1966, pp. 322-334.

20. Kline, Morris, *Mathematics: The Loss of Certainty*, New York, Oxford University Press, 1980.

21. Lakatos, Imre, *Proofs and Refutations*, Cambridge, Cambridge University Press, 1977.

22. Lauden, Larry, *Progress and its Problems*, Berkeley, University of California Press, 1977.

23. McKinney, Kay, "How the Experts Teach Math," *Research in Brief*, U. S. Department of Education, Office of Educational Research and Improvement, Washington, DC, November 1986.

24. National Council of Teachers of Mathematics, *An Agenda for Action: Recommendations for School Mathematics in the 1980's*, Reston, VA, National Council of Teachers of Mathematics, 1980.

25. National Council of Teachers of Mathematics, *Curriculum and Evaluation Standards for School Mathematics*, Reston, VA, National Council of Teachers of Mathematics, 1989.

26. National Research Council, *Reshaping School Mathematics: A Philosophy and Framework for Curriculum*, Washington, D.C., National Academy Press, 1980.

27. Nickles, Thomas, "What is a Problem That We May Solve It?" *Synthese*, Vol. 47, 1981. pp. 85-115.

28. Paulos, John, *Mathematics and Humor*, Chicago, University of Chicago Press, 1980.

29. Peters, Richard S., "Why Doesn't Lawrence Kohlberg Do His Homework?" in *Moral Education: It Comes with the Territory*, David Purpel and Kevin Ryan (editors), Berkeley, CA, McCutchan Press, 1976, pp. 288-290.

30. Polya, George, *Mathematical Discovery*, N.Y., NY, John Wiley and Sons, 1962.

31. Polya, George, *Mathematics and Plausible Reasoning*, Princeton, NJ, Princeton University Press, 1954.

32. Popper, Karl, *Objective Knowledge*, Oxford, Oxford University Press, 1972.

33. Scheffler, Israel, "In Praise of the Cognitive Emotions," *Teachers College Record*, Vol. 79, No. 2, 1977, pp. 171-186.

34. Siitonen, Arto, "Demarcation of Science from the Point of View of Problems and Problem Stating," *Philosophia Naturalis*, Vol. 21, 1984, pp. 339-353.

35. Sintonen, Matti, "Separating Problems From Their Backgrounds: A Question-Theoretic Proposal, *Communication and Cognition*, Vol. 18, No. 1/2. 1985, pp. 25-49.

36. *Tomorrow's Schools: Principles for the Design of Professional Development Schools*, The Holmes Group, Inc., East Lansing, 1990.

37. *Tomorrow's Teachers: A Report of the Holmes Group*, The Holmes Group, Inc., East Lansing, 1986.

38. Tymoczko, Thomas, "Making Room for Mathematicians in the Philosophy of Mathematics," *The Mathematical Intelligencer*, Vol. 3, No. 8, 1986, pp. 44-50.

39. Wertime, Richard, "Students, Problems and 'Courage Spans'," pp. 27-36, *Cognitive Process Instruction*, Jack Lockhead and John Clements (editors), Philadelphia, The Franklin Institute Press, 1979.

40. Worth, Joan, "Problem-Solving in the Intermediate Grades: Helping Your Students Learn to Solve Problems," *Arithmetic Teacher*, February, 1982, p. 16-19.

Appreciating the Humanistic Elements within Mathematical Content: The Case of Definitions

Raffaella Borasi
University of Rochester
Rochester, New York

1. Introduction

If we ask laymen, students, and often even *mathematics* teachers what they think about mathematics, the picture that is painted tends to be that of a discipline which is "cut and dried," rigid and impersonal where results are always either absolutely right or wrong, and where there is little space for creativity or personal judgment. If this is what mathematics is perceived to be, no wonder that even creative and able students are discouraged from pursuing its study!

The realization that most people's conception of mathematics is rather false and misleading, and that it may detract from their learning of the discipline, is not new. Conveying an appropriate image of the nature of mathematics was already recognized as an important curriculum goal by the authors of the many "new math"

Mathematics is in many respects similar to other disciplines and to other areas of human activity.

---

projects. Their concern, however, was mainly to communicate to students those characteristic elements of mathematics that make it unique—such as the emphasis on logical deduction and structure. On the contrary, the contributors to this book have assumed the position that what is mostly needed in order to open people to an appreciation of mathematics is the realization that mathematics is in many respects similar to other disciplines and to other areas of human activity.

This chapter will contribute to this overall goal by focusing on how students can be brought to experience, in the first person, the humanistic dimensions existing *within technical mathematical content*. Too often, in fact, there is the tendency to make mathematics more *humane* and personal simply by making the students acquainted with the life of great mathematicians and with some episodes in the history of mathematics, or by allowing the students to express and discuss their feelings about the discipline. If this, however, is not

accompanied with a change in the approach to learning the technical mathematical content covered in the curriculum, we can expect little effect on the students' conception of mathematics and their approach to its learning.

I certainly do not want to underplay the value of making the students aware of the historical and philosophical dimensions of mathematics. Realizing the difficulties and struggles encountered by mathematicians to achieve what are now widely accepted mathematical results, becoming aware of the proposal of alternative theories and the existence of some unavoidable contradictions, and furthermore recognizing the presence even today of conflicting positions about the very foundations of mathematics, could all greatly contribute to challenging the common notion of mathematical knowledge as absolutely predetermined and unquestionable. Similarly, I do not deny the crucial role played by affective elements in the learning of mathematics, nor the importance of helping students acknowledge their *math anxiety* as the first step towards effectively fighting it. However, I would like to point out that isolated efforts in these directions may be easily perceived by students as digressions, or at best as "icing on the cake," something that is not really relevant to the "real" essence of school mathematics—i.e., memorizing rules and solving problems. In order to have a real impact on the way students approach their study of mathematics and their mathematical activity, we need to make them explicitly appreciate the role that elements such as doubt, uncertainty, personal value, or context play when working with technical mathematical content; discuss whether the truth of the mathematical results studied as part of the school curriculum is always and absolutely determinable; make them aware that alternative interpretations or solutions to the mathematical problems with which they are presented may be possible and may need to be evaluated.

In this chapter, I would like to show that such an approach is possible, and I would like to open discussion on its considerable educational implications. I intend to do this by reporting and discussing a specific experience: a teaching experiment on the notion of mathematical definition, conducted with two sixteen-year-old students. The report of this teaching experience will be preceded by a brief mathematical/ philosophical analysis of the notion of mathematical definition, which will reveal how this notion is much more problematic and "humanistic" than it is usually perceived.

II. A Preliminary Inquiry
Into the Nature of Mathematical Definitions

How could the characterizing elements of a "good" mathematical definition be identified?

Most mathematicians and mathematics teachers probably would agree that a mathematical definition should at least:

1. **be non-contradictory and non-circular** — that is, the properties stated can all coexist, and the definition does not use essentially the term that it is trying to define;
2. **use precise and unambiguous terminology** — all the terms employed should have been previously defined, or be one of the few undefined terms assumed as starting points;
3. **"isolate" perfectly the concept in question** — so that all instances of the concept will meet the requirements stated in the definition, while a non-instance will not;
4. **be essential, non-redundant** — only terms and properties which are strictly necessary to distinguish the concept from others are mentioned.

The first criterion stated seems reasonable and obvious, if we want the definition to allow us to recognize instances of the concept in question.

It is easy to recognize the importance of criteria 2 and 3 when analyzing incorrect definitions of a very familiar concept, such as the following definition of "circle";

(a)   *A closed, continuous, rounded curve.*
(b)   *Something whose area is πr^2.*
(c)   *All the points equidistant from a single point.*
(d)   *Circle: $x^2 + y^2 = r^2$. Round.*
(e)   *Definition of a circle: a perfectly round, closed figure with a radius r and a circumference c where r is the distance from the midpoint of the circle to any outside point and c is the distance measured around the outside.*

For instance, how could we interpret terms like "rounded" in definition (a), in order to decide whether a certain figure (say, an ellipse or an 8-shaped figure) is or is not a circle? On the other hand, even if we may not object to the terminology employed in definitions (b) and (c), these definitions do not succeed in isolating circles from other objects (such as spheres in the case of

(c), and a much wider variety of plane figures in the case of (b)—as for example a rectangle of sides πR and R). One could criticize definition (d) because it describes only circles with centers at the origin—even if in this case it could also be argued that, given a circle, we could always verify the condition stated by choosing our coordinate system with origin at the center of the circle.

The length alone of definition (e) already sensitizes one to the disadvantage of redundant definitions. Still, the value of "essentiality" can best be appreciated when we try to use a definition in the derivation of some mathematical results. A very good example can be found by considering the following alternative definitions for "isosceles triangle":

(f)   *An isosceles triangle is a triangle with two equal sides AND two equal angles.*
(g)   *An isosceles triangle is a triangle with two equal sides.*

While both definitions allow us to precisely distinguish isosceles triangles from other figures, only definition (g) would allow the following simple proof for the theorem "An angle inscribed in a semicircle is a right angle".

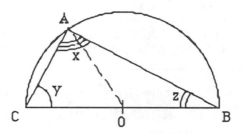

Figure 1

1.   OA = OC = 0B, since they are all rays of the same circle.
2.   Using def. (g) we can deduce that the triangles AOC and AOB are isosceles.
3.   Hence ACO = y and ABO = z.
4   The sum of the interior angles of triangle ABC then provides the equation 2y = 2z = 180° .
5.   Hence: x = ACO + ABO = y + z = 90° .

The crucial steps in this proof (steps 2 and 3 above) would not have been possible had we assumed (f) as our definition of isosceles triangle. In that case, in fact, to conclude that two triangles AOC and AOB are isosceles, we would have previously and independently verified that not only two sides, but also two angles are equal.

What has been said so far probably did not provide any new information for a mathematician or mathematics teacher—except perhaps for illustrating how errors could be used to make the students themselves more aware of these well-known characteristics of mathematical definitions. A more careful analysis of some of the previous examples as well as of some commonly "accepted" definitions could lead, however, to more surprising results, which may eventually challenge some of the four traditionally accepted criteria previously stated.

First of all, we may reasonably question whether it is realistic to expect totally precise and unambiguous definitions, even in mathematics. For example, let us consider the following typical formulation of the definition of circle:

    (h)"A circle is the set of all points in a plane that are at the same distance from a given point in the plane" [Jacobs, 1974, p 449]

Suppose we are in a city, where distance cannot be measured as the crow flies (since we cannot go across buildings!). If we interpret the definition of circle in this context, or in its simplified model consisting of a square grid (a context often referred to "taxigeometry"), we will obtain a set of points which will no longer correspond to the shape we usually associate with circles. In taxigeometry, for instance, "circle" of radius 3 will look like this, where each X represents a point on the "circle":

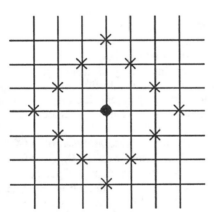

Figure2

There could be at least two different reactions to this rather surprising result.

The first one is to conclude that the previously accepted definition of circle [def.(h)] is not precise enough and needs to be improved. Our efforts in this case should be directed towards modifying such a definition in order to avoid interpretations in contexts other than the usual Euclidean geometry. This could be achieved, for example, by specifying the kind of distance to be considered.

There are some problems, though, with this solution. First of all if we really become picky we can find other

terms which may need a more "rigorous" definition than is usually the case. Though we may have already agreed about the necessity of assuming the notions of "point" and "plane" as undefined, what about words such as "all", "set", or "given"? Besides, even if it were possible to define every term in the definition precisely, we may wonder whether an extreme precision would unnecessarily complicate even the definition of such a simple concept as a "circle" and thus make it more difficult to understand it intuitively and to remember and use it.

An alternative reaction to these criticisms of def. (h) could be that of recognizing the values of accepting a certain level of ambiguity in a mathematical definition. Mathematicians have in fact found it useful to generalize the notion of "circle" as "the locus of points

---

Students can be brought to experience, in the first person, the humanistic dimensions existing *within technical mathematical content.*

---

equidistant to a fixed one" to contexts where "points" and "distance" may assume meanings quite different from their usual geometric ones. Think of the variety of metric spaces considered in advanced analysis [1]! Clearly, though, we cannot expect all the properties of the usual circle to transfer to these new situations. An excellent mathematical activity would be to investigate which properties do and do not transfer—an exercise which would make us realize better what characteristic elements of circles are strictly due to their metric property of equidistance from a center point. It is worth noticing that we will find ourselves in similar predicaments whenever we attempt to extend a certain mathematical notion to a wider domain. Indeed this happens often in mathematics, especially in the area of arithmetic, algebra and number theory, where the expansion of the set of numbers we are working with (going from natural numbers to integers, rationals, reals, etc.) may each time require a reexamination of all operations previously defined and their properties.

Once we come to examine more critically the third item of our list of criteria to evaluate mathematical definitions, we may find some problem even with the seemingly reasonable requirement of "isolation of the concept." If we have not yet defined the concept precisely, in fact, how can we expect to be able to analyze a tentative definition and decide whether a specific object belongs or not to the category itself?

The seeming paradox of having to know a priori what is

an instance of the concept which we are trying to define becomes more evident as soon as we move away from concepts as intuitive and familiar to us as the "circle" and consider instead the activity of mathematicians who define new concepts as a means of exploring a mathematical domain.

Lakatos (1976) has presented a beautiful historic example in this sense, when he analyzed the development of one of the fundamental theorems of topology—first suggested by Euler—which can be stated as follows:

> In a polyhedron the relationship among the number of faces (F), sides (S) and vertices(V) is: $V + F - S = 2$

Lakatos' analysis shows the interplay between successive refinements of this statement, its demonstration, and the definition of "polyhedron" itself. Suppose we start with the intuitive definition of "polyhedron" as "a solid whose surface consists of polygonal faces," then the "pathological instances" of polyhedra in Figure 3 could present counterexamples to the theorem as stated above:

A solution (though not the only possible one) to this problem could consist in a revision of our original definition of "polyhedron" so as to add conditions which would eliminate the pathological examples. For instance:

— the alternative definition, "A polyhedron is a connected surface consisting of a system of polygons" would allow excluding figure 3A as an example of a polyhedron (since this figure consists of two disjoint surfaces), so that it will no longer present a counterexample to the theorem; however, according to this definition Figures 3B and 3C should be considered as polyhedra, and thus would still disprove the theorem;

— the threat posed by Figures 3B and 3C can be eliminated by further refining the previous definition as follows: "A polyhedron is a

system of polygons arranged in such a way that (1) exactly two polygons meet at every edge and (2) it is possible to get from the inside of any polygon to the inside of any other polygon by a route which never crosses any edge at a vertex."

It is interesting to notice that even this last rather detailed definition will not be sufficient to solve all the problems presented by "pathological counterexamples" to the theorem. The value of the procedure described by Lakatos, however, should not be measured in terms of reaching a final determination of the definition of "polyhedron" nor even of providing a proof of a refined version of the Euler theorem—though either of these results may be reached in the end. Rather, the work around the definition of "polyhedron" may just become a valuable means to explore fundamental differences amongst solid figures, which may in turn motivate and lead to possible ways of categorizing them—a fundamental objective in the study of topology!

The situation described above is certainly not unique to the historical example examined by Lakatos, and similar contexts and activities could be easily created for students, dealing with even simpler notions than "polyhedron"—as it will be shown later when discussing the design of the teaching experiment. Reflecting on this kind of situation can help us appreciate how definitions are really *created by us*, even in mathematics where everything may seem so deterministic (at least to most students!). Yet, at the same time, the process is not totally arbitrary since mathematicians indeed try to define "useful" concepts, which attempt to fit our intuitive images, and provide a valuable tool to describe and identify properties of the situation studied (be it "real" or "mathematical").

The examples and considerations brought forth so far have supported the thesis that definitions in mathematics cannot always be as "perfect" as we may wish them to be, and have helped us uncover some of the limitations. I believe that this experience, besides being a humbling one for a mathematician, should also

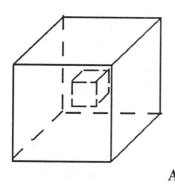
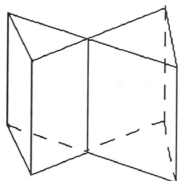
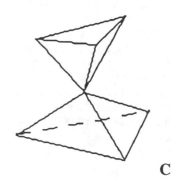

A   B   C

Figure 3

make us all reevaluate the power and value of intuition in every human being (including students!).

First of all, it may be important to realize that we do not really need absolutely rigorous definitions in all mathematical domains and contexts. For instance, we constantly use notions such as "variable", "equation", "area", "geometric figure", knowing what we are talking about sufficiently to *operate* with these concepts and *do* some genuine mathematical activities (such as solving algebraic equations, finding the area of certain figures, or discussing their equivalence with others). Yet, we may not be able to give "rigorous" mathematical definitions of these concepts, definitions which will allow accountability for all pathological instances, and exceptions in various possible contexts, until we acquire some sophisticated tools in advanced areas of mathematics such as abstract algebra, analysis, or differential geometry.

Secondly, relying on the intuitive definition of a notion can be valuable, not only because it makes it possible to work with the concept earlier in one's mathematical education, but also because it may allow more agile reasoning and applications even when a rigorous definition is available. However, as always when relying on intuition, we will also have to be aware of its limitations and possible pitfalls, and thus be constantly more "alert" in our manipulations—a healthy attitude to gain, in mathematics as well as any other domain.

While the previous analysis may have made us aware of the need for reconsidering some of the original expectations regarding the nature of mathematical definitions, it should not bring us to relinquish the four criteria stated at the beginning of this section as useless. Rather, we should recognize that these requirements are

---

This kind of situation can help us appreciate how definitions are really created by us.

---

essential when working within an axiomatic system, though not in all other mathematical contexts, and try to create different criteria, each appropriate for the evaluation of mathematical definitions in a specific context.

The conclusions stated above summarize the position I personally reached with regard to the role of mathematical definitions as a result of the arguments developed in this section. Starting from the consideration of some absolute criteria set by mathematicians to describe mathematical definitions, I subsequently moved to the realization that these criteria

may not be appropriate in all mathematical contexts. The confusion brought by this relativistic outlook was somehow resolved by reaching a position of personal and context-bound commitment, which respects the value of the characterizing elements passed along by the community of mathematicians within a specific context, but at the same time refuses to take them as an absolute standard or the ultimate criteria to evaluate mathematical definitions [2].

Though I cannot expect everybody to agree whole-heartedly with these conclusions, it is important that the reader recognize them as the background that motivated and led to the creation of the instructional unit whose implementation I am going to report and analyze in the rest of the chapter.

III. Exploring the Nature of Mathematical Definitions With Secondary School Students: Overview of the Teaching Experiment

The elementary nature of the examples examined in the previous section made me hope that high school students, too, could be actively involved in an inquiry into the nature of mathematical definitions along the same lines. As I also believed that such an experience would be highly beneficial to mathematics students, since it could make them reconceive some of their inappropriate and dysfunctional views about mathematics, I set forth to design and implement a didactic unit on this topic.

The first step in this direction consisted in the creation of a series of activities, mostly organized around the in-depth analysis of specific incorrect definitions, which I hoped would stimulate high school students to question and explore the nature of definitions and the criteria for their evaluation in mathematics.

I chose to focus on the concepts of "circle", "polygon", "equation" and "exponent", because they all dealt with fundamental yet elementary mathematical notions. Students in their last years of high school could be expected to have some degree of familiarity with these concepts, and a sufficient technical background to work with them. At the same time, from the point of view of mathematical definitions these concepts presented quite different characteristics.

"Circle" denotes a very familiar figure, which students can easily identify, and which can be described mathematically by a precise and simple definition in more than one way. I planned first to take advantage of these elements by asking students to evaluate a list of incorrect definitions of "circle", with the goal of having them recognize the need for elements such as the use of precise terminology, isolation of the concept, and so on. At the same time, by means of this exercise as well as the application of alternative definitions of "circle" in the derivation of some mathematical results, I hoped to make them aware of the possibility and value of having more than one way of defining the same mathematical concept. An interpretation of the common definition of "circle" in the context of taxigeometry would then show the importance of considering the context in which one is operating, in order to interpret or evaluate a mathematical definition.

The concept of "polygon" is also elementary, yet much less familiar than "circle" to high school students. I expected that, if I did not provide them with a definition of "polygon" to start with, the students would have only a vague idea a priori of what figures should or should not be considered as instances of this concept. Thus the activity of creating a reasonable definition of "polygon" for themselves would provide a context in which they might experience the situation of a mathematician who wants to define a new notion. After having let the students grope for a while with tentative definitions, I could provide some direction to their activity by suggesting the consideration of a tentative theorem "à la Lakatos". The statement "The sum of the interior angles of an n-sided polygon measures $(n - 2)180°$" could well serve this purpose.

Students are usually familiar with the notion of "exponent" as "the number of times you have to multiply a number by itself." However, they may not always realize that this definition applies only to exponents which are whole numbers. And yet, the notion of exponent can be extended to other numbers (such as negative integers or fractions) *at a cost*. This cost may include relinquishing the original intuitive meaning of the operation and dealing with some "exceptions", such as 0^0. An analysis of the dangers, as well as the advantages, of these extensions of the operation of exponentiation could bring new elements into the students' consideration of the nature of mathematical definitions, by illustrating the intrinsic limitations with which mathematicians sometimes have to deal.

I expected that the problems experienced while working with the definitions of "polygon" and "exponent", as well as "circle" in taxigeometry, would challenge some of the "absolutist" expectations that the students might have developed in the first activities focusing on the more familiar notion of "circle". In addition, I thought it would be worthwhile to direct the students' attention to notions which could be intuitively understood and used by them, yet be so complex that a precise definition would be out of their reach. I planned to use the concept of "equation" for this purpose, though many other notions would have served equally well. In fact,

in the actual implementation I ended up using the concept of "variable" instead, because its consideration could be introduced more naturally in that context.

The sequence of activities just described provided me with an initial structure which could be used as the starting point for specific implementations with high school students. I was aware of the need for extreme flexibility in the instructional design, not only because of the different mathematical backgrounds of the students who might be involved in the experimentations, but even more importantly because I wanted to allow the students to participate in genuine exploration, which might lead in directions different from those initially planned by the instructor.

The decision of conducting my first experience as a teaching experiment was motivated by several reasons. First, it would be quite difficult to implement such a long and articulated unit within a regular mathematics course. Second, to allow for the educational potential of the methodology to be fully exploited, I wanted to create an ideal environment, where the students would not feel the constraints (in terms of time, expectations, established teacher and student roles) of the traditional schooling environment. Finally, I needed a context that

---

This experience, besides being a humbling one for a mathematician, should also make us all reevaluate the power and value of intuition.

---

would allow for very careful monitoring of the experience, in order to be able to analyze the effect of the instructional unit on each student's conception of "definition" specifically and mathematics more generally, as well as on the student's ability to reflect and reason.

An ideal occasion to conduct the teaching experiment presented itself when two students, who for various reasons had missed many classes in an experimental high school course I had previously taught, asked for an opportunity to "make up" for these absences and thus be allowed to get credit for the course.

To interpret correctly the results of the experience reported here, it will be important to realize that the subjects (who will be called Kim and May) were far from being enthusiastic "mathematically gifted" students. Though both quite bright girls, neither of these students was especially interested in mathematics, and they had not always had positive experiences with their mathematics courses in the past. The following quotes, taken from an autobiographical essay assigned before the beginning of the teaching experiment, well illustrates these students' views of mathematics:

M.:   Mathematics has always been a sore area for me. Actually, when I was in my first years of elementary school, Math was fun. It was simple and precise and I have to (painfully) admit that I LIKED IT. In sixth grade things changed ... I failed the class and every math class since has been a trial for me. I consider myself LOGICAL, and I *can* understand things, but I'd rather avoid trying to get math.

K.:   I don't like math much. It does not stimulate me. Numbers just don't seem very significant in retrospect to everything else that my life involves. math skills are mandatory in dealing with money. Statistics are interesting ... still I have a hard time doing my math homework, probably because it all seems so silly and pointless. Mathematics just isn't as amusing as talking on the phone, isn't as interesting as reading a book and certainly not after a quarrel with my mother. The prospect of sitting down for two hours, trying to figure out the probability of picking one red chip if there are 2 red chips, 7 black chips and 1098 yellow chips in a bag seems ridiculous. ... Doing mathematics makes me feel frustrated, annoyed, repulsed—confident and competent when I understand. My good tests go on the frig.... The problem with mathematics is it is too impersonal, useless, I often don't understand how it will help me or relate to my life. Basic math is necessary but I do not understand why I continued in it.

The teaching experiment involved nine instructional sessions of about 40 minutes each, in which both students always participated. The sessions were audio-taped and transcribed, and detailed narrative reports of the same prepared, integrating the dialogue from the transcripts with other data and observations provided by the field notes of a non-participant observer. In addition, data to evaluate the effects of the experience were collected through various interviews with the students [3] and a final project.

It is quite difficult to summarize such a wealth of data in a few pages. In the following section I will try to give an idea of the depth reached by the students in their reflections on mathematical definitions by selecting and reporting a few significant episodes in the unit. An overall evaluation of the effects of the experience on the students' conceptions of "definition" specifically and of the nature of mathematics more generally will then follow. I refer the reader interested in further

information on this experience to my book *Learning Mathematics Through Inquiry* (Borasi, 1992) for the detailed narrative and commentary of each session as well as for a more thorough analysis of the experience.

IV. Episodes From the Teaching Experiment

To illustrate the variety of activities and discussions experienced by the students involved in the teaching experiment, as well as the quality of their achievements, I have selected three episodes, all dealing with definitions in geometry, yet presenting some interesting differences, in terms of mathematical context, aspects of definitions considered, and timing within the unit.

In the first episode I will report on the variety of questions and mathematical activities that were generated by the students' analysis of a list of incorrect definitions of "circle". This experience occurred at the very beginning of the unit, and took most of the 40 minutes allotted for our second instructional session.

The second episode occurred towards the middle of the experience, and dealt with the students' attempts to define "polygon"—a notion they did not feel very familiar with. This activity developed through our fourth and fifth meeting, taking about 45 minutes altogether.

Finally, I will report on the considerable challenges presented to the students by the examination of the usual definition of "circle" in the context of taxigeometry. About half an hour of our eighth instructional session was devoted to this topic.

All the dialogues reported in this as well as the following sections are excerpts taken verbatim from the transcripts of instructional sessions or interviews. The following abbreviations will be employed: K. = Kim; M. = May; R. = researcher/instructor.

Analyzing Definitions of 'Circle'

Our starting point in this activity consisted in the following list of definitions of "circle", previously selected from amongst those produced by the two students and by other students of similar mathematical background

A.   All the possible series of points equidistant from a single point (May)

B.   Circumference πr^2, = radius, an exact center, 360°. (Kim)

C.   Round—3.14—shape of an orange, coin, earth—Pi.

D.   Circle = something whose area is = πr^2.

E.   Circle: $x^2 + y^2 = r^2$. Round.

F.   A circle is a geometric figure that lies in a two

dimensional plane. It contains 360 and there is a point called the center that lies precisely in the middle. A line passing through the center is called the diameter. 1/2 of the diameter is the radius. I don't like circles too much any more because they look like big fat zeros but they can be fun because you can make cute little smiley faces with mohawks out of them.

G.   A closed, continuous, rounded line.
H.   I sometimes find myself going around in them...

When asked to comment on these definitions, May spontaneously started to criticize a few of them on the ground that they were describing also figures other than circles:

M.:   "Closed continuous rounded line": that could just be a spiral...a closed one.

Definition H was also immediately rejected by both students because it was "too vague". While both students seemed satisfied with May's definition (def. A), their belief was challenged by the instructor's observation that a ball would also satisfy that description. May, however, was quick in recognizing where her original definition was falling short, and quickly produced the following modification:

All the possible series of points on a plane equidistant from one single point.

The consideration of def. D, followed by Kim's own dissatisfaction with her definition (def. B), brought up more explicitly the issue of the difference between "definition" and "properties" of a concept, as well as a concern for "essentiality" in a definition:

M.:   whose area is πr^2! ...improper English
R.:   Would that be a definition of a circle?
M.:   No, it's an element of the circle...
R.:   Why would you exclude [this definition] then? Can you show a figure which satisfies this definition and it is not a circle?

(May suggests her spiral once again as an example of a figure with the same area of the circle, yet different from the circle)

R.:   Is there any other definition which you would would like to eliminate?
K.:   Yes, mine. (The response comes very quickly and forcefully).
R.:   Why? (laughing)
K.:   Because I just put anything I could possibly think of...And I was wrong. πr^2 is not the circumference, it's the area...I put 360 degrees but I didn't know...I was not able to put down a round answer, I just put what came to my mind...
R.:   That's also what definition F does. Why do you think we may not want to have a long list

of properties?
K.:   I am not saying that it would not be good, but...
R.:   Oh, you would like to have put more?
K.:   I just did not remember...
M.:   But for a definition...it should be stated as simply as possible...
R.:   So, we want a definition to be able to identify only circles. And a long list of properties would probably do that even better. Then why you would not like it?
M.:   Why? Because a definition is something you have to remember...you don't want to remember all the little things...the whole list...

Both the issues of essentiality and the distinction between "definition" and a set of properties of a given concept were too complex to be resolved completely at this point. We came back to these issues several times throughout the unit, especially in the third session, when we tried to use alternative definitions of "circle" to solve problems such as "Find the circle passing through three given points" and "Find the measure of the interior angle of a regular pentagon inscribed in a circle."

Different kinds of concerns were raised by an analysis of def. E, which was not as familiar to the students as I had expected (despite the fact that they had recently completed a unit on graphing!). Perhaps because of their lack of mathematical background, the students were able to point out several important pieces of information which had been left out by the author of the definition since such elements are usually taken for granted in the context of analytic geometry.

K.:   I don't know what x and y are...
R.:   You are right, we have to say what x and y are, or it doesn't make any sense ...
K.:   Like, in mine, if I should do it over, I should say what r means...
R.:   Let's say we were using graph paper...
K.:   Oh, that makes sense!
M.:   But this is not the full sense of what a circle is...because you do not always have graph paper...
R.:   With some work... This is a good definition though, because it will only give circles...
R.:   But how can you check if it does?

This question was first answered by trying to plot some points satisfying the equation, and then by checking with the use of a compass whether they all belonged to a circle. However, later it also motivated a more rigorous deduction of the equation of the circle based on the equidistance property. Since the students had forgotten what had been done in previous courses, this provided the opportunity for genuine mathematical activity on their part, while they were creating some geometric informal arguments to convince each other of the essential equivalence of the metric and analytical definitions of "circle".

In the context of this exercise the occasion for another interesting digression occurred because of an observation regarding the similarity between the equation of a circle and the formula contained in the Pythagorean theorem which was raised by Kim.

K.: But, How would you know...how could you say if that is the equation of the circle, but it is also the formula for the Pythagorean theorem for triangles...

This confusion initiated a discussion on the difference between "variable" and "constant", which was further developed in a later session when introducing the consideration of definitions of complex mathematical notions such as "variable" or "equation".

The increased level of sophistication and criticism reached in the analysis of the definitions by the end of the session is evident in the following concerns expressed by May with regard to our revised versions of both definitions A and E:

M.: How would you figure out if something is a circle, if there is no measurement for the radius?
R.: Ah! That is a good point!
M.: What if they just say "circle", "draw a circle" and you are... what's its r?... So it doesn't work, because you don't know a circle if you don't know its radius.
R.: Do you think you need also to know were it is placed? Where is the center of the circle?
M.: No...

This argument reveals an interesting expectation with respect to mathematical definitions; that is, a definition should allow us to construct the object in question. However, in this case this expectation should be reconsidered since "circle", as most other mathematical concepts, is an abstraction—of all possible circles, with different radii and placed in different positions in the space. So its definition can at most allow us to construct an instance of the concept, once we have given a specific value to some of the general parameters left open in the definition itself—such as the value of r in the case of "circle", for example. Unfortunately, I did not seize the opportunity of expanding on this point at the time this happened—a regret I felt more than once in this experience!

To conclude this section I would just like to report May's response when, at the end of the lesson, I asked whether having two different yet acceptable definitions of 'circle' was disturbing:

M.: No, they are just two different ways of looking at it. That's a very healthy attitude! (laughs)

Creating a Definition of "Polygon"

Defining "polygon" presented a real challenge to Kim and May, and brought to their attention some aspects and roles of mathematics definitions that they had not realized before.

From the very beginning, the students felt uneasy about the task of coming up with a good definition of "polygon", because the notion was not clear in their minds. Even the activity of analyzing and refining some tentative definitions, which had been so successful in the case of "circle", now proved rather frustrating. Our starting point in this case consisted of the definitions produced by the two students at the beginning of the unit.

M.: Polygon: a geometric figure of straight lines that have no sides of equal length.
K.: Polygon (quadrilater): sided geometric figure angles add up to be 360.

Various figures were considered, such as:

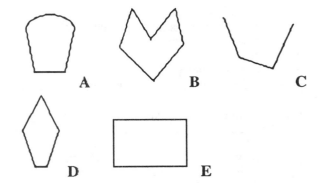

Figure 4

Deciding whether each of these figures should be considered an example of "polygon" brought the students to refine May's original definition by adding the constraint that the figure should be closed (to avoid considering Figure 4C as a possible polygon), and by eliminating the requirement that the sides should not be equal (so as to consider figures like the square as polygons). The refined definition now read:

Polygon: a closed figure of straight lines.

However, in this activity it was difficult for the students to motivate their decisions, as shown in this brief exchange between their instructor and May:

R.: Do we want this to be a polygon?
M.: No, let's kick'm out. Let's be elite?!
R.: What do we want to do with polygons?...so we

can decide if certain kinds of figures will be, or not be.

M.:   I know we have several different kinds of polygons. It doesn't really have a definite shape... (mumbling!)

The students welcomed with relief the suggestion of making the definition of "polygon" more meaningful by attempting at the same time to prove some interesting results about polygons. To this end, I proposed the consideration of the following statement:

The sum of the interior angles of an n-sided polygon measures 180(n - 2)°. [4]

Interestingly, this could be considered as a generalization of the property of quadrilaterals mentioned in Kim's definition.

R.:   How do you think we can prove something like this?

M.:   I don't know. Take a polygon as an example. (The instructor immediately draws one, a pentagon.) We never really got to the definition of a polygon. We think this is a polygon.

R.:   Right, so for the moment we think it's a polygon. How to figure out the sum of the interior angles?

M.:   Put a circle around it that meets all the points.

R.:   (draws a very "skinny" polygon) What if the polygon is like this?

K.:   (has a great insight) Make it into triangles.

M.:   Take a center point. But if it's really weird shaped you can't do it. Oh yes, you could do it.

R.:   (seizing another opportunity but refusing to get side-tracked now) That's an interesting question: "When can we draw a circle around a polygon?" But keep it aside. (Getting back to the last issue raised by May). It seems that whenever I have a polygon I can pick a point in the center, more or less. Does it matter which point I pick? Maybe we should ask Kim why it is that you wanted to break it down into triangles like this?

By developing Kim's insight, the students managed with some help from the instructor to find that the sum of the interior angles in this case was 3 x 180°, thus confirming our tentative statement. The same procedure was then followed to derive the sum of the angles in a hexagon, confirming once again the statement, to the student's surprise:

M .:   She's right.

K.:   I don't want you to be right.

M.:   It's too neat.

In the effort to find some counterexamples, May spontaneously suggested the consideration of a 5-pointed star:

Figure 5

She questioned, however, whether this kind of figure should be considered a polygon or not. Since we were now trying to prove a general statement about polygons, it was easier so decide this issue than before. In the case of the 5-pointed star, in fact, one cannot even decide which angles are to be considered as interior angles, thus it would be impossible even to know how to interpret the statement we were trying to verify. We did not, therefore, want to consider the 5-pointed star as a polygon.

This discussion also helped us justify some of the previous decisions made about what figures should be considered as polygons (see previous Figure 4), and at the same time resulted in a further refinement of our definition of polygon, which now read:

Polygon: a closed figure of straight lines which do not cross.

In order to deal somehow with the "star", however, May creatively suggested we consider only the outside lines:

Figure 6

With this modification, the star could then be considered a "normal" 10-sided polygon, and applying our procedure we could confirm once again the correctness of the formula to compute the sum of the interior angles—with great satisfaction on the part of the students.

The impact of this experience on the students' conception of mathematical definition is best reflected in the students' responses to a review/questionnaire administered towards the end of the unit. The page regarding their work about the definition of "polygon" is reproduced below.

When trying to find a definition of "polygon", we had some trouble starting because we did not really know precisely what we wanted a "polygon" to be—except that it was going to be a concept generalizing triangles, squares, pentagons, etc. So we started with a tentative definition, and "refined it" so that polygons

would have some interesting properties.
We ended up with the following definition:

> (i) A polygon is a closed geometric figure, with straight edges which do not cross.

Are you satisfied with this definition? How could we ever know if it is correct? Explain your answer.

M.: Not quite. Play w/ polygons. Test the theory, use examples of figures that meet the requirements for the def. but don't meet properties. OR just T&E (NOTE: meaning "trial and error")

K.: I am satisfied with this for the time being but I think possibly as I start to use them more end more, I may desire something more exact.

What are the main differences between this case, and the definition of concepts such as 'circle'?

M.: We already had a definition and a picture of the figure in our head before testing definitions and exceptions (with the circle)

K.: Circle is just one figure, one concept—a polygon is more general and includes many kinds of geometric figures.

It is possible that in the future we may want to further modify the definition of "polygon" or that of "exponent".

What is the value, then, of a preliminary, tentative definition?

M.: It's like a theory that hasn't been fully tested yet and therefore is open to modification.

K.: getting acquainted with a new concept—have something to base further defining on. You have to start with something.

Surprises with Circles in Taxigeometry

One of our last meetings was devoted to a discussion of the meaning of the usual definition of "circle"—"All the points in a plane equidistant from a given point"—in a new context, that of taxigeometry. I believe that this episode well illustrates the remarkable level of reflectivity, independence and critical stance assumed by the two students towards the end of the unit.

In preparation for this lesson, the students had first been introduced to the new situation by means of a story. The instructor's review of the different interpretation given to the notion of "distance" in this situation (i.e., as the length of the minimal path along the streets of a city, or the lines of a grid), initiated an interesting discussion which reveals once again the critical attitude now assumed by the students, as well as their clear understanding of the notion of "distance":

R.: It seems reasonable to consider this (the shorter path along the lines of the grid) as the distance...

M.: But if you were an airplane, then the distance

would just be (and indicates a straight line connecting the two points)...

R.: Right. But an airplane could not go through the buildings...

M.: No... but a helicopter could!

R.: (laughing) Okay! But we are human beings, we have to...

K.: (jokingly) But what if we are... "super hero"!

M.: Yeah!

R.: (laughing) That would give you a different distance...

Once the constraints of the situation with which we wanted to work had been clarified, I asked the students to draw "all the points at distance 5 from a given point of the city".

At these instructions, Kim immediately recognized the definition of circle:

K.: May be a circle?

R.: Sounds like a circle, right? Let's try to construct it. Would it look something like this (She draws a "usual" circle with radius 5).

Meanwhile May had started figuring out her own version of the question. She had drawn all the points at distance 5 from the given point, using "taxidistance", and commented:

M.: I think it should be a diamond.

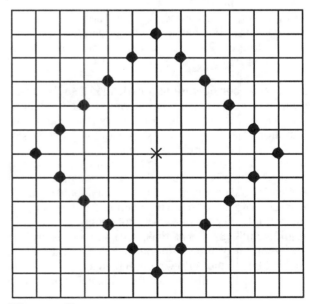

Figure 7

This difference of opinion gave rise to a very lively and interesting discussion between the two students.

R.: Now we have two theories.

K.: But... you want all the points that are 5 away?

R.: Yes. Why is that not a circle?

M.: Because you have to cut corners. There is no way to go up and down the buildings.

K.: So what? She said "5 points away".

To exploit this disagreement, I attempted to initiate an explicit discussion of the situation:

R.: Can it be that both of you are right?

M.: (without hesitation) Sure! It depends on what you want "distance" to be. It depends on whether you are on a graph or in a town. Uuh! It's fun, I like this! (She continues to embellish her graph paper)

R.: Can we call them both circles? They both suit the definition!

Since Kim was still in disagreement with the last statement, May explained to her once again why she thought that her "diamond" represented indeed all the points at the same 'taxidistance' from the center point. However, at this point, May herself brought up a new concern. She noticed that even if her "diamond" satisfied the definition of circle we had agreed upon, it did not satisfy many of the other properties of circles— an excellent point, which shows once again May's increased understanding of the notion of definition.

M.: You could call that a circle, but if you went over it with properties, it wouldn't be a circle, because it's not 360 degrees, is it? (pointing at the interior angles of her diamond).

R.: (instead, pointing at the angle at the center of the diamond) This is 360.

M.: But, other properties may not work in this case. For example, the circumference is no longer $2\pi r$, or the area is not πr^2. What others?

These new considerations challenged May's initial acceptance of the diamond as a 'circle', and brought us once again to consider explicitly the nature of definitions and their relationship with the properties of an object. The search for a solution to this conflict brought May to an interesting compromise. She started to analyze the original definition of circle in the attempt to identify either a vague term (which should be more precisely defined) or some misinterpretation on our part. Interestingly, her attention was not caught by the word "equidistant" (as I had expected) but rather by the interpretation of the condition of being "on a plane". This led her to the following arguments which is quite creative and logically sound (though some details may need refinement).

M.: I don't think so. We have *not* made a circle here. This is not... In the definition we said it was on a one-dimensional plane.

R.: Isn't this a plane?

M.: Yeah, but not "one-dimension" because of the buildings, this is more than one dimension.

R.: The situation is not the same.

M.: Yeah, I'm sure ...in our definition we said "A geometric figure on a plane"...and this is not a plane...(becoming silly) this has chimneys... going way up. These are all like buildings that are coming up in 3D. So [the diamond] is not a circle. (Pointing to Raffaella's) that's a circle!

What a nice way to describe the Euclidean plane: a plane with no obstacles!

May had clearly grasped the notion that the context may modify the meaning of a definition, and had resolved her initial conflict by finding the "mistake". There did not seem to be, however, an acknowledgment that for the same reason the definition could be considered correct and would describe either figure depending on the context—a realization that would probably require a much more radical departure from the common view of the nature of mathematics.

Interestingly, this activity spontaneously suggested to May a desire to define precisely "diamond"—perhaps still in the effort to contradict the disturbing fact that her "diamond" and a circle could share the same definition. Both students got very involved in the task, and with minimal intervention from my part, they proceeded along the same lines as initially done with the definition of "circle" and that of "polygon"—that is, they attempted to refine successively a tentative definition by considering many examples and counterexamples based on their intuitive notion of "diamond". In my opinion, this episode provides the best possible evidence of what the two students had internalized about the nature and role of definitions, since they themselves posed the problem (that is, they felt the need for a definition of a concept which they had clear in their minds) and chose a reasonable procedure to approach a solution.

V. What did the Students Gain from the Experience?

The episodes reported in the previous section already provide valuable data to illustrate what the two students got from their participation in the teaching experiment. In this section, I would like to complement those data with an overview of what the students learned with respect to the notion of mathematical definition specifically, and with respect to the nature of mathematics more generally, on the basis not only of their behavior in the teaching sessions but also their performance in the final project and their explicit reflections as expressed in a series of interviews at the end of the experience. Once again, the reader is referred to the detailed report of the teaching experiment (Borasi, 1992) for further information and evidence to substantiate the conclusions reported here.

Learning about Mathematical Definitions

As the episodes previously reported well illustrate, the organized activities brought the students first of all to realize several aspects of mathematical definitions, such as: the importance and justification for requirements like essentiality and precision in the terminology used in definitions of terms such as "circle"; the difficulties involved in the creation of the definition for a "new" notion, as they experienced in the case of "polygon"; the unexpected results that can occur when a definition is interpreted in a different context, as when they worked with "circle" in taxigeometry and "exponent" with numbers other than the whole numbers.

The ability to transfer principles and procedures learned in some specific activity to new situations was sometimes unexpectedly shown by the students. The half-serious discussion regarding the interpretation of "distance" in taxigeometry, and May's desire to define "diamond" precisely (both briefly reported in the third episode), are very good examples in this sense, especially because in both cases the task was not set by the instructor but generated by the students themselves.

One cannot conclude, however, that the students gained a clear and explicit understanding of the different roles and uses of mathematical definitions, thus reaching the position stated by the author at the end of section II of this chapter. At the end of the unit, in some cases it was still difficult for them to decide which criteria would be the most appropriate in the evaluation of a specific definition requested by the instructor. In other words, though they had been made aware of the diversity

---

Contrary to the general objectives and expectations of mathematics teaching, I would like to consider this confusion as a positive achievement.

---

existing among mathematical definitions, they had not yet been able to generalize, from the specific examples analyzed, criteria to distinguish between different mathematical contexts, and consequently to set reasonable expectations about the characteristics of definitions employed within that particular context.

As a result, at the conclusion of the experience the students still experienced some confusion and frustration, as May explicitly mentioned in one of the interviews:

M.:   Well, it's good to have...I guess...I'm glad I found out about the exceptions. But I think

there should be someone out there that is going to perfect the definitions, because it would be nice to know that, in math at least, that you can always go by it. Because if a definition is what you'll go by in all your work and then you find an exception and you can't do anything with it, I imagine that'll be frustrating. But it's like, where do you go when the definition written "in blood" isn't right?

Contrary to the general objectives and expectations of mathematics teaching, I would like to consider this confusion as a positive achievement. It would have been too ambitious and unreasonable to expect the students to go much further within the limits of the teaching experiment; reaching an explicit clarification of the characteristics and requirements of mathematical definitions, depending on their role and use in different mathematical contexts, should rather be a goal of the whole high school mathematics curriculum! More importantly, I believe that a state of confusion may be a necessary premise to a more sophisticated and differentiated understanding and use of mathematical definitions. The doubts and conflicts this confusion may generate would in fact make the students more cautious and reflective in their future use of mathematical definitions, paying an increased attention to the context they are operating in and to the various facets that mathematical definitions may show.

Changing Conceptions of and Attitudes towards Mathematics

As expected, the most remarkable effects of the experience were not limited to the students' improved understanding of the complex notion of mathematical definition.

Throughout the instructional sessions, but especially during the interviews conducted at the end of the unit, the students explicitly commented on the different approach to learning mathematics implemented in the unit. They recognized and were appreciative of the opportunity to discover and analyze mathematical issues on their own. They found the material more interesting, complex, and challenging than they had ever thought possible in a mathematics class, and were surprised at their own achievement and ability in a subject where they had always felt at a disadvantage.

The following excerpts from the final interviews illustrate the impact of the experience on these students' conceptions of and attitudes towards mathematics.

K.:   I thought math...everything that was going to be discovered in math has already been discovered and being a mathematician would be a really stupid thing to do because everybody already knows everything that there is to know....But even the smallest thing is questionable. The same definition could get you something completely different, like with taxigeometry....I learned a new method of doing math.

M.:   When it cannot be figured out by me, or by the next person, it just reminds me that this was all invented by people. It's not something like we were born and there is a tree and it has been there for ever. It's like we invented this, out of our minds.

It seems like a totally different way of looking at math...instead of being taught something, we were like, learning it and discovering it, as we went along. I start to see patterns in it [math]. And all the separate rules, all the rules are pretty regular...and there are exceptions just like in everything ... Still, it holds a definite shape and it does have a purpose, and it can be figured out, even by me! But I didn't think I could figure out math before. I used to think that the material was to blame, but I really think it is the way people were told to teach it. And I think that it is the way they were taught.

Contrast these quotes with the thoughts expressed by the two students before the beginning of the teaching experiment (see section III) and furthermore with the following statement May made earlier in the unit, referring to the exercise of writing definitions for a few concepts assigned by the instructor:

M.:   When writing definitions...I felt it was going to be like right or wrong, according to what the classic idea was. You [the instructor] didn't make me feel this way, but I felt that way myself.

However remarkable these passages may seem (especially considering the age and background of these students!), one may also want to be alert to the danger of rhetoric whenever trying to elicit people's beliefs. It is important, therefore, to complement the students' explicit comments with some implicit evidence gathered from their behavior during the instructional experience.

As may be partly evident from the episodes reported, a gradual change in the students' mathematical behavior was observed throughout the unit. It was evident, as the unit developed, how the students' critical attitude and independence increased, making them more active and in control of their mathematical experience. First of all, they became more and more cautious with regard to both the results of their own mathematical activity, and the results suggested by Authority—consisting, in this case, of the instructor's input and what they had been taught in previous mathematics courses. The students also took advantage increasingly of the opportunity of shaping their own mathematical

explorations, by generating new questions rather than simply executing the tasks imposed by the instructor.

I would like to report an episode which I find extremely revealing of the new attitudes towards learning mathematics reached by the students as a result of this experience. This episode concerns the definition of the task for their final project—which was supposed to replace the usual test expected at the end of any instructional unit.

I had suggested at first that they write an essay about mathematical definitions which would synthesize and communicate to other people what they had discovered through our inquiry. The students, however, were disappointed by this decision, and while explaining politely their reasons, they were also able to state what their expectations had been. This comes out clearly in the following excerpt of the conversation:

K.:   Is this something...

M.:   Is this the project you meant? (both surprised)

R.:   Yeah.

M.:   Mmm, because I had a totally different idea, I was...

R.:   Okay, what did you think of?

M.:   (Almost talking over R.) Hum, I was just... last night, in fact (laughter)...I was talking with (her boyfriend) about how much work I will have coming up, next week, you know, we were talking about doing some things. I said, I've got a big project coming up, and he said "Well, what do you have to do?", and I said "Well, I think what she is going to do is give us a figure, or something that needs to be defined, and then we have to define it and give all of its properties, by figuring out ... And I thought, that would be fun, but hard work, and I'm going .. it will probably be something really complex, and I will have to work on it for hours and hours and hours (laughs)...

(R during this time has been giving just some encouraging sounds)

K.:   I thought something like that, too, or else like ... that you would give us graph paper or something, and we would have to construct a little town or something, and tell a little story about how long it would take to get... (everybody is laughing supportingly)...I don't know!

R.:   Would you prefer something like that?

K.:   I don't know.

M.:   Yeah! This (meaning R.'s idea for the project) is interesting, just...but it's such...I don't know, kind of review...it would be like...

K.:   It would be neat to take...this will be just like rounding everything which we have learned, and just like putting it on paper.

R.:   Right.

K.:   But instead, doing something...taking everything which we have learned and actually applying it to something else...

R.:   Yes! Okay .

M.:   That's why I thought you would take all this and we would have to apply it to something new...

As a result of this discussion, I indeed changed my original idea for the project, which finally resulted in the task of proving the equivalence of three alternative definitions of "perpendicular bisector", and in the discussion of the differences between ellipses, parabolas and perpendicular bisectors in the contexts of Euclidean geometry and taxigeometry, respectively.

In the dialogue reported above, the students implicitly showed a remarkable understanding of what could be a good evaluative experience for what they had learned—that is, showing the ability to "transfer" it to a new situation. At the same time, their observations show an appreciation for the possibility of creative activity offered by a mathematical task.   Most importantly, they took a stand in an aspect of the mathematical learning experience which is usually considered the ultimate discretion of Authority—that is, students' evaluation .

This event is even more remarkable because the students seemed totally unaware of the significance of their gesture, and brought forth their criticisms and alternative suggestions very simply and spontaneously!

The impressive effect of this short unit on the students' conceptions of and attitudes towards mathematics was certainly due to many elements of the teaching experiment besides the topic studied—for example, the opportunity for flexibility and individual attention allowed by the small number of students, the attitude and role assumed by the instructor, the personality of the two students involved.   However, I believe that the results of the teaching experiment confirm that an analysis of the nature of mathematical definitions, along the lines suggested here, has the potential to involve students in activities that may have a considerable impact for their overall approach towards mathematics.

VI. Concluding Observations

It is my hope that the experience reported in this chapter has provided sufficient evidence for the claim that pre-college students can be brought to appreciate the humanistic dimensions existing even within technical mathematical content, and that such an experience can have considerable impact on their overall conception of and approach to mathematics.

Focusing on the notion of *mathematical definition* well

served this purpose, since it allowed the students to work with mathematical concepts and tasks that were mostly familiar to them and typical of the school mathematics curriculum.   At the same time it brought them to recognize the possibility and value of having alternative definitions for the same concept, to realize the importance of taking the context (mathematical and/or practical) into consideration when interpretating and evaluating a specific definition, and to appreciate the

> ## It's not something like we were born and there is a tree and it has been there for ever.   It's like we invented this, out of our minds.

interplay between values, purpose, and personal judgment occurring when making a "mathematical decision".   Yet, the notion of definition is by no means unique for achieving the realizations of these humanistic aspects of mathematics.   For example, appropriate learning activities could have been similarly designed to make students appreciate the more humanistic dimensions of the concept *of mathematical problem* [5], or of *proof* [6].   Even the study of more typically "technical" mathematical concepts—such as *infinity* (see Borasi, 1984 and 1985) or *equation* (see Raman, 1989)—could provide rich opportunities for students to realize the presence of contradictions, ambiguity, alternative interpretations, and so on, within the very core of mathematics.

Similarly, the specific learning activities organized in this teaching experiment can be further generalized so as to provide ideas for the design of other experiences geared at making students appreciate the humanistic aspects of mathematical content.   Among the didactic principles embedded in the design of this teaching experiment, I would especially like to highlight the following ones:

- *concreteness of the tasks*: note how the exploration of the different facets of mathematical definitions was made possible as the students engaged in concrete activities around the definition of a few specific concepts and not by means of purely philosophical or general discussions;
- *building on conflict and doubt*: a challenge to the students' deterministic expectations about mathematical definitions was often initiated by presenting situations which were in contrast with such expectations, or which allowed for contradictory yet equally plausible interpretations; often the analysis of real or

perceived errors played such a role (for a more explicit discussion of the strategy of using errors as springboards for inquiry, see Borasi [1986a and 1987]);

- *integration with "traditional" mathematical activities*: the uncovering of the humanistic aspects of mathematical definitions went hand in hand with the use of specific definitions in typical mathematical tasks such as solving a problem, proving a theorem or exploring a topic, so as to make the students immediately realize the impact on everyday mathematical activity of their new philosophical realizations about the nature of the discipline;
- *providing opportunities for reflection*: the set-up of the teaching experiment provided plenty of opportunity for the students to voice and discuss the more philosophical doubts, concerns and reflections raised by specific "technical" mathematical activities; in larger classes, similar opportunities should be provided, by setting aside sufficient time for class discussions and perhaps also by encouraging personal introspection and reflection by means of appropriate writing tasks.

I hope that the experience described in this chapter, as well as these more general considerations, will provide sufficient interest and inspiration for other mathematics teachers, at all levels of schooling, to try to uncover the more humanistic aspects embedded in the mathematics they are expected to present in their courses, and communicate those to their students.

(This study was partially supported by a grant from the National Science Foundation, #MDR8651582; all the opinions expresssed here, however, are solely the author's. The instructional episodes discussed in this chapter have also already been reported and analyzed, though in more detail and from a somewhat different perspective in my book *Learning Mathematics Through Inquiry* (Borasi, 1992))

Footnotes

[1] Interestingly, mathematicians have continued to maintain the usual geometric and intuitive meanings for "circle" and "sphere", and to use instead a new term— that of 'ball of radius r'—for the generalized notion of "all the points at distance equal to or less than r from a given point in a metric space." However, this choice has to do mainly with the *name* assigned to the mathematical concept, rather than its *definition*—a distinction which may be worth some reflection, especially when discussing the "arbitrary" nature of mathematical definitions.

[2] It is interesting to notice a strong parallel between the various stages I went through in the argument developed in this section and the scheme suggested by Perry (1970) to describe the developmental changes in the conception of knowledge that he observed in several students throughout their undergraduate education. Perry summarizes his scheme in three parts starting from an absolutistic right-wrong outlook that relies on Authority for the true answers, then moving through a stage of relativism where the possibility of diverse but equally valid positions is recognized and finally teaching some sort of personal and context-bound commitment amongst these possibilities.

[3] A group interview was conducted by the researcher/instructor with both students at the end of the experience. An individual interview with each student followed, conducted by the non-participant observer. In the case of May only, I also had the opportunity to gather an audio-taped conversation on the differences between definitions in mathematics and those in other domains.

[4] It may be worth reporting that the statement initially suggested by the instructor was the following: "The sum of the interior angles of an n-sided polygon measures 180 x n degrees". The mistake had not been part of the instructor's design, yet it turned out to be beneficial since it made the justification and the discussion of the statement more lively and interesting and reinforced the students' critical attitude towards the instructor's input as well as their own work.

[5] See Brown [1982], Kilpatrick [1982] and Borasi [1986b] for a theoretical analysis of the notion of *mathematical problem* which could be used as a basis for the design of a unit illustrating the variety of mathematical problems, and the role of context, value, and personal judgment in the definition and solution of them.

[6] Lakatos [1976] and Kline [1980] can provide the historical background for a "humanistic" analysis of this fundamental yet controversial notion which is often considered as one of the ultimate characteristics of mathematics.

References

BORASI R.: 1984, "Reflections and Criticisms on the Principle of Learning Concepts by Abstraction," *For the Learning of Mathematics*, 4, 3, 14-18.

BORASI R.: 1985, "Errors in the Enumeration of Infinite Set," *FOCUS: On Learning Problems In Mathematics*, 7, 3-4 77-90.

BORASI R.: 1986a, "On the Educational Roles of Mathematical Errors: Beyond Diagnosis and Remediation," Ph.D. Dissertation State University of New York at Buffalo.

BORASI R.: 1986b, "On the Nature of Problems," *Educational Studies In Mathematics*, 17, 2,125-141.

BORASI R.: 1987, "Exploring Mathematics Through the Analysis of Errors," *For the Learning of Mathematics*, 7, 3, 1-8.

BORASI R.: 1992, *Learning Mathematics Through Inquiry*, Heinemann Educational Books.

BROWN S.I.: 1982, "On Humanistic Alternatives in the Practice of Teacher Education", *Journal of Research and Development in Education*, 15(4), 1-12.

JACOBS H.R.: 1974, *Geometry*, Freeman and Co., San Francisco.

LAKATOS I.: 1976, *Proofs and Refutations*, Cambridge University Press.

KILPATRICK J.: 1982, "What is a Problem?", *Problem Solving.* 1, 4.

KLINE M.: 1980, *Mathematics: The Loss of Certainty*, Oxford University Press.

PERRY W.: 1970, *Intellectual and Ethical Development in the College Years. A Scheme.* Holt, Rinehart and Winston, Inc.

RAMAN M.L.: 1989, Beyond the Solving of Equations," Proposal for an Ed.D Dissertation, University of Rochester Rochester, NY.

Mathematical Orchards and the Perry Development Scheme

Larry Copes
Augsburg College
Minneapolis, Minnesota

Part I

The orchard

> Imagine an orchard, with trees evenly spaced in rows and columns. Now extend that orchard infinitely in all directions. Finally shrink all of the trees until they are infinitely thin. Cut one down and stand on its stump. (You'll have to become infinitely thin, too.) Can you see out of the orchard?

What do we do as mathematicians when we first encounter the orchard problem? I have done some limited research, completely lacking in statistical validity. Our first reaction is to smile with delight. We enjoy finding problems that challenge our intuition. We see that the size of the orchard makes us think that any line of sight would hit a tree, but that the smallness of the trees helps balance that out. We're intrigued at the tension.

Almost immediately, then, we "mathematize" the problem. We restate it in terms of the Cartesian coordinate plane: Is there a line (or ray) through the origin that passes through no other point both of whose coordinates are integers? In idealizing the situation, we often implicitly assume answers to some questions that

---

> ## We don't stop with an answer. We reflect. We return to our original enjoyment of the problem's tug on our intuition.

---

are obvious to beginners: Can I look up? Does the orchard bend as the earth curves? How far can we actually see? What does "out" mean if the orchard is infinite? What Robert Davis (1990) calls the "mathematical culture" has taught us something of what a problem is, and what kinds of information to ignore in mathematical modeling.

We then reason. We ask ourselves what we could say

about a line through the origin that did "hit another tree," and we decide that the slope would be the ratio of two integers. That is, the slope must be a rational number if the line hits a tree. Of course, not all slopes of lines are rational. In fact, any line with irrational slope would miss all of the trees. So it is possible to see out of the orchard. But we don't stop with an answer. We reflect. We return to our original enjoyment of the problem's tug on our intuition. And we retrace our logical argument to see how our intuition can be reconciled with it.

Finally, we start asking other questions. If we had a spinner that determined the line of sight randomly, what's the probability that the line of sight chosen would hit a tree? What if the trees were not in regular rows and columns, but more random? How far could one see if the trees had some thickness?

In this chapter I want to step back a little and consider what skills and attitudes we use while engaged with mathematics, and to apply those observations to various notions of humanistic mathematics. In the process I'll introduce two delightful students and an organizing construct. I'll start with the organizing construct, and leave the best for later.

The Perry Scheme

The Perry Development Scheme grew out of an attempt to describe patterns in the ways some college students interpreted their experiences. From longitudinal studies in which Harvard and Radcliffe students were interviewed came a description of a "journey" that many American college students seem to take as the intellectual, ethical, and identity aspects of their lives change. The research effort was coordinated and first described by Harvard counselor William G. Perry, Jr., so the result bears his name.

The Scheme consists of nine "positions" from which students view their worlds, but I shall oversimplify these by grouping them into four major categories. For more details, please see Perry (1970 or 1981). Students moved from one position to the next, with some backtracking, but none moved through more than a few of the nine positions over their years in college, and apparently nobody skipped any of the positions.

Young people initially interpret their experiences from a point of view that Perry calls **Dualism**. This is not mind-body dualism, but dualism between right and wrong. To these folks, things are pretty clear-cut. The morally good people are the ones who have the intellectually correct answers. They are Authorities—teachers, religious leaders, maybe even parents—with whom I'm supposed to identify and whom I am to emulate.

As you can see, the Perry Scheme is not merely a description of intellectual development. Rather, it shows interactions among the intellectual, ethical, and identity aspects of a person. Observe the mix of these three lenses in Dualism: getting answers right in school makes me a morally right person. I identify with those Authorities who know the Truth, and I am good if I strive to learn how to read those same Truths from tablets in the sky. As one student said,

> Yeah, it wasn't in the book. And that's what confused me a lot. Now I know it isn't in the book for a purpose. We're supposed to think about it and come up with the answer! (Perry, 1970, p. 78).

This same mix will appear to varying degrees in the other three categories.

Although many beginning college students see at least some aspects of life Dualistically, peer pressure for change usually prevails fairly quickly. We don't make friends and influence people by telling them that they're

---

We want our students to be able to distinguish between intellectual and ethical correctness.

---

damned to hell just because they're different from us. Students often find great relief in adopting a new *zeitgeist*, which Perry calls **Multiplicity**. As one interviewee said, "I mean if you read them [critics], that's the great thing about a book like *Moby Dick*. *(Laughs)* Nobody understands it." (Perry, 1970, p. 98) Diversity becomes not only legitimate but required. Everyone has an *intellectual* right to any opinion whatsoever, and anyone trying to insist on standards is *morally* wrong. The ideal teacher is the Ph.D. in blue jeans whose class sessions consist only of discussions. Lacking a justification for grades in terms of truth, grades can only be a function of the amount of effort expended.

This Multiplistic attitude of students clearly presents a conflict with what we as professors want. We want our students to be able to weigh evidence. We want them to be able to distinguish between intellectual and ethical correctness. We want them to be able to rule out those opinions to which nobody has an intellectual right—since those opinions ignore some of the data or are logically inconsistent. We want students, nevertheless, to be tolerant in the face of the realization that there still remain, and always will remain, conflicting opinions that are equally valid. So we ask them to compare and contrast, to evaluate. And we are frustrated when they don't do it.

Perry calls this lens **Contextual Relativism**. It goes beyond philosophical relativism, which is more like Multiplicity, since within Contextual Relativism ideas are related to context. Contextual Relativists see the *ethical* as a special case of the *intellectual*, something to be reasoned about. For the first time students can look at ideas from a variety of perspectives, and can analyze their own thinking from "outside." They now see how the authorities really have been thinking, and they can think that way and see themselves as fellow searchers with their teachers.

But *is* the ethical merely a special case of the intellectual? Can I really live a meaningful and moral life within the realization that there are several conflicting but equally valid religions, political philosophies, ethical systems, careers, relationships? Perry calls the fourth category **Commitment**. The capital C indicates that the term refers to more than a commitment to go to the ball game Friday night; it's Commitment within Relativism, a Commitment that comes from saying, "Yes, but I have to decide on something for me, to make my life meaningful. I must make some leap of faith." At first my identity is in my Commitment: I am a college teacher. Then, as Commitments to career and relationships and ethics begin to conflict, my identity is in my balancing style. All the while I live with the realization that I might have made other Commitments with equal validity, and may occasionally have to reconsider all of my Commitments. Polanyi (1958) calls the accompanying epistemology "Personal knowledge."

That's a brief and oversimplified summary of the Perry Development Scheme. Before going on, I want to point out that a person probably will view different kinds of experiences from different positions. I, for example, am pretty well into Commitment with respect to my career and some human relationships, but I'm not ready to make a "leap of faith" in religious areas of my life, and with respect to child-rearing I'm pretty Multiplistic.

The Perry Scheme and the orchard

A person who viewed mathematics Dualistically would probably be impatient with the entire orchard problem endeavor. What good is this problem? What's the answer, or what's the best way of getting the answer? Perry tells of a mathematics professor who showed his class three ways of solving one particular problem and refused to say that one way was the best. One student reported to the board of trustees of that religious college that the professor was "ungodly." Persons who think of mathematics Dualistically find it difficult to believe that mathematics is a human enterprise.

Persons viewing mathematics more Multiplistically might be happier with the process, but would not be comfortable with the tension between intuition and logic that is inherent in this, as in most, mathematical endeavors. They would not be proficient at viewing the problem both intuitively and logically, and might be quite disturbed by the way that the logic changes their intuition. My students say that their intuition is part of their identity, as if neither ever changed. Moreover, Multiplists would not see any reason to reflect or pose new problems.

The mathematical encounter with the orchard problem with which I started was from a stance of Contextual Relativism. Mathematicians must be Relativists to tackle problems from a variety of perspectives, to move between intuition and logic, and to reflect on their own thinking.

The difficulty is that most college students are Multiplistic with respect to most areas of their lives, and most students are Dualistic with respect to mathematics. For our students to be able to do mathematics, they must move beyond thinking of mathematics as a collection of formulas to be memorized.

How do we get students to develop?

Those of us who teach students beyond first grade rely on "show and tell" even more than the teachers of younger students. Our attitude toward student difficulties seems to be: "If they can't do it, show them how; if they don't know it, tell them.'

It follows that all we need to do for students who are not viewing our discipline Relativistically is to tell them to do so. Adding one lecture on the Perry Scheme might save us immense amounts of time haranguing students to consider conjectures from both intuitive and logical perspectives.

The only difficulty with that approach is that it doesn't work. By definition, students who are not Relativistic do not switch perspectives easily. More significantly, a person's Perry position is closely tied with that person's identity. Changing that way of interpreting experience means changing who the person is. Such change is not always considered necessary, and is almost never pleasant. That is, we can't "get" students to develop in any direct way.

There is hope, though. In investigating development on the Kohlberg model of moral development, Turiel (1966) found a phenomenon similar to what Perry researchers have discovered: people did not change when in an environment more than one stage later than their own. But persons in an environment just one step beyond their current stage moved toward their environment. In Perry's scheme, Multiplists in a Relativistic environment would be more likely to move than Dualists in the same situation.

Growth is more complex than that, however. Adopting a new way of looking at the world takes time. When first moving into a new position, one is not ready to be challenged to move on. For example, those who have recently begun to adopt Multiplistic ways of looking at

> For our students to be able to do mathematics, they must move beyond thinking of mathematics as a collection of formulas to be memorized.

things need time to consider the consequences and expand the areas to which they can apply their new viewpoint. Only after they have become comfortable with Multiplicity can they be challenged effectively to start interpreting their experience Relativistically.

A major challenge, then, is in finding ways to teach to all the different students in a course. Of course, in smaller classes the methods and assignments can be tailored to the students present. What happens in larger classes, though?

Fortunately, complexity is diminished somewhat at the undergraduate level. Most college students view most academic areas Multiplistically, and most students think of mathematics Dualistically. Since many experiences that challenge Dualism also support Multiplicity, Multiplistic experiences can do double duty.

When planning a course, I like to increase the overall Perry position of the course as it progresses. For freshmen it will begin fairly Dualistically, with a high degree of structure, and I'll gradually introduce more choices (at first in assignments, and later in approaches to the ideas at hand and in their intuitive feelings about those ideas). A course for more experienced students might have a more Multiplistic structure, with lots of choice and perhaps even a contract system of grading. Within that structure students will be asked increasingly to act more Relativistically—to defend their intuitive opinions, compare and contrast ideas and methods, and examine their own reasoning processes. (For more details on how that planning might go, see Copes, 1982.)

How can we encourage the development mathematically?

The preceding comments concern courses in general. How do they apply to mathematics courses? In a *mathematical* context, what could I mean by "different approaches to ideas," "intuitive feelings about those ideas," "comparing and contrasting ideas and methods," and "defending intuitive opinions?"

Let's imagine introducing a class of thirty calculus students to the orchard problem. Elsewhere (Copes, 1974) I have discussed the appropriateness of several teaching models for encouraging development on the Perry Scheme. Here are two of those possible approaches:

1)   Write the problem carefully on the board, making sure to state the problem precisely to avoid ambiguity. (That is, we'd say that the orchard is on a plane, that we are to look horizontally, etc.) Then we'd carefully restate the problem in terms of Cartesian coordinates. We'd probably state and prove a lemma such as: "If a line through the origin goes through a point other than the origin both of whose coordinates are integers, then the slope of the line will be rational." Finally, we'd point out that the contrapositive of this lemma shows that any line with an irrational slope will not pass through any other point both of whose coordinates are integers. Amazing, we'd say. It would take ten minutes at most.

And students would not have seen different approaches to any mathematical ideas, would not have encountered different intuitive feelings or approaches, would not have compared or contrasted ideas or methods, would not have defended their intuitive opinions. Teaching and telling are not the same thing.

2)   State the problem with all the ambiguity used at the beginning of this chapter. Ask students to brainstorm about it, expressing any intuitive ideas without fear of evaluation. Encourage every response that comes up if it's an honest statement of the proposer's ideas, even if the idea is somewhat immature right now. Point out as needed the variety of ways of thinking about the problem. As students make

suggestions about looking upward or ask about the effect of the curving earth, slip in a point about refining problems when modeling. As much as possible, let students respond to each other. Come back to the problem repeatedly for short periods of time over several class periods. When one student suggests using a coordinate plane, see how many others can go along. Don't push. Find where their levels of understanding are, and work with them. Remind them of definitions of slope, rational number, and irrational number as needed. Don't put students down for their level of understanding of something they should have "had" before; after all, why should it be "theirs" if they hadn't

---

A person's Perry position is closely tied with that person's identity. Changing that way of interpreting experience means changing who the person is.

---

grappled with it? Where possible, mention explicitly how their intuition and logic are in conflict, and how intuition can and does change.

From this teaching method students will have seen different approaches to mathematical ideas. They will realize that people have intuitive feelings about these ideas, and that intuition can inform and be influenced by logic. They will have deepened their understanding of some previously-encountered mathematical ideas. They will have had some valuable experience doing mathematics. They will have experienced what Buerk (1993) calls the "private voice" of mathematicians. And they will have been challenged to move into a more Multiplistic view of mathematics.

But if we take all of this time on the orchard problem, how will we ever get to all of the topics of the course? Simply put, we won't. That is, we won't have time to *tell* the students about every detail in the textbook. But if this approach to the orchard problem, to help review some "old" ideas, is woven into similar approaches to "new" ideas, which are then overlapped and used to have students encounter the most basic ideas of the course, students will see the body of knowledge in the course in the context of how mathematics is done rather than as something to be memorized for an exam. In later courses and through their lives they will have a deep understanding of the fundamental ideas into which other notions can be fit. Is it such a loss if we don't tell them about some of the more superficial ideas? As Clarence Stephens has said, one of the impediments of implementing a high-quality mathematics program such as that at Potsdam is "excessive faculty concern about the subject matter to be covered." (Gilmer and

Williams, 1990)

In short, we can encourage development in mathematical maturity as well as movement on the Perry Scheme by getting students engaged in doing mathematics. I don't mean "doing mathematics" as in, "I did my math, and now it's time to do my English." I mean "doing mathematics" as mathematicians "do it"—toying with ideas and looking for patterns and balancing intuition with logical arguments. Like mathematicians, students may focus on answers; but our focus as teachers is no longer on the body of knowledge, but rather on the mathematical process: on students' acquiring the reasoning abilities, the critical thinking attitudes, and the maturity to look at things from a variety of perspectives.

The orchard problem and humanistic mathematics

I have been addressing most directly how to improve students' performance by teaching them to do mathematics as mathematicians do rather than only by memorizing facts. There's an underlying message here that could be called "humanistic": that mathematics is done by human beings.

I remember that even as a senior in college I was amazed when one of my professors said that people doing mathematical research "discovered theorems." I don't know where I thought theorems had come from; I suppose the question had just never occurred to me. Our own students should find out where theorems come from before they are college seniors.

Not only should letting our students do mathematics help them become better mathematicians, however. It should attract more students, with more diversity, to mathematics. Buerk (1981) has found that women who were repelled by the Dualistic conception of mathematics were quite attracted to mathematics taught more Relativistically, as it is done by human beings.

Part II

Linda

Linda is in my calculus class. She's older than traditional-aged college students and works as a programmer for my college. She is articulate and willing to tell me how she reacts to what I do—a rare opportunity for me.

Linda took calculus when in college, but she asked whether she could sit in my course. She has participated fully, spending many hours on assignments and taking all exams. From many conversations two major reasons have emerged for her dedication.

One reason was that she had difficulty thinking as well as she wanted to. She felt that her "mental circuits were blocked," probably for various emotional reasons, and that she could "do better." As a programmer, for example, she would hammer out a new code that worked pretty well, but she couldn't do the up-front thinking and planning she thought she should be doing.

A second reason for Linda's taking calculus is that she had abandoned some areas of interest to her in the past, and she wanted to go back and begin to explore them. Mathematics and music were two of these areas. She's dealing with the math first; when she tried to do both, she became overwhelmed.

Linda sees these two reasons as intimately related, but I haven't yet understood her attempts to explain how. Since she knew of no generic "roto-rooter" for "cleaning

> As Clarence Stephens has said, one of the impediments of implementing a high-quality mathematics program such as that at Potsdam is "excessive faculty concern about the subject matter to be covered."

out the pipes," she is "tackling the circuits a piece at a time." Now, she says, "I'm doing something I want in a safe environment."

During the first term of calculus, she reports, she "freaked out" whenever I asked the class to think. She wasn't confident that she could do it. She had always succeeded in math courses by memorizing, and she was quite good at it, so that's how she approached Calculus I. Instead of memorizing algorithms and formulas, though, she now had to memorize explanations and concepts, because that's what I emphasized explicitly. She knew that memorizing wasn't the kind of thinking she was trying to get better at, but she knew no alternative.

So she found herself in what she calls a "lose/lose" situation. Failure on an exam would have indicated not accomplishing her goal of "getting into" math. But she was upset with her success on exams because it meant that she had memorized sufficiently, thus failing at her goal of thinking better.

Finally the amount of time required for memorizing got to be too much. She had to find an easier way, and she was forced to let go of the old way and try a new one—do more thinking. Why did she do that instead of abandoning the course? She kept on, as she says, because "That's why I'm here." She saw it as consistent with both of her goals.

Now we're in second term, and she's somewhat more comfortable. "The thought of thinking about things is not so unpleasant," she says. She and her supervisor report that she is more comfortable "mulling over" programming ideas. I find that she's much more willing to think about mathematical ideas, too, and to engage in the mathematical process. Her fascination with orchard problems seems to have helped.

Brown on humanistic mathematics

In an important piece on humanistic mathematics Brown (1973) examines how "aspects of mathematics learning could have an impact on the way one conducts his life and views the world in other domains as well." He writes, "I am asking .... whether reflection about mathematical experiences might be used as a reasonable

---

If I can't have perfection, I will at least have the satisfaction of honesty.

---

starting point (in conjunction with other areas) in order to encourage a kind of self consciousness that is frequently ignored especially by those scholars who intentionally select that discipline as a means of avoiding the kind of confrontation with self I am suggesting."

Mathematics is not an "hermetically sealed body of knowledge," says Brown. Doing mathematics really involves posing problems as well as solving them. Moreover, conflicts between intuition and logic result in a great deal of doubt.

Brown suggests that mathematics educators not only engage students in mathematical explorations, but also take time to have students reflect on those experiences. Thus discussion of the orchard problem with students would reflect all aspects of the process described above, including reflection on the methods used. Students would think about their thinking, about how their intuition had changed, and about the degree to which it felt threatening to them.

As Brown puts it, "One point of extreme irony for us has been that the more we focus on mathematics as a tool for dealing with the world in humanistic terms, the closer we seem to come to a kind of training that if appropriately applied might also produce first rate mathematicians." I would add that they would gain

insights into themselves and the human condition in general.

Linda, Brown, and the Perry Scheme

Brown's idea of analysis of one's own processes is given support by Linda's experience. Her frequent self-examinations have been the driving force behind her adopting new ways of coping both with calculus and with her more general thinking problem. What she has learned in, or through, the calculus course has transferred to other domains of her life.

This transfer took place even though I was not consciously doing much to encourage this kind of reflection in these students. It happened because Linda was predisposed to engaging in it.

It was part of her reason for taking the course in the first place. If I had conscientiously asked students to talk or write about their thought processes, would more students have increased their self-consciousness in Brown's sense of the term?

Probably, but not so much as might be ideal. Perry found, after all, that college students could not think about their own thinking very well until they could view it from perspectives other than from "inside"— i.e., until they entered Contextual Relativism. Most college students are Multiplistic. Although many of them can and should be challenged with Relativistic thinking, hopes that a class consisting mostly of freshmen will grow much from this kind of exercise are probably unrealistic. Linda's Relativistic approach to life in general allowed her to, even demanded that she, engage in this kind of reflection.

Linda and humanistic mathematics

Does the discipline of mathematics belong in the Humanities rather than the Sciences? It is not my place here to analyze this question. But I must point out that, to the extent that the Humanities are disciplines that study or engage these "deeper" aspects of human beings, Linda's experience with mathematics is similar to what we expect from courses in poetry or history. One could argue, on the other hand, that, even if her experiences had been a result of conscious efforts on my part, that would only show that education, not mathematics, is part of the Humanities.

Nevertheless, Linda has demonstrated how experience with mathematics can reach much further down into one's humanity than having to miss a party to study for an algebra exam.

Part III

Dawn

Dawn is a junior mathematics major from our local metropolitan area. She has performed at average to high levels in traditional mathematics courses, and plans to pursue a mathematical career.

One of my colleagues has been asking the students in her analysis course to read articles about mathematics outside their textbook. Recently she asked students to read and respond to some pieces (from Campbell, 1976 and Guillen, 1983) on the nature of mathematical truth. They read about the philosophical implications of non-Euclidean geometry, Russell's paradox, and Gödel's theorems. Dawn's response opened with this paragraph:

> How can it be that in a few pages, a mathematician's idea of truth can be so shattered? Yes, it's true (or is it?). My concept of truth, my very core thoughts about math have been radically changed. I thought math was (naive as I am) something that was relatively unquestionable. I came away from these readings feeling as if my world and future career have a totally different meaning. I am no longer as satisfied with math as I once was. I was forced to gasp at Russell's and Godel's discoveries. How can so many years of the unshaken ground of math suddenly erupt into an earthquake of uncertainty? I still wonder, even shudder, that I haven't suddenly decided to change my major. Math no longer has the unquestioned appeal of certainty that it had had for me since I first learned that 2+2=4. I probably did wonder why, but I never really stopped to question the awesome power of MATHEMATICS. I guess I just thought, 'How can so many unopposed truths be made from a few assumed truths if we didn't make the right assumptions to begin with?' (Yotter, 1990, p. 1)

After several pages of summarizing the articles, Dawn turned inward again, to conclude the paper:

> At any rate, I feel that although math is no longer as steady as I had once thought it to be, I am inclined to continue to believe and have faith in a system in which some of its participants are aware of the uncertainty that exists.... It now makes me feel better to at least know that mathematics makes no claims of being any more superior than any other field of study. I know. It's sour grapes. But if I can't have perfection, I will at least have the satisfaction of honesty. (Yotter, 1990, pp. 3-4.)

The world of academic philosophers is not the only one that has been shaken by mathematical uncertainty.

Dawn and the Perry Scheme

One attraction of the Perry Scheme is that it doesn't pretend to capture all of the complexity of human beings. Dawn's writing is a good example of how, as Perry (1976) puts it, "People are much bigger than any boxes into which we might want to put them." It is nevertheless insightful to consider Dawn's essay from the perspective of the Perry Scheme.

Dawn, like many of us, was apparently attracted to mathematics as a major because it appeared to be an island of Dualism in a sea of Relativism. She now has had to integrate mathematics with the rest of her world, which seems to be Relativistic. And, as her concept of mathematics changes, so does her identity as a mathematician. Her dreams for the future, for a career, "have a totally different meaning." Her old world is lost.

When Relativistic students' worlds are shattered, Perry found that some of them "retreat" into Dualism or Multiplicity. Most of them go on, though. And what gives them the strength to go on is often a mentor. As Perry (1976) puts it,

> I think that what they mean by faculty-student contact is: "You look real. You look as if you care about what you're studying and doing and teaching. How do you care in a world like this? This is a terrible world, and you know it. What are you caring for? You must be kidding yourself! I want to get near enough to you so I can bang on you and bang on you and bang on you and find out if you're real. Because if you're real then there is hope. Then I might live with zest and care about something."

From the last paragraph of Dawn's essay we can see that she has already started processing the grief that accompanies such a loss. She has turned to "participants" in the system who "are aware of the uncertainty that exists." If we more experienced mathematicians can continue on, even knowing that we won't be finding any Absolute Truths, then she can, too.

Must our students be Relativistic?

Dawn and Linda are atypical in their Relativism. Must our students be Relativistic in order to understand mathematics more humanistically?

I believe not. Several times I have had the privilege of teaching our course entitled Mathematics for the Liberal Arts. Each time I have taken the attitude that the goal was to engage students in mathematical processes and to let them in on ways that mathematics has grown out of and, in turn, influenced the development of civilization.

Toward the second goal I have organized the course chronologically. They have read about and discussed how the "axiomatic game" grew up with Greek goals for finding absolute truths. They studied how projective geometry grew out of attempts to make paintings more realistic in the Renaissance. They learned about how the success of The Calculus influenced attempts at axiomatizing virtually all areas in the eighteenth century.

And then they encountered the same ideas that shattered Dawn's conception of mathematics. These students, not many of whom were very Relativistic in general, faced some of the same shock. As a student once burst out, "Is it really *true* that the sum of the angles of a triangle is less than 180 degrees? Do you really *believe* that?" We can learn a great deal from Dawn's articulateness about how less articulate students are feeling.

Dawn and humanistic mathematics

As a human endeavor, mathematics does not take place in a vacuum. Even though the problems of "pure" mathematics are not influenced by direct applications, mathematics itself is not separate from cultural influences. And the surrounding culture certainly has been affected by mathematical results, not only those that led to atomic weaponry but also those with deep philosophical implications.

Should these kinds of connections be made known only to "liberal arts" students majoring in areas other than mathematics? Dawn's situation demonstrates that teaching mathematics can be a powerful tool toward development as described by the Perry Scheme. Perhaps we could encourage the development of mathematics majors toward the Relativism so necessary for doing mathematics if we remembered that they, too, are liberal artists.

Part IV

An ideal

These days it seem unfashionable to dream. Even young people are more cynical than idealistic. If they're not enthusiastic about saving the world from us older folks, then who will do it? Toward that end, I share a dream with you.

It can be stated rather simply, as a summary of what I've already said: I want students in mathematics courses to do mathematics as mathematicians do, to reflect on the processes in which they are engaged, and to realize how mathematics grows out of and influences the cultures in which it is embedded. As a result, more students will enjoy and be attracted to mathematics;

students will become better mathematicians; and they will be further along on the Perry Scheme, and thus better able to handle the complexities of modern society.

There. I've said it and I'm glad. But is it possible? Can we have it all?

A proposal

Imagine a course like this: The teacher poses a problem for students, perhaps in an historical context. Working alone or in groups, students tackle the problem, sharing ideas with each other. When the students have acquired "ownership" of the problem, the teacher directs them to various sources and helps them interpret what they read or hear about solutions to the problem. The teacher then poses other problems that can be solved in similar

Should these kinds of connections be made known only to "liberal arts" students majoring in areas other than mathematics?

ways, and guides students' practice. They also discover the historical influences of the mathematical ideas they read about. At various points they think and write about the processes they used in approaching the problem, and they pose other problems that arise from this one. The teacher selects one or several of those problems for future exploration.

Now imagine a lot of courses like that. Mathematics majors graduating from a college curriculum filled with such courses might have seen fewer of the mathematical trees than those leaving now. But they would know the topography of the mathematical orchard, and they would be able to read well enough to fill in the gaps. More importantly, they would be able to do mathematics. And, most importantly, they would be better human beings because of their education in mathematics.

What more could we ask?

References

Brown, Stephen I. (1973). Mathematics and humanistic themes: Sum considerations. *Educational Theory* 73,3, 191-214.

Buerk, Dorothy (1981). *Changing the conception of mathematical knowledge in intellectually able, math avoidant women.* Doctoral dissertation, SUNY Buffalo.

Buerk, Dorothy (1993). {chapter in this book}

Campbell, Douglas (1976). *The Whole Craft of Number*. Boston: Prindle-Weber-Schmidt.

Copes, Larry (1974). *Teaching Models for College Mathematics*. Doctoral dissertation, Syracuse University.

Copes, Larry (1982). The Perry Development Scheme: A metaphor for learning and teaching mathematics. *For the Learning of Mathematics* 3,1: 38-44.

Davis, Robert (1990). Untitled talk at Conference on Humanistic Mathematics, Louisville, KY.

Gilmer, Gloria F. and Scott W. Williams (1990). An interview with Clarence Stephens. *UME Trends* 2,1: 1.

Guillen, Michael (1963). *Bridges to Infinity*. Boston: Houghton-Mifflin.

Perry, William G., Jr. (1970). *Forms of Intellectual and Ethical Development in the College Years: A Scheme*. New York: Holt, Rinehart, and Winston.

Perry, William G., Jr. (1976). *The Scheme from Within*. Talk at Perry Development Scheme Conference, Ithaca, New York.

Perry, William G., Jr. (1981). Cognitive and ethical growth: The making of meaning. In Chickering, A., *The Modern American College*. San Francisco: Jossey-Bass.

Polanyi, Michael (1958). *Personal Knowledge: Towards a Post-critical Philosophy*. Chicago: University of Chicago Press.

Turiel, E., (1966). An experimental test of the dequentiality of developmental stages in the child's moral judgment. *Journal of Personality and Social Psychology* 3: 611-618.

Yotter, Dawn (1990). Writing assignment #2. Unpublished manuscript, Department of Mathematics, Augsburg College, Minneapolis, MN 55454.

Getting Beneath the Mask, Moving Out of Silence

Dorothy Buerk*, teacher
Jackie Szablewski**, student
Ithaca College

Mathematics has a public image of an elegant, polished, finished product which obscures its human roots. It has a private life of human joy, challenge, reflection, puzzlement, intuition, struggle, and excitement. Mathematics **IS** a humanistic discipline, but the humanistic dimension is often limited to its private world. Mathematics students see the elegant mask, but rarely see this private world, though they may have a notion that it exists. Their exposure to mathematics, limited to its public image, often leaves them silent and insecure. Mathematics is a combination of both its public image and its private world. Students of mathematics need to experience both of these domains.

This chapter presents the story of Jackie, who knew mathematics' public image and was "successful" in high school mathematics, including a calculus course. She felt disconnected from mathematical ideas, however, and considered herself "math phobic." In 1985 Jackie came

---

Mathematics has a public image of an elegant, polished, finished product which obscures its human roots. It has a private life of human joy, challenge, reflection, puzzlement, intuition, struggle, and excitement.

---

to Ithaca College, a private, co-educational, four-year comprehensive college in upstate New York, as a freshman. The College, with a strong commitment to developing students' writing skills, places freshmen in writing courses based on their writing abilities. The Writing Seminar Program offers writing courses in content areas for better than average writers. Jackie was placed in Dorothy's course, "Writing Seminar in Mathematics." Jackie, a dual major in psychology and English, enjoyed writing and wrote competently. The other students in the class were humanities and social

science majors who, for the most part, would have preferred to avoid mathematics. Dorothy's goal in the course was to help students like Jackie to come to know mathematics as a humanistic discipline by experiencing its private world, and to write about that experience.

In this chapter Jackie will tell her story of her experience in this course and include excerpts from the journal that she wrote during the course. Dorothy will discuss her teaching strategies for the course and present the mathematical activities used to engage the class in the private world of mathematics.

Jackie: Math Phobic

I was one exposed only to the public image of mathematics. To me, there seemed no room for interaction with the content, no possibility of connection with the ideas. Mine was:

> ... the role of the tourist who merely looks out at the sights that surround [her] as they travel past in a blurred rush.
>
> Journal Entry #1—Writing
> Seminar in Mathematics, 1985

There was no chance of stopping, of touching those mathematical concepts that lay out there.

It was true, I had taken Calculus in high school and had even done well. One would think I had to be connecting with math. But I viewed my mathematical experiences much like building a house that was supposed to be for me, according to someone else's design:

> ...[About] my math experiences, I mean it always turned out to be "the house that Jack[ie] built," but according to someone else's step by step instruction. I never really liked what I was building and never felt comfortable with the tools I was using. Meaning I never felt

sure of my mathematical background. If I was using defective tools or materials who was to say that the house would not fall in at any moment. Thus though I "built the house," it was for someone else. I was but the [contractor] attempting to follow the blue prints accurately enough to please the owner, 'cause after all, I certainly never felt I "owned" any mathematical ideas, or understood them well enough to call them my own, until this course.

Journal Entry #18—Writing
Seminar in Mathematics, 1985.

I did not feel that I could interact with math on a personal level and therefore did not feel as though I owned or could own any of the mathematical ideas and concepts with which I was presented. I found math simply an exercise of going through the motions and, as Carlyle (1830, p. 965) put it, "believing for" someone else, rather than for myself. I was frustrated. Later in the semester I began to understand some of that frustration. I said in a Letter to a Future Math Teacher:

> I realize that in order to help us realize all that already exists in the world, in order to guide us through all the worlds of mathematics, you must keep to a strict itinerary. If you didn't, we would not be exposed to all we must be exposed to in order to reach the destination of "mathematician," "chemist," "well-rounded person." But don't you see that in your well intended efforts to show us all the "landmarks" of those worlds, you are not allowing us to touch? How can we come to say that we believe in a thing, a concept, an idea, if we ourselves do not know it is real?

Journal Entry #28—Writing
Seminar in Mathematics, 1985.

I wanted to touch, to believe for myself. But I knew that interaction required process and what kind of process could there possibly be with an already finished and perfected product. It didn't matter that I had taken Calculus, or any other course for that matter. I still felt estranged from math, aggravated by it, afraid of it. I was math phobic. I wrote in a 1985 journal entry:

> As in the cases of other phobias; though you would rather be afraid of heights than face the challenge of a roller coaster, you might find the roller coaster appealing after gaining the

courage to try it. In the same way, one may find the concepts of math exciting and exhilerating after building up enough courage to at least attempt to open the mind.

Journal Entry #1—Writing
Seminar in Mathematics, 1985.

The question remains, how do we go about "building up enough courage"? What does it take to "at least attempt" to try? The first step to conquering fear is naming it. Yet how is it possible to name a thing that renders you voiceless? Unlike English class, math was not a place for ideas in process. You could not say or share something you were thinking about. You could only share with the class completed, perfected thoughts and I simply had no such thoughts concerning math:

> We would just get out of our English class where I never shut up and then we would come right to Math and there I would sit dumfounded, lost, unable to speak. I was intimidated by those of my classmates who could rattle off formulas and understand them by the time our teacher finished writing them on the board.

Journal Entry #6—Writing
Seminar in Mathematics, 1985.

Because of my inability to connect and interact with math in the way it was presented to me, I was left voiceless and because of this I could say "I hate math," to anyone 200 times over. I could say, "I am no good at math," but I saw no possibility of saying, "I am afraid of math," and then talking about why. For one, no one had ever asked how I felt about math. Why should they? Either you can do it or you can't; either you like it or you hate it; like all else in math, it's black and white. You don't talk about it. You don't write about it. You just do it. And even if I could talk about my feelings concerning math, that did not automatically provide me with the environment in which to do so. I, like most people, need a supportive, understanding and encouraging milieu where I feel at least somewhat comfortable before exposing vulnerabilities like math phobia. I didn't know at the time there was another whole world of mathematics hidden away.

Then one day someone asked how I felt about mathematics, someone in a teacher's position of authority in a freshman level course at Ithaca College called Writing Seminar in Mathematics. And the discovery of another whole world of mathematics began.

Dorothy: The Course Itself

My goal in the Writing Seminar in Mathematics course is to change my students' conceptions of mathematics as a discipline and of themselves as learners of mathematics. For me mathematics is creative, dynamic, and evolving. I value the personal, formal, intuitive, logical, and reflective dimensions of mathematics. I enjoy the process through which mathematics is created. This process of developing mathematics involves conscious work, unconscious work, intuition, conjecture, false starts, collaboration, new conjectures, reflection, redefining the question, asking new questions, and finally a degree of certainty, and a formal, logically correct and consistent presentation of the completed idea. This is a very human process. As you finish one problem or proof you are often left with many new questions that might be pursued at another time. This is precisely the process used by mathematicians, by educators, by students, by me—in fact, by all inquisitive persons as they approach **ANY** question that is new to them, not just a mathematical question. True, we are not always able to reach the last step of a formal, logically correct presentation. However, this universal process is indeed what makes mathematics a humanistic discipline for me. I want my mathematics students to experience this process of learning mathematics and of getting in touch with their own intuitions and questions as they learn mathematics. I want them to see that mathematics requires the same process that they use to resolve questions in other disciplines and in their personal lives, and then I want them to develop their competence in using this process. I want them to experience the private world of mathematics.

Many of my students, like Jackie, see only the product of the mathematical inquiry in its final, formal, carefully crafted form. They have trouble, as Jackie has indicated, making sense of it and bringing it into their own world of thought. I want these students to stop being only tourists who observe the fine works of others, and to begin to touch, and experience for themselves, the mathematical ideas that they study. To do this I create an environment to help my students to see and experience the private world of mathematics and to listen to themselves and to their ideas and gradually to develop their mathematical voices.

Everybody Counts: A Report to the Nation on the Future of Mathematics Education, released by the

National Research Council in January, 1989, is one of many recent reports documenting concerns about mathematics education in our country. It presents a plan for radical change in mathematics education from kindergarten to graduate school. The report stresses that "students learn mathematics well only when they *construct* their own mathematical understanding." (p. 58) In the Writing Seminar I try to create mathematical situations to help students construct their own understanding. I encourage my students to explore the situations, generally in small groups. Through these situations I help my students become aware of their own experiences with mathematics, accept that they can and do have ideas and intuitions about mathematics, and validate these ideas and intuitions through the use of mathematical theory. Specific examples of these mathematical situations appear later in this chapter. Others appear in Buerk 1990.

Students write about their experience with these events first through a process called "freewriting" developed by Peter Elbow (1973). This particular type of writing encourages the students to listen to their own ideas by putting on paper anything that comes to mind. Elbow's process further entails rereading those ideas, sorting them out, and organizing the relevant ones into a coherent whole. This whole then appears in the form

---

But don't you see that in your well intended efforts to show us all the "landmarks" of those worlds, you are not allowing us to touch? How can we come to say that we believe in a thing, a concept, an idea, if we ourselves do not know it is real?

---

of journal entries which I read and comment on every other week. My comments support the students' ideas and their struggle to make sense of the material, raise questions of clarification, try to lead students to see their misconceptions and consider alternative points of view, and invite them to think more deeply about mathematics. Thus their informal writing in their journals allows me to participate in this process by offering support and encouragement to their struggle to recognize their own thinking, and by challenging their thinking when that seems to be appropriate. I try to affirm the students where they are and to respect and encourage each student as a thinking person. I find that students respond to that respect and affirmation by courageously deepening their thinking and reflecting. While I suggest journal topics to students, I also encourage them to use the journal in their own ways. Each student writes two two-page entries per week. I try to develop dialogues with the students urging them to respond to my questions and comments regularly. (For an account of one student's experience in using a journal in this course, see Buerk 1986.) Well-developed essays in which their ideas are expressed more formally evolve as a result of this process and include at least one draft, followed by a revised version.

Students seem to go through a developmental process in changing their view of mathematical knowledge. A first step is to listen to their own ideas and to acknowledge them and own them. This requires a supportive environment where a student feels safe enough to try out untested ideas. It requires a group of classmates who will be supportive. In this setting students can begin to speak and can begin to own the ideas they have spoken. In addition, in mathematics this seems to involve acknowledging that mathematics was and is made by people. This happens before students become comfortable with the theory, the algorithms, or the procedures that are so necessary in mathematics.

I have found William Perry's theory of intellectual and ethical development in college students very helpful in understanding my students (Perry 1970, 1981.) Larry Copes' chapter in this volume will help you understand Perry's theory in relation to mathematics teaching and learning. See Buerk (1982) for a description of my work with able, math-avoidant women using Perry's scheme. More recently I have also found the work of Mary Belenky, Blythe Clinchy, Nancy Goldberger, and Jill Tarule, in *Women's Ways of Knowing* (1986) insightful as I struggle to understand my students, including students like Jackie, more deeply. They note: "We found that women repeatedly used the metaphor of voice to depict their intellectual and ethical development; and that development of a sense of voice, mind, and self were intricately intertwined." (p. 18). Jackie uses the voice metaphor often, and will continue to do so as she struggles to define her relationship to mathematics, to gain a sense of herself as person and female, and to speak her mind in a public forum.

Many students, like Jackie, enter my course believing that mathematics is something that they can have no ideas about; that every question has an answer; that every problem has a solution. Teachers know these answers and teach methods to reach them. Students learn mechanically, following the rules precisely, without questioning or reflecting. This view is termed "received knowing" by Belenky, Clinchy, Goldberger, and Tarule. It is what Perry called "dualism". The "received knower" feels confident about her abilities to absorb and store truths, but these truths are received from external authorities or experts. This person might say, "I have listened and taken down your words. There is nothing that I can add or say differently; you have said it all." Students with these beliefs are precisely the ones I want to help to see a broader view of mathematical knowledge.

I create an environment in this course that helps the students become aware of both diversity and uncertainty in mathematics and gain a more personal, subjective view of mathematics, by working in its private world. The students begin to realize that their own experiences

---

Because of my inability to connect and interact with math in the way it was presented to me, I was left voiceless.

---

are worth considering. They still believe, however, that knowledge is the product rather than the process of the inquiry. They can now learn from the experience of peers, for peers as well as authorities (teachers, textbooks) have ideas worth hearing. They begin to mold ideas, but often without being willing or able to support them. According to Belenky et al. the subjective knower is still dualistic in the conviction that the right answer exists, but now believes that truth resides within the person rather than in the external world.

Subjective knowing is particularly important since it does offer students the opportunity to look inward, to trust their own inner voice and to begin to acquaint themselves with their own needs and desires. The internal voice present in subjective knowing marks the emergent sense of self and sense of agency and control, so important to learning.

Students become aware that intuitions may deceive; that some "truths" are truer than others; that theories can be shared and expertise respected. They become aware that it is the process of constructing knowledge that is important. They move away from subjectivism, learn to respect established procedures and their own insights, listen carefully to other points of view while evaluating their own. This course helps students move through this process in their view of mathematics. Students become empowered and then become less fearful of the procedures and techniques needed to do mathematics.

Jackie: On Writing about Mathematics

I definitely found Dorothy's course empowering. Through it I was able to develop my private mathematical voice, as well as gain the courage to believe my private voice was valuable enough to integrate into the public sphere. Yet neither of these two things happened simultaneously nor without much resistance. For example, the writing seminar in itself seemed a unique idea, but to combine writing and mathematics seemed to me absurd. Writing meant thinking, talking about, reflecting upon all that has been said, and somehow integrating it. In mathematics it is different. All there was to say had already been said. All possible answers had already been etched in stone. How then, was I to write about mathematics? Still I had nothing to say and no voice to say even that. But Dorothy asked the "right" questions, different questions which no math teacher had asked or ever seemed to care about before. Among them were: "How do you view mathematics?" and "What is it you would like to gain from this course?"

These were questions the thirteen of us in the course talked about with Dorothy; we also wrote about them. Dorothy introduced us to Elbow's process of freewriting which I found especially useful in beginning to develop my voice because of its inclusiveness of personal thought and ideas. I did not have to immediately pass judgment upon or edit out something I was pondering. This was extremely important for me who would have, as in past experience, edited out everything, believing no thought was worthwhile, of value, or made sense. By getting something down on paper, there was a distinct possibility that there was some "gem buried amidst the dirt." I found I really did have things to say and now needed to decide whether or not to try to say them.

Since freewriting was not a thing to be shared, I still had an option out. Dorothy would read and respond only to our journal and the more formal essays that were to grow out of our journal entries. The very fact that she was willing to read thirteen journals and take them seriously enough to respond to them was a definite indication that she cared about what I had to say. She had after all, asked those questions and

listened to what we had to say in class thus encouraging us to listen to each other.

So I wrote and I shared my phobic feelings about mathematics, my hopes and fears about the course and Dorothy read and responded with questions and comments and challenges. In response to my first journal entry, noticing that I used only male pronouns, she wrote, "Mm, about you and it's HIM." (This is why the "her" in entry #1, quoted earlier, is in brackets.)

Either you can do it or you can't; either you like it or you hate it; like all else in math, it's black and white. You don't talk about it. You don't write about it. You just do it.

I began to notice that in my writing I was speaking about myself as if I was someone else; not only that, I was speaking of myself as if I was male. I could not use the pronoun "I" nor was I using the pronoun "she." I could not claim my experience as a female nor could I claim my experience as a person. I began to notice that I was claiming nothing for myself and began to notice as well that somebody cared.

Dorothy became someone I could trust with my struggle in mathematics. She told me, she told the class, that there was another whole world of mathematics which she wanted to share with us. But she didn't want us to "believe for" her. She wanted us to believe for ourselves, to accept our intuitions about mathematical concepts, to test them out against mathematical theories, and to own our ideas. And that was the key! In order for me to claim my experience as both female and a person, I had to feel I was entitled to it. I had to own my experience and my ideas.

Adrienne Rich, whose writing speaks powerfully to me, states in "Claiming an Education," (1979) that we each have a responsibility to ourselves that will not be fulfilled through passivity. I needed to interact with the mathematical concepts and content presented to me; I needed to connect with them to find something in them for myself. I needed to have such a relationship with mathematics in order to claim my mathematical experience. Dorothy offered such opportunities through mathematical activities and situations in class.

Dorothy: Mathematical Situations

Classroom time was used primarily for the development of mathematical ideas and the study of mathematical situations. The writing and reflecting on the activities occurred outside the classroom. My individual encouragement of my students' thinking happened within the classroom through discussions and questions, in our conversations outside of class, and as I responded to the students' journal writing once each two weeks.

I designed the mathematical activities for the course to generate mathematical thought, questioning, discussion, and reflection. Since I believe that mathematical thought is stimulated by instances of surprise, of contradiction, of believing, of doubting, and through the

---

every question has an answer; every problem has a solution. Teachers know these answers and teach methods to reach them.

---

collaborative sharing of ideas, I try to create situations with these in mind. See Larry Copes' chapter in this volume for a discussion of the Orchard Problem, which is an example of a situation that fits these criteria.

We studied the Fibonacci sequence and the golden ratio, discussing the historical context and the connections between these ideas and their occurrences in art, nature, and mathematical questions. I began with several seemingly unrelated problems that each resulted in the Fibonacci sequence. This surprised the students. Finding the relationship also occurring in nature, by counting spirals on pine cones, artichokes, and pineapples for example, was another surprise. Some students commented that if mathematics is in nature, then mathematics must be very different from what they had thought. They wanted to know more about this part of mathematics. Jackie found that the study of the Fibonacci sequence helped her to begin to see mathematics as a developing process, rather than just a polished product. It helped her break down the notion of mathematics as "magic."

We studied prime numbers, continuing our search for patterns and learning to make conjectures based on the patterns that we saw. Prime numbers provided us the opportunity to ask a variety of kinds of questions with a variety of kinds of answers. We found that

counterexamples were sometimes hard to find. We found that verifying the truth of a statement was also often difficult. The study of primes provided the opportunity to study the process of mathematics, including testing specific examples, making conjectures, trying to verify our conjectures, searching for counterexamples, sharing our examples and counterexamples, and formalizing what we were certain about. For Jackie the primes gave her more experience with the private process of mathematics.

We talked about mathematical proof, what constitutes a proof, and what makes a proof believable. We began with the Pythagorean Theorem and, through the use of four identical right triangles and the appropriate square, showed that the pieces formed the square on the hypotenuse of the right triangle. Then the same pieces were used to form the sum of the squares on the other two sides of the triangle. We discussed the Greeks and the geometric relationship inherent in the Pythagorean Theorem. We also discussed the knowledge of this relationship in earlier cultures and the tendency of Western scholars to name theorems for their Western discoverers. But did our work constitute a proof? Was our demonstration a convincing argument? We returned to this question later after we *seemed* to turn an 8 x 8 square into a 5 x 13 rectangle. (See note at the end of this chapter if you are not familiar with this "paradox.") Here something was clearly wrong. We had two demonstrations making different figures from the same pieces. In each case we should obtain equal areas. Why should we believe one of these demonstrations if we could not believe the other?

We discussed a proof by contradiction of the statement, "The square root of 2 is irrational." Jackie could accept the logical sequence of ideas that took us step-by-step through the proof, but she still was not convinced that the statement was true. She was reassured by reading in David Henderson's paper, "Mathematics and Liberation," (1981, p. 13):

> How do we view mathematical arguments? When do we call an argument good? When do we consider it convincing?—When we are convinced!—Right?—When the argument causes us to see something we hadn't seen before. We can follow a logical argument step by step and agree with each step but still not be satisfied. We want more. We want to perceive something.

I asked my students to look carefully at each step, come up with a reason for each step of the proof, and to write down any questions that they had about the step or the reason. Then I asked them what part of the proof they still had questions about. I wanted them to think about what they did believe and what they still were not ready to believe. Jackie found by looking carefully at the proof she could come to believe it. She wrote:

> Dorothy, you knew this would happen, didn't you? Looking back at it, step by step and writing out my reasons why, I can actually *see* the proof step by step and *believe* the entirety of it, which was really a surprise to me since I kept telling myself over and over again that I didn't believe it. As I said in my explanation for step 5, I was just not looking closely enough at it which goes back to my all time narrow minded assumption that "I just can't understand." I'm really thrilled that this makes sense to me.

Journal Entry #9—Writing
Seminar in Mathematics, 1985.

Then we went through the "proof" that all triangles are isosceles. Here Jackie could say she understood the logical sequence of the steps, but knew that the statement could not be true. She needed to doubt what was on the page and she found out what was wrong with this incorrect proof—and she found it out **FOR** herself.

We spent time discussing non-Euclidean geometry. Three weeks before this unit began, I brought a large rubber ball to class and asked the students wheter there could be straight lines on the surface of this ball? A long discussion ensued. What did "straight" mean? Could a straight line have curvature? If a line "appeared" to the eye to be straight was it really straight?

Students then each responded in their journals to the question, "Can there be straight lines on the surface of a sphere?" I read and responded to these entries before we began to discuss Riemannian geometry, talk about great circles, or raise questions about the similarities and differences between Euclidean and non-Euclidean geometries. We began our discussion about great circles using rubber bands on the same large rubber ball. Before we got deeply into this discussion, I put on the ball, without comment, a "triangle" that clearly has each of its three angles equal to 90°. I watched the students' expressions as they noticed it, and as some indicated their disbelief, "No, no! That can't be!" We then were ready to look more closely at the similarities and differences between the Euclidean and Riemannian geometries. I used Douglas Campbell's book, *The*

Whole Craft of Number (1976) for this unit, but we did not begin the readings until we had the experiences mentioned and all the students had written in their journals and received comments on that writing. (See Buerk 1990 for a fuller description of this unit and other writing assignments used in this course which are not presented here.)

Another unit, on Chromatic Triangles, is designed to have the students work more deeply in the private world of mathematics. I want them to think about ways that mathematics might be developed and experience developing mathematics for themselves. I do not tell them in advance, however, that that is what I am asking them to do. The question comes from Niven's *Mathematics of Choice* (1965, pp. 122-123). I ask the question in the following way:

Consider n points in the plane, no three of which are collinear (on a straight line.)

Draw all lines connecting these points. Color these line segments in any way using one or two colors, say red and blue; all segments may be red, all may be blue, or some may be red and the rest blue. Each segment may contain only one color.

Say any triangle connecting three of the points is *chromatic* if all of its sides have the same color.

Consider the existence of chromatic triangles for values of n from zero to seven.

The students are to indicate, for each value of n, one of the following: every drawing will contain chromatic triangles, some drawings will contain chromatic triangles, or no drawing will contain a chromatic triangle. They are to support their answer by giving a drawing with at least one chromatic triangle and a second drawing without any chromatic triangles, if they say that there are sometimes chromatic triangles. If they answer that there are always chromatic triangles, they are to try to explain why it is that they cannot avoid chromatic triangles for that particular value of n.

Working both collaboratively and independently, students carefully develop their own responses to the Chromatic Triangles question and then struggle to explain why, after a certain value of n, they can no longer avoid chromatic triangles. This is a journal assignment. As part of this assignment I ask:

> What do you think that you are doing that is like what a mathematician would do in pursuing this idea?

To that question Jackie responded:

> I feel that in pursuing the idea I have functioned as a mathematician first in that I did not just assume something did or didn't work. After finding an example of a case that did work, I looked for a counter example to prove the validity or predictability of a particular statement. Secondly, I functioned as a mathematician in the way I went about looking for the counter example. I derived a method I felt was credible and could be used effectively in each case. I was also alert in observation and in noticing what events occurred as a result of the process. I was utilizing, and still was able to come up with questions, which is really what mathematicians do: They come up with one answer that produces a thousand more questions.
>
> Journal Entry #24—Writing
> Seminar in Mathematics, 1985.

Jackie, who entered the course believing that mathematics was all known and completed by someone else, now saw herself doing mathematics for herself. She found herself experiencing the mathematicians' private world, and able to develop a convincing argument to explain why she could not avoid chromatic triangles for $n = 6$.

After I had read and responded to each student's journal entry on the Chromatic Triangle question, we discussed as a class our various approaches to the problem, the conjectures we each made, and how we tried to resolve each of them. We discussed our arguments for our inability to avoid chromatic triangles until we each had an explanation that was convincing for us.

But we did not leave Chromatic Triangles once the problem was solved. I wanted my students to have experience asking new mathematical questions as well as answering ones that were given to them. Stephen Brown and Marion Walter present some powerful strategies to generate new questions in their book, *The Art of Problem Posing* (1990, 1983). Their first strategy presents a situation, without a question, for people to consider. The emphasis is on "accepting the given" and raising questions and concerns that come to mind. To use this strategy on Chromatic Triangles you would present the situation as I did on page 18, omitting the last instruction: "Consider the existence of chromatic triangles for values of n from zero to seven.", and ask instead, "What are some questions you might ask?"

To look more deeply at Chromatic Triangles, I used Brown and Walter's second problem-posing strategy, called "challenging the given" or "What-If-Not." First we listed the attributes of the Chromatic Triangle problem. For example:

> No three points are collinear.
> The problem looks at triangles.
> We may use two colors.

Then we asked, "What-If-Not" for each of the attributes we listed. For example:

> What-if-not "no three points are collinear"?
> What-if-not "triangles"?
> What-if-not "two colors"?

Next we asked, "But if this attribute is not true, then what **could** be true instead?" For example:

> Suppose three points could be collinear.
> Suppose we look at quadrilaterals.
> Suppose we used three colors.

Then from our "what-if-not" work we posed specific new questions including the following:

> Will we need to change any other rules if we allow three points to be on the same line?
> Discuss the existence of chromatic quadrilaterals for values of n from zero to seven.
> If we use three colors, will we still reach a point where we can no longer avoid chromatic triangles?

Each student could choose one of the new questions that the class generated to pursue in his or her journal.

Jackie: A Private Mathematical Voice

Along with these opportunities to interact with mathematics came the chance to write about and make sense of my experience in my own terms. Accompanying the activities Dorothy describes were substantial journal entries where I discussed how I felt about the process, my reflections on it, my questions about it, and the mathematical ideas that I developed. Metaphor was a device I used throughout the journal. Indeed, it had been a device I had used often in my poetry and in my other writings. It had been and continues to be a technique I use to establish and maintain connections and meaning in my life.

Salvador Minuchin (1984) speaks of the use of metaphor as a way to own ideas and experiences by deriving meaning and developing voice. "To describe life we need metaphors. Objectivity will not suffice." (p. 111). Objectivity does not communicate what the meaning is for us. My use of metaphor in regard to mathematics indicated that math had become a part of my life. The particular metaphors I chose and used demonstrated the ways I interacted with mathematics and the meaning I had given to it in my life.

As is evident through my writing and my mathematical experiences, I had been introduced to, and had even become involved with, the private world of mathematics. I had made meaning of it for myself and believed my ideas and conceptions were valuable enough to test against mathematical theory. We were close to the end of the semester when I noticed that my perception of math was different from that of the "math phobic" who had entered the course in September. I had experienced mathematics for the first time, as something graspable, something with which I could interact and even make connections. My metaphor had changed from one of passivity, from the "tourist who merely looks out at the sights that surround [her] as they travel past in blurred rush," to one of action, interaction, and ownership. In responding to the question, "Was mathematics invented or discovered?," I wrote in my journal:

> I had heard of the Fibonacci Number sequence before and I knew some guy had made it up, but when I discovered the pattern in the pinecone, and the tiles, and the bee question solution, for myself—that first time, I really felt as if I had "stumbled upon something big" accidently. Maybe to the world the "Jackie Szablewski number pattern" would go unnoticed since for them (the rest of the world), Fibonacci had invented it. But for me, I would have never really believed in such a pattern if I had not developed it for myself. So though Fibonacci may have invented the "real pattern," to me it was my invention because I had developed the process it took for me to understand it—it *was* my invention.

In such experiences I found the two options of discovery and invention pulling at each other, and in this realization, the thesis statement of that first paper stood out in my mind. Math was indeed invented, by the Pythagoreans, Fibonacci, Pascal and also *myself, my classmates*—for such inventions as Pythagoreans' etc., led us to discover the potentiality for understanding within us, allowing us to be inventors also.

Journal Entry #10—Writing
Seminar in Mathematics, 1985.

I was the inventor of my ideas. I owned them. Yet did I have the courage to put them out there in the public sphere?

The struggle that lay ahead was that of integrating the voice I had gained and developed in private in my journal, into the public sphere and is illustrated most lucidly in the further development of my response to Dorothy's "Invent/Discover" assignment.

Dorothy: The "Invent/Discover" Assignment

Students in the seminar struggled to develop their own individual responses to the question, "Was mathematics invented or discovered?" The first day of class, after introducing the students to each other and to the course and doing an introductory mathematical activity, I asked the students for a sample of their writing in the following way:

> People both inside and outside of mathematics often speculate about whether mathematics was discovered or invented.

> Here are some crude definitions of these words:

> invent—to think up, to contrive by ingenuity, to originate something that did not exist previously.

> discover—to find and bring to the knowledge of the world something that has always existed.

> Look at these definitions and think about mathematics as you see it.

> **For your eyes only:**

> Freewrite for at least five minutes on the question: "Was mathematics discovered or invented?"

> Read over your notes.

> **To hand in at the end of class:**

> Write three carefully developed paragraphs that respond to the question: "Was mathematics invented or discovered?" Use specific examples to support your choice.

About halfway through the semester, I asked them to think about the same question with their experience in the course in mind. After working in their journals, talking with their classmates, and rereading what they wrote the first day, they wrote a draft of their "Was Mathematics Invented or Discovered?" paper.

At this point Jackie was clear in her view and could clearly express it through her private voice in her journal. She wrote an extensive entry including the two paragraphs which appear above as entry #10. In that entry Jackie conveyed the excitement of her own inventions. She talked about her own ideas and her discovery of a way to understand the mathematics with which she was working. Her mathematical voice was very clear and very strong.

Jackie: Acquiring a Public Voice

Yet, the journal, the place where I personally thrived was to be pitted against the "Formal Well-Developed Essay." I questioned whether the mathematical ideas, concepts, notions I had developed and explored in my journal were "good enough" for such an arena. The journal was one thing; I felt comfortable there. It was supportive and warm. My voice had just grown from a whisper. It was too vulnerable to just sit out there by itself. I still lacked the authority to say anything in polished, perfected mathematical form. I may have been

---

Since I believe that mathematical thought is stimulated by instances of surprise, of contradiction, of believing, of doubting, and through the collaborative sharing of ideas, I try to create situations with these in mind.

---

able to live in the journal but I wasn't the type of third person authority required in these formal essays. And so again, I disguised myself. I attempted to express the ideas of my journal in the formal *public* essay without claiming any of them, without admitting any of these experiences were my own. The following are two paragraphs from that essay:

> Though many feel their role in mathematics is limited to that of a Christopher Columbus, an explorer upon the seas of education, discovering that which has always existed, one cannot deny the continual presence of a Mathematical Revolution and the participation

of humankind as inventors in it—Eli Whitneys whose interchangeable parts are numbers. Yet, in mathematics, it is the mind that draws up the patent for invention. For only the individual mind knows, through not recognizing, those ideas it has newly contrived.

> In examining the Pythagorean Theorem, Fibonacci's Numbers, and Pascal's Triangle, it is often assumed that all one can do is land on the shores of such concepts, simply discovering their existence, tracing the paths of footsteps made by those who came before. "After all, these ideas carry the name of their creator who had to have been sure of exhausting all natural resources themselves, leaving only complete developments and no materials for future building." However, one must keep in mind that the Mathematical Revolution is fed by the fire of thought. The necessity of man to answer his own questions creates more questions and so the revolution thrives.

> —excerpt from "Invent/Discover"
> paper draft #2, 1985

Looking at the journal and essay excerpts side by side, it seems impossible that both could have been written by the same person and that this same person was me. My extensive use of the passive voice and the metaphors I chose disguise very well both author and meaning. I again talk about "man answering his own questions," clearly excluding my own experience, the very experience that had helped to make these mathematical concepts meaningful to me.

Dorothy was puzzled and perplexed that in that essay I had lost all of my powerful, personal voice. She asked for, and received, my permission to share both my journal entry and my essay with a colleague. Dorothy returned the draft of the essay to me, clearly indicating through her own written comments and those of her colleague that it was **MY** experience in which she was interested, in which I needed to believe, and which I needed to share in order to speak effectively to others through my writing about mathematics. Dorothy and her colleague encouraged me to read the journal entry and my essay side by side and ask myself, "Which has a clearer voice?, a clearer sense of audience?, a clearer thesis?" "You are *you* in that journal as you are not in your paper. Try being the *same you* in your formal papers!", they wrote. I used the journal entry for courage, as I heard again, my experience, my thoughts, my words here speaking to me. I closed my eyes, took a deep breath, and, pretending it was but a revised version of a conglomeration of journal entries, dove in. It was terrifying, difficult. But at this point I felt it almost less exerting to just trust it and see what happened than to keep fearing and fighting. The result was my final "Invent/Discover" paper which follows and of which I am very proud:

Mathematics as a Personal Experience: Invented or Discovered?

In responding to the question of whether mathematics was invented or discovered, most people look past themselves into a universe characterized by spheres, triangles, rectangles, and squares that seem always to have existed. They thus assume that since these fundamental elements of mathematics prevailed prior to human existence, humans could do nothing but discover them. Yet what are now known as mathematical concepts were not recognized as such until a system capable of defining and explaining such occurrences was invented.

Similarly we, in our mathematical experience encounter such concepts as the Pythagorean theorem, prime numbers, and the Fibonacci sequence, which appear to us as fundamentals, age old and "invented out." Most feel, "We can do nothing more to develop these concepts because someone else has invented them. All we can do is discover what someone else says about these ideas." It is just such an attitude that fosters dissociation between ourselves and our mathematical ideas. Until we derive our own system to define the mathematical experience in our lives, the concepts which these experiences consist of will go unrecognized, no matter who tried to explain them in the first place.

Only when we invent something for ourselves can we grasp its essence, for it is only then that we know what it is made of and how it works. It is only through this individual invention process that we can come to claim and understand our own mathematical experience and ideas. Therefore, in this way, mathematics was and continues to be invented. As Thomas Carlyle (1830) suggested, the merit of originality is not novelty, it is sincerity. The believing man (woman) is the

original man (woman); he (she) believes for himself (herself), not for another.

Unfortunately, it is precisely this attitude of "believing for another," nurtured by today's society, that leads to the feeling of detachment from mathematical experience common to so many of us. Oftentimes, in early mathematical encounters mathematical ideas are presented with the expectation that we, the students, will see it the teacher's way and adhere to the teacher's methods to arrive at a specific outcome designated as acceptable to the instructor. So, what is our natural course of action? If we are at all interested in preserving the social order of the classroom, it is of course, to attempt to see things the teacher's way. In such cases one of two things may occur: we may either receive a grade of 75% or above on a test or quiz having arrived at the desirable conclusion by way of the desirable method, or we will receive a failing grade and engage ourselves in a seemingly endless struggle with mathematical concepts that will eventually breed the frustration and hostility later to surface in our attitude toward math. In either situation, we are not believing for ourselves but we are instead, believing for the teacher and thus cannot claim these ideas as our own, because we have put no bid in to gain their possession. It is almost as if we have bought these ideas for someone else with someone else's money.

In this are also evident the reasons why we may choose to "hold on" or "let go" of these ideas. If we have bought something for another with the money they had given us for it, we may be sure to keep it in good condition because they may call for it at anytime. On the other hand, if the article (or mathematical idea) is lost or becomes tarnished, though we may feel badly, it won't matter very much, because it didn't really mean anything to us in the first place. The same is true in mathematics. Until we have put some stake in owning mathematical ideas, until we can recognize them for ourselves, they have no value to us and "for us" may as well not even exist.

Therefore, to the individual, mathematical concepts are not recognized as such until that individual derives their own system of explaining and defining those particular experiences. In doing so, the individual will be inventing their own system of mathematics.

In my own experience, though I had heard of the Pythagorean theorem, Fibonacci sequence, and prime numbers, they meant virtually nothing to me because I had never understood them and had never taken a "stake" in these concepts. Up to the time of a Writing Seminar in Mathematics course I am now taking, I just assumed that somebody had invented them and through the ages they had become an important part of our mathematical system.

In my Writing Seminar in Mathematics class, I was given the opportunity to experiment with a cut out square and four cut out triangles that I discovered could be put together in certain ways to form a larger square. Through discovering such a pattern, I also discovered in myself the potentiality for invention. I could break down the squares in such a way that the side "c" of the triangle could be multiplied by itself and be equal to the side "a" of the same triangle times itself, added to the side "b" of the same triangle times itself. Now to someone else this may have been the Pythagorean theorem, but to me it was the "Jackie Theorem" because it became a concept to me only as I invented it. I had put a "stake" in it and made my claim. I was "believing for myself."

Notice, though, that through experimenting with the cut out figures, I did not discover the Pythagorean theorem. I had only discovered within myself the potentiality to develop a system to explain and define such occurrences. The system I had invented to define what was happening with certain formations of the square and triangles was what was to me the "Jackie Theorem," otherwise known as the Pythagorean theorem. This concept now carried the status of being a new invention in my mind, an idea I had contrived through my own ingenuity.

A similar experience was most likely shared by the Pythagoreans when they first invented the theorem and is one, I feel, we must all, at some point, experience if we are to acknowledge our place in the mathematical world. Thus, it matters not who developed a thought for the first time. If we believe something because we have come to understand it by inventing our own conceptual process, the thought or idea is "patented" as our own. In view of this, mathematics can be perceived as a continuum of invention.

Dorothy: Epilogue

Jackie, who "closed her eyes, took a deep breath, and dove in," could claim her mathematical experience and present it clearly, powerfully, and personally in her finished essay. I admire her courage and her persistence; I know the depth of her struggle. Jackie had come to my course able to "do" the mathematics asked of her in her high school courses, but considering herself "math phobic," feeling disconnected from the mathematical ideas she studied. Although an enthusiastic, reflective, outgoing person in most settings, she found herself voiceless in a mathematics classroom.

Through this course she became involved in mathematical activities and learned to listen to her own intuitions, to make conjectures, and to share her incomplete thoughts with her classmates and with me. Using freewriting and journal writing she began to develop her private mathematical voice in creative ways, to reflect on her own thoughts, and to listen to my questions, comments, and suggestions, as well as those of her classmates. She experienced the private world of mathematics, became excited about the mathematical ideas that she studied, and gained confidence by developing those ideas and discussing them in the private world of her journal.

With great difficulty, she kept her voice and spoke in the public domain. Her struggle is the humanistic struggle. She made meaning of the mathematics she studied *for* herself and learned to share that experience— sharing it now with all of us.

For me this experience was exhilarating, but often very difficult. Jackie's determination and her growing ability to trust her developing mathematical ideas empowered me. But I was unprepared for the stark contrast between the draft of her Invent/Discover paper and journal entry #10. How could Jackie, who wrote such a powerful, confident, personal journal entry, remove her voice and herself from her formal paper? Fortunately my colleague, Susan Laird (of whom Jackie speaks), understood my distress and Jackie's struggle, and helped me respond to Jackie's paper. I needed to share my disappointment, my disbelief, and my frustration with

Susan before I could respond to Jackie with support, while encouraging her to try the paper once more. Intellectual growth and change are difficult. In this course I am continually reminded of that difficulty, and struggle with the balance between support and challenge. I am often exhilarated as I see students like Jackie meet these intellectual struggles head-on.

Jackie is a special person, articulate and bright, but she is not unique in feeling disconnected from and apprehensive about mathematics. I taught the Writing Seminar in Mathematics course ten times to groups of

---

Objectivity does not communicate what the meaning is for us. My use of metaphor in regard to mathematics indicated that math had become a part of my life.

---

eight to seventeen students. In each class two or three, male and female, have had experiences as powerful, as personal, and as reflective as Jackie's. Each class has had more who come away with new confidence and joy in their ability to think mathematically.

True, to bring the private world of mathematics into the classroom and really listen and share in the intellectual struggles of our students takes courage. But the rewards are empowering for the teacher and the students alike. One of the greatest rewards is the chance, now six years later, to work collaboratively with Jackie on this paper and to share our experience with you, our readers.

NOTE

If the reader is not familar with the "paradox" mentioned in the middle of this essay, see *Fascinating Fibonaccis* (Garland, 1987, pages 84 and 86.) Garland indicates that it was a favorite of Lewis Carroll, the author of *Alice in Wonderland*. It is also discussed in *The Divine Proportion* (Huntley, 1970, pages 48-50.)

* This work was supported, in part, by an Exxon Education Foundation grant entitled, "Reclaiming Intuition in Mathematics."
** Jackie graduated from Ithaca College in 1989 and received her Master's Degree from Harvard University in 1991 in preparation for a career in the mental health field.

REFERENCES

Belenky, Mary., Blythe Clinchy, Nancy Goldberger, and Jill Tarule. (1986). *Women's Ways of Knowing: the Development of Self, Voice, and Mind.* New York: Basic Books.

Brown, Stephen I. and Marion I. Walter. (1990, second edition; 1983, first edition). *The Art of Problem Posing.* Hillsdale, New Jersey: Lawrence Erlbaum Associates.

Buerk, Dorothy. (1982). An Experience with Some Able Women Who Avoid Mathematics. *For the Learning of Mathematics*, 3(2), 19-24.

Buerk, Dorothy. (1986). Carolyn Werbel's Journal: Voicing the Struggle to Make Meaning in Mathematics. Working Paper # 160. Available from the Wellesley College Center for Research on Women, Wellesley, MA 02181 for $3. (ERIC Document Reproduction Service No. ED 297 977, microfiche only.)

Buerk, Dorothy. (1990). Writing in Mathematics: A Vehicle for Development and Empowerment. In Andrew Sterrett (Ed.), *Using Writing to Teach Mathematics*, MAA Notes #16, (pp. 78-84). Washington, D.C.: The Mathematical Association of America.

Campbell, Douglas. (1976). *The Whole Craft of Number.* Boston: Prindle-Weber-Schmidt.

Carlyle, Thomas. (1830). From "Characteristics." In M. H. Abrams (Ed.), *The Norton Anthology of English Literature*, 1962, Vol. 2, 4th edition, pp. 964-975. New York: W. W. Norton.

Copes, Larry. (1993). Mathematical Orchards and the Perry Developmental Scheme. (Chapter in this book.)

Elbow, Peter. (1973). *Writing without Teachers.* London: Oxford University Press.

Garland, Trudi Hammel. (1987). *Fascinating Fibonaccis: Mystery and Magic in Numbers.* Palo Alto: Dale Seymour Publications.

Henderson, David. (1981). Three Papers. *For the Learning of Mathematics*, 1(3), 12-15.

Huntley, H. E. (1970). *The Divine Proportion: A Study in Mathematical Beauty*. New York: Dover.

Minuchin, Salvador. (1984). *Family Kaleidoscope*. Cambridge: Harvard University Press.

National Research Council. (1989). *Everybody Counts: A Report to the Nation on the Future of Mathematics Education*. Washington, D.C.: National Academy Press.

Niven, Ivan. (1965). *Mathematics of Choice: How to Count without Counting*. New York: Random House/Singer.

Perry, William G. Jr. (1970). *Forms of Intellectual and Ethical Development in the College Years: A Scheme*. New York: Holt, Rinehart and Winston.

Perry, William G. Jr. (1981). Cognitive and Ethical Growth: The Making of Meaning. In Arthur Chickering (Ed.), *The Modern American College*, (pp. 76-116). San Francisco: Jossey-Bass.

Rich, Adrienne. (1979). Claiming an Education. In *On Lies, Secrets, and Silence: Selected Prose, 1966-1978*. New York: W. W. Norton and Co.

When Is Calculus Humanistic?

Robert B. Davis
Rutgers University
New Brunswick, New Jersey

For nine years I was associated with a very special course, one that met every day, Monday through Friday, and every one of those days was an experience I looked forward to with pleasurable and excited anticipation. Class lessons seemed more like a matter of getting together with friends for the most gratifying kind of mutual inquiry. This is what I think education ought to be, and once in a while it actually is. With all of our present national attention on the teaching and learning of mathematics, it seems important to look more closely at this particular example.

The course was a two-year calculus program, offered to juniors and seniors at University High School, the lab school of the University of Illinois, located in Urbana, Illinois. The mathematics offerings at Uni deserve to be recorded in more detail, for at least two reasons. First, the Uni program has pursued a direction quite different from most of the proposals that are popular nowadays, and the possibilities of alternative directions for the evolution of mathematics courses needs to be discussed and studied. Second, Alvin White's creation of the Humanistic Mathematics Network gives us a new and more promising perspective for analyzing the course at Uni. When we ask whether a course of study is *humanistic* we ask a very different kind of question from those that are usually posed, and the answers can point us in quite different directions. Given that the most common approaches usually fail (or cause the students to fail), the existence of significant alternatives must surely be taken seriously.

What Makes Mathematics "Humanistic"?

The recent meetings of the Humanistic Mathematics Network have included valuable discussions of what "humanistic" means when it is applied to a course in mathematics. For the purposes of analyzing the program at Uni, an excellent list of criteria is the one proposed at the Louisville MAA meeting by Reuben Hersh. A course is "humanistic" in the sense of Hersh if its main goals include helping students to develop:

independence
skepticism
intellectual honesty

self-reliance
an appreciation of some of the great achievements of the human mind, and if the course does not limit its evaluation of student performance to merely test scores on the ability to carry out memorized algorithms.

As a result of working with the students at Uni, I want to add to Hersh's list at least the following goals for students:

learning to help others to understand a mathematical situation or problem more deeply;
being able to talk about, and write about, mathematical situations;
developing a lively sense of intellectual curiosity.

But there is also another kind of criterion: I personally wanted to pass on to the students at Uni some of the almost palpable culture or world view that I believed I had acquired from my own teachers, including Dirk Jan Struik, Norman Levinson, Witold Hurewicz, Mark Kac, W. T. Martin, Raphael Salem, and Felix Browder. The two-year sequence in calculus at University High School seemed to offer an appropriate setting for doing something of this sort.

It is important to point out that this discussion is emphatically *not* merely an attempt to define the word "humanistic." The process that Alvin White and his colleagues have started is far more important than that. Confronted by generous evidence that mathematics instruction in the United States fails more often than it succeeds (see, for example, Douglas, 1986; McKnight et al., 1987; MSEB, 1989; Voytuk, 1989), we urgently need ways to do better. Achieving this depends in part upon powerful analyses of *what* is wrong and *how it can be changed*. Far too many of the early responses to this need have been marked more by desperation than by perspicacity, and have often focused on either the "more" theme (more days of schooling, more time spent on mathematics, etc.), or on the "testing" theme (make more use of tests, and make the tests count for more in the lives of students and teachers). In the real world, both of these directions for improvement are likely to turn out to make matters worse (see, e.g., Whitney, 1985). The approach to analyzing the needs of mathematics students that is offered by the "humanistic" criterion seems to be one of the most promising now available. As such, it is one answer to a very serious need. This implies that we, in turn, must try to see what a humanistic approach can possibly have to offer.

The Mathematics Sequence at Uni

University High School is a school for gifted youngsters, and the students there, despite their youth, are people with whom one can enjoy exploring mathematical ideas. (In a recent four-year interval, three graduates of Uni were awarded Nobel Prizes, one each in economics, medicine, and physics; how remarkable this really is can be seen when compared with the size of the graduating class of Uni, which may number about 25 students each spring; for one description of the program at Uni, see Driscoll, 1987; for another, see Davis, 1984.) The normal six years that most schools devote to grades seven through twelve had long ago been abbreviated to five years at Uni, roughly speaking, by eliminating grade 8. In the case of mathematics, the students themselves quickened the pace even further, so that the usual six-year math sequence was completed in less than three years. It is important to emphasize that this faster pace was introduced by students, not by faculty—it was simply the case that students would arrive at class already having worked successfully on the

---

When we ask whether a course of study is _humanistic_ we ask a very different kind of question from those that are usually posed, and the answers can point us in quite different directions.

---

next chapter, or the one after that. The instructors merely went along at the pace that the students determined. [Note 1.] In the end this resulted in a program where the students began the study of algebra in grade seven (during which they also studied the computer-programming language BASIC, and did some work in geometry). If we adopt the Uni idea that there was no grade eight, then the year following seventh grade should be called grade nine; at this point the students studied more algebra, more geometry, and some trigonometry. The following year—grade ten—they studied Euclidean geometry from three points of view: the Euclidean synthetic approach, a coordinate-based (or "Cartesian") approach, and a vector approach.

This left two years for a leisurely and rather deep consideration of calculus. Although it is this two-year calculus sequence that is the main topic of this note, it

will still be useful to look a bit more carefully at the three years of mathematical studies that precede it.

Mathematical study at Uni seeks to give students many different views of mathematics, including:
  mathematics as a description of reality;
  mathematics as an imperfect description of reality—the use of ideal elements, or abstractions;
  mathematics by proclamation and implication—the role of axioms and theorems;
  mathematics as a process of seeking patterns and making conjectures;
  mathematics as a matter of reconciling seemingly irreconcilable requirements;
  mathematics as a process of achieving things that seem impossible;
  mathematics as a study that sometimes seriously challenges our naive intuitions;
  mathematics as a field for wonderful opportunities in scheming and planning (especially including the powerful use of **heuristics**); mathematics as consisting of _statements_ that may be true, or may be false, or may (for the moment) be of unknown truth value;
  mathematics as a study that often requires the recognition and resolution of ambiguity (this includes "mathematics as the gradual explication of intuitive ideas");
  mathematics as a collection of fundamental **concepts**;
  mathematics as a collection of procedures or algorithms.

Only the last of these—or possibly the last two—are prominent in most pre-college mathematics studies. Hence the program at Uni is already unusual in its choice of ways in which students should come to see mathematics.

Perhaps a few of these need some clarification. The idea of "mathematics as a description of reality" means just what it seems to mean, and is intended as an antidote to the kinds of meaningless rote that students often encounter. But it is also seen as important for students to recognize that mathematics is frequently a partly fictional account of reality. We say that we lift a 5 pound weight—from the floor, up to a table, say—by exerting on it a force of five pounds. Were we really to exert a force of five pounds, the weight would not move. It is initially at rest, and with a gravitational force downward of 5 pounds and our upward force of 5 pounds it would be in equilibrium, and would remain at rest. If we want the weight to stop at the table top, we must allow it to decelerate, by exerting on it, toward the end of its journey, a force somewhat less than 5 pounds. So our claim that we lift it with a force of five pounds is something of an oversimplification, though a very helpful one when we try to compute the work that we do in lifting the weight up to the table top.

This process of simplification begins to become important already in the seventh grade course, where students are asked to draw triangles with angles of 90°, 45°, and 45°, using whatever lengths of side they wish. They are then asked to measure the sides, and to compute the ratio of the length of a leg, divided by the length of the hypotenuse.

Of course different students get different answers, but not *very* different answers. Class discussion is then used to bring up the question of what might be gained if we were to *imagine* a "perfect" 45-45-90 right triangle (even though we could not construct it with our measuring tools), and to define the ratio of leg to hypotenuse (even though we can, and do, show that we could never write out its actual numerical value in decimal form).

(One 12-year-old student summarized this aspect of the work by saying: "Ah! I see what science really is. It's seeing the world the way it really isn't" He did, in fact, have essentially a correct understanding of abstractions vs. specific concrete examples, as his more extended remarks made clear.)

"Mathematics as the reconciling of seeming irreconcilable requirements" also deserves some explanation. Some psychologists refer to this as "oblique thinking," the recognition that there is often a third way possible then the two obvious possibilities have been ruled out. When we seek to find the *instantaneous velocity* by dividing the distance covered by the elapsed time, we find the divisor would be zero, and we could not proceed. If we carefully guarantee that the divisor is *not* zero, we can divide, but we have computed the wrong thing. A third possibility *does* exist, but it takes considerable effort to work it out; we can use non-zero divisors to get some answers to *other* questions that *can* in fact be used to shed light on the instantaneous velocity we are seeking. Of course we have to work out the theory of limits in order to accomplish this—so this is what the students are asked to do. It is significant that this is accomplished by group discussion. The students are not *told* what to do, nor how to do it. Of course, the teacher provides help in two ways: first, by assigning problems (instantaneous velocity is one; finding areas and volumes are others) where "improbable" approximations are appropriate and make intuitive sense; second, as the students suggest ways to define "the limit of a sequence", by guiding class discussion, by offering hints, by posing questions, and by suggesting counterexamples. The students are not left without guidance, but the responsibility for suggesting definitions, criticizing the suggestions, and making

improvements lies unmistakably with the students, not with the teacher.

The importance of being able to see mathematical statements as true or false, and keeping track of implications, is recognized by most mathematicians as an essential aspect of mathematics. This is one of the points of view that the Uni course wants students to develop. Achieving this has turned out to be far more difficult than one might have expected. Even though instruction at Uni does *not* focus on the "stimulus-response" approach to mathematics—"whenever you see **A**, you should do **B**"—it seems that many students nonetheless tend to interpret mathematics in these terms. This was revealed dramatically in an episode when Barbara, an exceptionally bright senior, attempted to deal with a problem in simultaneous equations by looking at what seemed to be implied [Davis, 1988], and was able to demonstrate an apparent contradiction.

Barbara's suggestion was creative, and demonstrated a kind of thinking that is certainly valuable in mathematics. As it happened Barbara made an error, which was in a way surprising, for she made very few; but the real surprise lay in the response of the other members of the class, for it revealed the strong attachment that the other students felt to "doing the orthodox thing, without thinking much about it."

The lesson dealt with three-dimensional geometric interpretations of algebraic equations. (In fact, it was review, prior to some new work in vector geometry.) Class discussion established that the equation
$$Ax + By + Cz = D$$
might represent the empty set (if A=B=C=0, and D≠0), or the entire 3-dimensional space (if A=B=C=D=0), but would "typically" represent a plane. The simultaneous equations
$$\{Ax + By + Cz = D$$
$$\{Ex + Fy + Gz = H$$
would, similarly, in general represent a line (although in special cases they could represent a plane, the entire space, or the empty set).

At this point in the discussion, Barbara suggested that one could find a number K such that D + K = H. By adding K to both sides of the first equation, one would have
$$\{Ax + By + Cz + K = H$$
$$\{Ex + Fy + Gz = H$$
from which, Barbara suggested, one could deduce the equation
$$Ax + By + Cz + K = Ex + Fy + Gz$$
But, Barbara complained, the two simultaneous equations represented a line; the last equation represented a plane. "How," Barbara asked, "can a *line* be the same thing as a *plane*?"
The teacher asked the members of the class to write out

their answers to Barbara's question. Barbara herself, being relatively comfortable with logic, immediately recognized the flaw in the inferential fabric of her question, and wrote out a complete discussion of which statements implied which others. None of the other students were able to get anywhere near the crux of the matter, and nearly all sought to explain the situation by saying that Barbara "had not used the procedure that you were supposed to use."

But this, of course, represented a denial of the legitimacy of the idea of *implication*. It was a surprising retreat into the all-too-common view of mathematics as "the carrying out of some procedure that you have memorized on some previous occasion." In this view, if you do not carry out the procedure "the way it's supposed to be done," then you are in error. But the concepts of "right" and "wrong" are not adequate substitutes for *truth*, *falsity*, and *implication*. This episode revealed to the faculty at Uni that getting students to feel really at home with the various views of mathematics is not easily accomplished, and is not always completely successful. (For details, see Davis, 1988.)

Mathematics as a challenge to our naive expectations appears from time to time in the Uni sequence of courses, for example when we look at Cantor's transfinite arithmetic, and prove that there are the same number of points on a long line segment as on a short one (or, more distressingly, the same number of points on a *whole* line segment as there on one half *of that same segment*!).

The use of heuristics is an important aspect of the coursework at Uni, but one we have never learned to describe successfully. Indeed, we have never even been able to decide whether it would be wise or foolish to attempt to *list* all of the heuristics that we commonly use. (The problem is that we cannot devise a systematic way to *name* the various heuristics, so that many would appear under several different labels. There is also the question as to whether the collection is really well-defined, or whether it *should* be defined and thereby limited.) Among those we certainly do use, however, are the following: (1) Do *something*; the one strategy sure to fail is to sit there and do nothing! (2) If you can't do something right, then do something wrong—in the process of figuring out *why* it's wrong, you may get a deeper insight into the problem; (3) Try to solve *part* of the problem; (4) Try to make up a simpler, but easier, problem; solve it, and see what you can learn from observing how you do it; (5) If you can't seem to solve the given problem (call it Problem A), then solve some different problem (call it Problem B). See if you can change the solution of Problem B into a solution of Problem A. (6) Relax some of the constraints, and see if you can then solve the problem. (7) *Try some easy*

cases! (Sometimes this comes down to making numerical replacements for some of the variables.) (8) Try to identify the specific thing that is making the problem difficult—then try to: (a) eliminate it; or (b), somehow work around it. (9) Can you change the *form* of the problem? Or the form of some part of the problem? (10) Think about the *meanings* of the various symbols, constraints, etc.

Perhaps surprisingly—given the academic excellence of Uni students—many of them find the idea of using heuristics quite difficult. Even such seemingly simple heuristics as "trying it out in some simple cases" or "making numerical replacements for the variables" do not, at first, get much use by most students. [Note 2.] Over time, however, students come to use these heuristics in an impressively powerful way. Some excellent examples of student work appear from time to time in *The Journal of Mathematical Behavior*. (See, for example, Pandharipande, 1986, where he proves the inequality

$$(c - a^2 + b^2)^2 + (d - 2ab)^2 > 4$$

provided that

$$a^2 + b^2 > 4$$

and

$$c^2 + d^2 < 4$$

by reinterpreting some of the quantities as components of vectors, so that this becomes a matter of a certain point lying within the interior of a certain circle. (The inequality was needed by Lionel Shapiro, another student in the class, for some computer work that he was doing.))

The Calculus Course

The Very First Day. On the first day of the calculus course, the students are asked to compute the work done in compressing a spring one foot, if a force of one pound will compress it exactly this much. They know that work can be computed by multiplying force times distance, provided that the force is constant. They also know that in the present problem the force will be proportional to the distance that the spring has been compressed. Hence the force is not constant, and the "force times distance" law would not be applicable.

The problem is solved by class discussion. If no student suggests using the "product" law, to see what can be learned from so doing, then the teacher will ultimately suggest it. But even to use it at all, one must select some mythical "constant" value for the force. What value should be used? The students always opt for the average value, in this case one half of a pound. This is of course an eminently reasonable choice, but the teacher dissuades its use, primarily on the grounds that it is often more important to know the *sign* of your error than it is to get as close as possible to the correct answer. (Examples would be used in class; e.g., if you were going into Chicago to watch a Cubs game, is it more important to estimate very accurately how much money you will need, or is it more important to know that you will have enough?) Hence, under teacher influence, the class does the computation twice, once using the largest possible value, and once using the smallest possible value. This

Mathematics instruction in the United States fails more often than it succeeds.

gives the two answers O and 1, and the realization that the correct answer must be larger than zero, and smaller than one.

But pretending that the force is constant for the entire compression—for the entire movement of one foot in length—was clearly foolish. Would we do better if we made such pretenses for only half of a foot? Maybe, but maybe not. The class tries. This time the answers are 1/4 and 3/4, so the correct answer must be greater than 1/4 and less than 3/4. (By now most students suspect that they can guess what the answer should be, but of course they realize that they do not yet *know*.)

Assuming the force to be constant for a distance of one-fourth of a foot leads to the answers 3/8 and 5/8. Now, as all of this is being carried out, there is an on-going stream of mathematics: graphical representation of the force as a linear function of distance, and graphical interpretation of these upper and lower sums; algebraic computations that involve determining the sum of

$$1+2+3+ \ldots + n$$

(and, in subsequent problems,

$$1^k + 2^k + 3^k + \ldots + n^k).$$

This last computation, in turn, leads to an informal use of finite difference methods, and to careful proofs by mathematical induction.

But of course the main result of all of this activity has been for the students to find themselves confronted by a new kind of mathematical object, unlike any that they have previously encountered. We who are already familiar with the territory would call these new things *infinite sequences*, but of course the students at first have no name for them. They realize that what they have is a potentially endless collection of better and better "approximations" that seem to suggest a clear "correct answer"—but by this point in their education the students have seen enough instances where intuition leads one astray, and they realize that these new objects

must be dealt with carefully.

That becomes the students' next assignment: make up as many different kinds of sequences as you can think of, and try to classify them as "looking promising" or else as "looking useless." For each sequence that looks promising, state what number you would associate with the sequence. (This is called the *associated number*, because the word "limit" would have undesirable connotations, as in the 55 MPH "speed limit." Of course, *after* the students have worked out their own correct theory of convergence of sequences, the standard names are introduced, being harmless once the students possess correct concepts.) Of course, you need to try to formulate a precise definition of "associated number," and each suggested candidate is tested against the examples that have been collected.

This can be a fascinating discussion. One can expect that some student will give the example of .9, .99, .999, .9999, ..., and suggest that the "associated number" should be 1, justifying this by saying that "1 is the number that the terms in the sequence are going towards." If no student points out the weakness of this definition, then the teacher steps in and suggests that 1984 should also be considered the associated number, because the terms in the sequence are going toward 1984. To be sure, they don't get very close, but they surely are *going towards* 1984.

This may lead to the revised definition "the smallest number that the terms are going toward." But this, in turn, should be met with the example

$$2.5, \ 2.05, \ 2.005, \ldots$$

or with some similar example. This discussion is accompanied by graphical representations of sequences, and attempts to define the "associated number" at several different levels of formality (for more details, see Davis, 1984, Chapter 15).

Students are asked to show that *the limit of a sequence* is a very different idea from an "approximation," by proving that, in the case of the limit of a sequence, *if the limit exists, it is unique*. Clearly, this is not at all the way approximations work.

A major question has to be: How is it possible to determine an absolutely precise limit, by using terms in a sequence which are only approximations. Clearly, one could not do this by what might be called "numerical" methods—but it *can* be done by using *logic*. Given the Law of Trichotomy, if one can find approximations good enough to prove the falsity of two of the three possibilities, then the third alternative must be true. If, for example, one can prove that, for any A less than 1, it must be false that the limit of the sequence .9, .99, .999, .9999, ... equals A, and similarly for any B greater than 1, then the Law of Trichotomy tells us that the limit *must be exactly equal to one.*

Mathematics and Reality. A considerable effort is made to get students to see mathematics as potentially related to the real world. Current events are often used for problems in the class. A few years ago a batter hit a home run completely out of Detroit Tiger's Stadium, an unusual event because of the great height of the upper portions of the stands. This suggests an excellent problem: how fast was the ball traveling when it left the bat?

This is in many ways a problem typical of the course. For one thing, the teacher has posed it in a deliberately ill-stated form. Hence the first task of the students is to discover that this formulation makes no sense; then the second task is to reformulate the question so that it does make sense. Clearly, if some speed v would be sufficient for the ball to sail over the top of the stadium, then any greater speed would also suffice. Hence the best that one can say is that the speed must have been *at least* some particular value.

It is also typical in that it was taken seriously. One student telephoned a local sportscaster to get the dimensions of Tiger Stadium. The sportscaster did not know, but he knew a phone number at the Tigers organization, and so the student was able to get some realistic dimensions for Tiger Stadium.

Further, solving the problem requires some careful thought. This is not the typical "maximum distance" problem, familiar from physics courses, because the question is not how far away will the ball hit the ground. The requirement is that the trajectory have at least a certain specified height at a certain specified distance. There are also interesting implications. How did the speed compare with the fastest pitches that are known to have been thrown? Could Nolan Ryan stand at home plate and *throw* a baseball out of Tiger Stadium?

"Taking Charge" in Problem Solving. Students are encouraged to feel that they can propose lines of inquiry, invent appropriate notations, suggest definitions, and so on. In one noteworthy instance, some students reformulated a mathematical problem as a competitive game, in order to work out the necessary implications, rather as one move in chess might imply some required answering move (Davis, 1987B).

Student Initiative. Students play an even more generative role in the progress of the coursework. In one case, a student made an error in computing the mean of n numbers, then used this incorrect value in computing the covariance of a set of ordered pairs of numbers, obtaining the *correct* value for the covariance. Most of us might have dismissed this as some sort of accident. One student did not; Rahul Pandharipande thought about the matter, and finally proved that a covariance will be computed correctly if *one* of the means is correct, no matter what value is assigned to the other. (Pandharipande, 1987.)

In another noteworthy instance, a student posed the question: "You buy a billiard table in the shape of an ellipse. You put a pool ball at one focus, and hit it sharply with the cue stick. No matter where the ball hits the elliptical bumper, it will be reflected to the other focus. But there's nothing there to stop the ball, so it keeps on rolling. When it hits the bumper, it is reflected to the first focus. Suppose there is no energy loss. The ball keeps on rolling forever. Will the path ever close up and start retracing itself?" By way of response, many students tried to make some drawings. Everyone's drawing seemed to show the same result: the path seemed to converge to the major axis of the ellipse. Was this in fact the only possible outcome?

The next day a student had worked out a simulation on a computer, but the computer's graphics were not fine enough to offer much of an improvement on the hand-made drawings. The teacher went out and bought a Macintosh, which had better graphics (indeed, he had not previously bought a computer; it was this episode that finally made him take that step)—but the matter was still not really settled. Another student—Lee Imrey—made what he claimed was a proof. In its original form it was so elegant that no one else, the teacher included, could understand it. Imrey claimed that he had proved convergence to the major axis. But at this point, yet another student got into the act, with the argument that convergence to *any* "final" configuration was impossible. The laws of reflection are symmetric, so that any path that could be followed in *one* direction, could also be followed *in the opposite direction*. This clearly involved a contradiction. Students continued to argue about the problem, to propose alternative solutions, to find purported errors in Imrey's proof, and so on. Lee himself gave a simpler proof that began to convince people, although it did not eliminate the argument that any path could be followed in either direction. The matter was finally settled *on the afternoon of Christmas Day!* (Davis, 1987A; Imrey, 1987.)

Independence from the Textbook. Where appropriate, students were encouraged to work out a theory, or a method, on their own, *before* reading the relevant portions of the textbook. As one example, the topic of curvature was approached as follows: The class drew various curves (many rather like the draftsman's "French curve"), and began a discussion of "how curved they were" at various points. Clearly, some were bent very sharply, whereas others were nearly straight, with many falling somewhere in between. The class was then asked to invent a theory of curvature—create a mathematical way of making sense out of "how curved" a curve might be at some specific point along it.

The first suggestion from the class was this: "When we wanted to find the slope of a curve, we approximated the curve by that straight line that fitted the curve the most closely at the point in question. For curvature, why don't we fit the curve with that *circle* which fits it most closely at the point in question?" The teacher was forced to admit that this was a very shrewd interpretation of the task, and acknowledged that precisely this approach could be used, and in fact *had* been used by some of the mathematicians who had dealt

---

You buy a billiard table in the shape of an ellipse. You put a pool ball at one focus, and hit it sharply with the cue stick.

---

with this problem. However, the teacher thought that this approach might be too difficult for the class to carry through, so he asked them to continue thinking about it, and see whether they could find any other ways of proceeding. To help matters along, the teacher suggested a possible context: "You are up in a helicopter, looking down on the flat prairie land of Illinois, at night. There is a car driving along a road. You can see the direction of its headlights. How curved is the road?"

Since this was a calculus course, the students immediately suspected that this was either an accumulation/integration situation, or else a rate-of-change situation, and it surely seemed to be the latter. That meant that one needed to deal with some derivative, giving the rate of change of some variable (call it "u") with respect to some other variable (call it "v"). It was also clear that the thing whose rate of change needed to be considered was the direction of the headlights. But, once again thinking back to the earlier study of *slope*, there was a neat way to specify the direction of the tangent line, namely, by using the angle $\emptyset$ that the tangent made with the positive direction of the x-axis. Hence an appropriate choice for the variable u seemed to be $\emptyset$.

One still needed to settle on a choice for the variable v. The class made the obvious (and entirely reasonable) suggestion that v be taken to be the *time*, t. This had the result of making a road "curvier" when one drove along it faster—and, indeed, there is more than a little truth in this—but it was agreed (some teacher

intervention here!) that it would be better to try to make up a theory of curvature that depended only upon the road itself, and not upon the car or person driving along it. So t was out.

The class then made the next most obvious (and reasonable) suggestion: let v be the horizontal coordinate x. After all, x had worked well in a very large number of previous studies.

At such a point it was the teacher's job to give some telling counterexample, in this case the example of a circle, which (under this rule) was bent more in some places than in others. Clearly a deficient theory—if anything deserved to have the same curvature everywhere, it was surely a circle!

With x and t disposed of, the class next suggested that perhaps v should be the *arc length* along the curve. With this choice, they were able to use their knowledge of the derivative of the arc tangent to find a formula for the curvature. All of this had been accomplished without recourse to the textbook!

How Should One Describe this Course?

It can be argued that what "humanistic mathematics" is mainly about is the way we *think about* mathematics courses, and therefore also the way we *describe* them, because these, in turn, will gradually influence the kinds of experiences that we choose to provide for our students. So it is not in fact idle chatter when we ask whether, and why, the calculus course at Uni deserves to be described as humanistic.

Well, is it humanistic? And if so, why?

For one thing, it is a group activity that both teacher and students *enjoy*, that involves: creativity; two-way communication with others; dealing with doubt, uncertainty, and judgment; carrying out meta-analyses of what you are doing, how appropriate it is, how likely it is to succeed, etc. The content is tied to major philosophical questions, to major practical human achievements (like designing bridges, buildings, and airplanes), and to major themes in human history. The whole activity is intimately related to how humans think, how people invent new ideas. It involves exploring how far you can take ideas, as when a law for constant forces is applied, in an ingenious fashion, to a situation where the force is not constant. It involves "oblique thinking," as in the case of resolving seemingly unreconcilable contradictions in the analysis of the path of a pool ball on an elliptical billiard table. It gives students an opportunity to invent their own methods and interpretations and definitions, and to compare their results with the work of professionals (as in the matter of creating a theory of the curvature of plane curves). It involves a thoughtful consideration of alternatives, as in the case of the various definitions of the limit of an infinite sequence.

Perhaps one should focus on what a student is *doing*. The student is *not* merely following explicit directions, not merely "doing what one is told to do," which we would argue is *not* humanistic. The student is thinking

---

Twelve years of being told how to perform largely meaningless operations on largely meaningless symbols! This is not humanistic education—it is not really education at all.

---

seriously about new situations, or about situations with some important novel aspects, and is trying to adjust old knowledge to be helpful, or to come up with some appropriate new ideas. Most of this *thought* is about *meanings*, and not merely about symbols.

It is also *thought* that is shared with others. It is *thought* that is created in the presence of others, and with the help of others. A New York artist, who entered his profession through the door of graffiti on the sides of subway trains, recently spoke on TV about the challenge of graffiti when he was a teenager. As he traveled around New York city, meeting other graffiti artists: "It didn't matter where you were from, whether you were from Brooklyn or the Bronx. It didn't matter whether you were black or white. It just mattered if they had seen your art. It [the community of graffiti artists] was a wonderful family that you could belong to." I think that this may have been quite a bit like the feeling of the calculus class at University High School.

Implications for Mathematics Education in General

At the present time, mathematics instruction is threatened by the very importance—and particularly the *usefulness*—of the subject itself. So many people want to know how to **do** something, be it "how to add fractions," or "how to factor polynomials," or "how to integrate products of trigonometric functions," that those who teach mathematics are often tempted to answer this need directly, and only this specific need. Telling someone how to do some very specific thing is *not* humanistic. It may be appropriate, especially in a setting where the main goal is not mathematics, but rather one in which mathematics plays only a momentary instrumental role—perhaps we want to

compare the magnitude of fractions, in order to choose a drill of a suitable size, say.

But think of what twelve years, or more, of this kind of experience adds up to—twelve years of being told how to perform largely meaningless operations on largely meaningless symbols! This is not humanistic education—it is not really education at all. Such information, in such large doses, cannot be remembered; it cannot be understood; it surely cannot be enjoyed.

It may do something even worse; it may give students the idea that *the world cannot be understood and is not worth thinking seriously about*—or, at least, it gives students the idea that *mathematics* cannot be understood and is not worth thinking seriously about.

And perhaps that points to one of the greatest values of a humanistic approach—properly implemented, it should give students an unmistakable realization, based on their own personal experience, that the world is worth thinking about, that creative thought is possible and powerful, and that some degree of understanding and problem solving is within their reach, if they work hard enough to achieve it. This could make a real difference in anyone's life, even for someone who does not go on to more advanced study in mathematics.

NOTES

1. The *pacing* of mathematics instruction deserves more discussion, and more study, than is possible here. In general, observation of mathematics classes in elementary school, high school, and college (including some observations by Andrew Porter, and others by Uri Treisman) show a fairly clear general pattern, of a *very* slow pace in pre-college mathematics, and a very rapid pace in many college courses. Both seem to go to extremes. (For one mention of the college-level problem, see Tucker, 1990.) It is, however, probably impossible to separate questions of *pace* from questions of pedagogical approach. Many precollege programs assign little attention to *meanings*, so that students are expected to learn largely meaningless manipulations on largely meaningless symbols primarily by the method of rote repetition. Approached in this fashion, learning even a small amount of content can indeed require a large amount of time. The program at Uni never sought "acceleration" for its own sake, but did always strive to deal with a thoughtful and meaningful approach, and to adjust the pace, on a day-by-day basis, to what the students found appropriate. Depth of understanding was seen as a more important goal than merely "covering more material in less time." But, given understanding, the students did in fact move ahead very quickly.

2. Or perhaps it is not surprising. A student trying to

follow through some sequence of steps, not yet well understood, does not wish interruptions, even by himself or herself, and even for so good a reason as the matter of relating the procedure to underlying concepts and explanations; it is a bit like being interrupted when you are trying to count.

REFERENCES

Davis, Robert B. (1984). *Learning Mathematics: The Cognitive Science Approach to Mathematics Education.* Norwood, NJ: Ablex Publishing Corp.

Davis, Robert B. (1987A). Mathematics as a performing art. *The Journal of Mathematical Behavior,* vol. 6, no. 2 (August), pp. 157-170.

Davis, Robert B. (1987B). "Taking charge" as an ingredient in effective problem solving in mathematics. *The Journal of Mathematical Behavior,* vol. 6, no. 3 (December), pp. 341-351.

Davis, Robert B. (1988). The interplay of algebra, geometry, and logic. *The Journal of Mathematical Behavior,* vol. 7, no. 1 (April), pp. 9-28.

Douglas, Ronald G., Ed. (1986). *Toward a Lean and Lively Calculus: Report of the Conference/Workshop to Develop Curriculum and Teachina Methods for Calculus at the College Level.* MAA Notes, No. 6. Washington, DC: Mathematical Association of America.

Driscoll, Mark (1987). *Stories of Excellence: Ten Case Studies from a Study of Exemplary Mathematics Programs.* Reston, VA: National Council of Teachers of Mathematics.

Imrey, Lee (1987). The key lemma in the elliptical billiard table problem. *The Journal of Mathematical Behavior,* vol. 6, no. 2 (August), pp. 241-245.

Mathematical Sciences Education Board (1989). *Everybody Counts: A Report to the Nation on the Future of Mathematics Education.* Washington, DC: National Academy Press.

Pandharipande, Rahul (1986). A Proof of Shapiro's Inequality Theorem. *Journal of Mathematical Behavior,* vol. 5, no. 1 (April), pp. 103-105.

Pandharipande, Rahul (1987). An interesting result concerning covariance. *The Journal of Mathematical Behavior,* vol. 6, no. 3 (December), pp. 369-371 .

Tucker, Thomas W. (1990). Calculus reform: Preliminary report from the field. *UME Trends,* vol. 2, no. 1 (March), pp. 1-3.

Voytuk, James A. (1989). A challenge for the future. *UME Trends,* vol 1, no. 5 (December), pp. 1-6.

Whitney, Hassler (1985). Taking responsibility in school mathematics education. *The Journal of Mathematical Behavior,* vol. 4, no. 3 (December), pp. 219-235.

PART V

Contemporary Views of Old Mathematics

A Mathematics Seminar from the National Endowment for the Humanities

William Dunham
Muhlenberg College
Allentown, Pennsylvania

On two occasions in recent years, I have had the opportunity to direct five-week seminars titled "Great Theorems of Mathematics in Historical Context" and supported by the National Endowment for the Humanities. These were held on the campus of The Ohio State University as part of NEH's "Summer Seminars for School Teachers" (SSST), a program designed to provide a select group of U.S. teachers with the opportunity to examine classic texts from the humanities.

Of course, NEH traditionally supports seminars in such disciplines as history, literature, and philosophy, and their standard summer offerings come with titles like *Mozart: The Man, His Music, and His Vienna,* or *Thomas Hardy and T.S. Eliot: Literature and Landscape.* For many people, the prospect of NEH's funding a seminar on great mathematical theorems seems as improbable as their funding a seminar on transmission repair.

But I have no doubt that Plato would have felt comfortable classifying mathematics among the humanities. Fortunately, so did NEH. The Endowment proved quite receptive to the idea of a seminar that considered mathematical masterpieces not for their practical utility but for their logical beauty Mathematics, after all, was among the first of the liberal arts; it has always placed a premium on intellectual creativity; and it provides an unsurpassed arena for the exercise of the human imagination. Even the terms mathematicians use to discuss their favorite theorems—terms like "beautiful," "elegant," "powerful"—are precisely those used to describe a Mozart concerto or an Eliot poem.

Thus, I proposed studying a collection of classic results—the "great theorems"—from the long and glorious history of mathematics, while paying attention to the creators of the theorems and the historical context in which they appeared. NEH accepted this proposal, even though no mathematics seminar had ever before been funded as part of its SSST program.

The Endowment initially expressed some concern as to whether sufficient interest existed among U.S. teachers, a concern that in no way diminished when I imposed a prerequisite of calculus for all applicants. To say the least, our worries proved groundless. In 1988, our first year of operation, 482 teachers wrote letters of inquiry and, of these, 151 undertook the time-consuming task of preparing an application. In fact, ours was one of the

> **Even the terms mathematicians use to discuss their favorite theorems—terms like "beautiful," "elegant," "powerful"—are precisely those used to describe a Mozart concerto or an Eliot poem.**

most heavily subscribed of the Endowment's 53 seminars that summer. Unfortunately, we could select only 15 seminar participants from this vast pool. Our rate of selectivity—accepting just one out of ten—is the sort normally associated with Ivy League student bodies or Broadway auditions. During the second year, 1990, the inquiries fell a bit to 375 and the applications to 68, but we still had an enormous amount of talent from which to choose.

As can be imagined, the associated paperwork nearly buried us beneath a mountain of file folders. It was here that the seminar's administrative assistant, Penny Dunham, came to the rescue. She proved to be an organizational genius who soon brought order out of chaos and allowed us to avoid the Frankensteinian fate of being destroyed by our own creation. Even so, our postage budget was a whopper.

The large applicant pool guaranteed that individuals selected into the seminar were amazingly well-qualified. Most, but not all, taught high school mathematics. Among the 30 participants in our two seminars were residents of 22 states and territories. Seventeen were women, and thirteen were men. Collectively, they held undergraduate and graduate degrees from 50 different U.S. colleges and universities, with concentrations in such diverse fields as mathematics, philosophy, English, American civilization, history, psychology, Latin, French, engineering, and theatre. Finally, all were gifted teachers.

When we convened at Ohio State, our goal was to examine, in a collegial fashion, a smorgasbord of mathematical landmarks ranging from the ancient Greeks to the modern era and spanning such subdisciplines as algebra, geometry, number theory, analysis, complex variables, and the theory of sets. We tried to analyze a theorem as we would analyze a painting—probing its aesthetic qualities and formal structure, considering its historical antecedents, investigating its impact on future developments. It was an ambitious but highly rewarding enterprise.

The choice of theorems was at once critical and yet unavoidably anachronistic, for there surely exist as many potential seminars as potential seminar directors. I began with Hippocrates of Chios and his quadrature of

---

Never had I seen a discussion where this kind of fervor could be generated by mathematical ideas.

---

the lune from 440 B.C. and proceeded through a collection of mathematical landmarks until reaching Georg Cantor in the latter years of the nineteenth century. Along the way, we delved into such classics as Euclid's *Elements* (ca. 300 B.C.), Archimedes' *Measurement of a Circle* (ca. 225 B.C.) and Cardano's *Ars Magna* (1545), as well as individual papers by Bernoulli, Euler, and others. (A daily seminar schedule and complete list of theorems appear at the end of this paper.)

I would like to think I chose wisely from a centuries-old mathematical menu. One consideration that went into my choice was the necessity of introducing key personalities from the history of mathematics. Thus, I had to insure that Euclid, Archimedes, Newton, Euler, and Gauss had starring roles. Fleshing out the cast were Heron, Cardano, the Bernoulli brothers, and more. Their lives—sometimes triumphant, sometimes tragic, and sometimes bizarre—gave a rich biographical counterpoint to the mathematics.

Still, it was the mathematics itself that held center stage. Coming to grips with Euclid's proof of the Pythagorean theorem, or Archimedes' determination of the surface area of the sphere, or Gauss' construction of the regular 17-gon provided unmatched, almost breathtaking, mathematical experiences. Likewise, it

was both instructive and fascinating to compare Cantor's first (1874) proof of the uncountability of the continuum—firmly planted in the field of real analysis—to his second (1891) argument—one that introduced the diagonalization process and was, in his words, "...remarkable...because of its great simplicity." In these and other cases, there could be no doubt that we were in the presence of genius.

Our typical seminar day revolved around a single theorem. Usually I would take some time to set the stage with comments of a mathematical, historical, or biographical nature, after which a participant would discuss the proof itself. It was here that their teaching talents were so strikingly evident. As I watched them in action, I could only envy those students who, a few weeks after seminar's end, would be fortunate enough to walk into their classrooms in New Hampshire or Texas or Guam.

Their presentations always generated comments, and these sometimes carried us down quite unexpected paths. For instance, one seminar member (who held a Ph.D. in classics) enlightened us on some of the finer points of Euclid's Greek; another introduced us to the dynamic dissections of Leonardo da Vinci; and yet others veered into topics as diverse as Lady Lovelace, the paper-folding of geometric solids, Voltaire's Dr. Pangloss, or the olde English letter "thorn"(¥). This willingness to share interesting morsels with colleagues proved to be one of the seminar's least predictable but most rewarding features. Needless to say, it was never possible to guess just where we might end up.

The participants displayed an infectious enthusiasm. In this regard, let me relate an incident from early in our study of the *Elements*. I had wondered aloud whether Euclid should be praised for the general excellence of his ancient text or condemned because of its occasional logical imperfections (one of which appears in the very first proposition of the *Elements'* very first book). I thought this was a fairly innocuous question, but the participants went after it with a vengeance. They took sides, both pro- and anti-Euclid, and defended their positions with gusto. There were even a few raised voices. For a time it appeared we would be lucky to emerge without an exchange of gunfire. Never had I seen a discussion where this kind of fervor could be generated by mathematical ideas. I left shaking my head in amazement at folks who discussed mathematics with a passion usually reserved for such emotionally-charged topics as politics, religion, or one's favorite brand of personal computer.

Occasionally our diversions could be of a more technical nature. While reading Euclid's Book I from its beginnings through his concluding proofs of the Pythagorean theorem (Proposition I.47) and its converse

(Proposition I.48), we noted that Euclid *used* the former in the proof of the latter. Of course there is no logical necessity to follow this route, so I issued a challenge to design an argument by which Euclid could have inserted the converse of the Pythagorean theorem immediately after Proposition I.46 but before the proof of the Pythagorean theorem itself. In short, I wanted to see how the converse could have appeared as Euclid's Proposition $1.46\frac{1}{2}$.

In response, the participants generated a number of proofs, and we spent a lively time checking whether they had met the challenge or had inadvertently used results not proved among Euclid's first 46 propositions (an error easier to commit than one might imagine). Perhaps this seems a subtle, even irrelevant, point of logic, and I admit that it may not be to everyone's taste. But those who show up at an NEH seminar on great theorems are prone to attack such problems as a hungry puppy attacks a plate of cheeseburgers.

Each day of the summer, I distributed a set of exercises. These focused on the day's theorem, often developing alternate proofs or extending key ideas in different directions. For instance, after looking at Heron's wonderful proof of his formula for triangular area, participants were asked to provide the details of two or three other routes to the same end. Or Gauss' construction of the regular 17-gon, which surprisingly required the algebra of imaginaries, suggested a host of related problems about the complex numbers. Work on the problem sets, although entirely voluntary, was the order of the day, and participants organized evening sessions to solicit hints or discuss solutions. By summer's end our list had grown to over 150 problems inspired by the great theorems.

These, then, were our continuing seminar activities. But there were a few extra flourishes thrown in for good measure. One summer afternoon we assembled in the Special Collections area of Ohio State's library to examine such holdings as a 1566 copy of the *Elements* or Jakob Bernoulli's 1713 masterpiece *Ars Conjectandi* (in which one of our theorems appeared). On another day, we welcomed as guest lecturer the noted mathematics historian Fred Rickey, who presented a spirited account of the trisection problem from classical times to its ultimate resolution by Pierre Wantzel in 1837.

In its conception of summer seminars, NEH envisioned a significant writing component focusing on the analysis of original texts. In this spirit, I asked each participant to choose a "great theorem" and prepare a paper describing its proof and importance. Working either individually or in small groups, they descended upon the library with their typical enthusiasm. The writing project gave participants the chance to approach mathematics not pre-digested, summarized, or even translated. Although some chose to write on theorems of Archimedes, Newton, or Agnesi, most dipped into the incredibly vast *Opera Omnia* of Leonhard Euler. The fact that his writings were in Latin, French, or German proved no obstacle for such a highly educated group.

It was clear that the participants found this writing assignment to be intellectually challenging, for they had to grapple with the terminology, notation, and language of our mathematical predecessors and in the process experience both the frustrations, and the joys, of exploring original mathematical texts. When their papers were submitted at seminar's end, I reproduced and bound them into a document that was both a very nice monograph and a tangible reminder of our summer together.

By the time our five weeks were behind us, there was no doubt that the seminar had been a revitalizing force for participants and director alike. Looking back, I think it is fair to describe us as people being stretched intellectually and enjoying ourselves in the process. I am very proud of all that we accomplished and regard the NEH seminar as the most exciting, most rewarding experience of my teaching career.

I think others may wish to consider this approach to the history of mathematics. Obviously not every group of students has the diverse background, language skills, or intellectual temperament of our NEH teachers. Yet the intrinsic beauty of the mathematics and the fascinating lives of its creators make this subject a "natural" for those willing to approach our discipline with an aesthetic, not just a practical, eye. The overwhelming response we got from across the country testifies to the fact that there is an intense and, I would suggest, largely unfulfilled desire to treat mathematics in a more humanistic, and humane, fashion. I would encourage one and all to give it a try.

APPENDIX A
Daily Schedule, NEH Seminar

Day 1— Introduction and general orientation. Sketch of ancient mathematics and the work attributed to Thales of Miletus.

Day 2— The notion of quadrature. Hippocrates' quadrature of the lune. EPILOGUE: squaring the circle from ancient to modern times.

Day 3— Greek mathematics in the time of Euclid. Careful development of Book I of the *Elements*, Prop. 1.1 - 1.46.

Day 4— *Elements,* Prop. I.47 and I.48 - the Pythagorean theorem and its converse. EPILOGUE: non-euclidean geometry.

Day 5— Survey of later books of the *Elements.* Euclid's proof of the infinitude of primes. EPILOGUE: some landmarks of number theory.

Day 6— Archimedes' life and work. His approach to circular area and approximation of π from *Measurement of a Circle.*

Day 7— Archimedes' determination of the surface area of the sphere from *On the Sphere and the Cylinder.* EPILOGUE: Archimedes and calculus.

Day 8— Later Greek mathematics. Heron's formula for triangular area from *Metrica.* EPILOGUE: modern proofs of Heron's formula.

Day 9— Islamic mathematics and transition to Renaissance Italy. Cardano's biography and the struggle to solve the cubic.

Day 10— Cardano's solution of the cubic from Ars *Magna.* Transition to the seventeenth century. EPILOGUE: Abel and the quintic equation.

Day 11— Newton's life and work. Newton's generalized binomial theorem and applications.

Day 12— Newton's clever approximation of π and Newton's Method. Seventeenth century mathematics on the Continent. Leibniz and the Bernoullis.

Day 13— The Bernoullis and the harmonic series. Biography of Leonhard Euler.

Day 14— Euler's summation of $\sum_{n=1}^{\infty} 1/n^2$. Euler's proof of the little Fermat theorem. Euler's refutation of a conjecture of Fermat.

Day 15— Visit to Special Collections at the Ohio State Library to examine rare mathematics books.

Day 16— Euler's (partial) proof of the fundamental theorem of algebra. Biography of Gauss. EPILOGUE: the fundamental theorem of algebra from a modern viewpoint.

Day 17— A "short course" on complex numbers and the roots of unity.

Day 18— GUEST LECTURE: Prof. V. Frederick Rickey of Bowling Green State University speaking on the trisection problem from classical to modern times.

Day 19— Gauss' construction of the regular 17-gon. Trends in nineteenth century mathematics. Biography of Georg Cantor.

Day 20— Cantor's theory of the infinite. Algebraic and transcendental numbers. Transfinite cardinals.

Day 21— More on Cantorian set theory. The Schröder-Bernstein theorem and applications. EPILOGUE: the continuum hypothesis.

Day 22— Cantor's theorem and implications. The contributions of Kurt Gödel. Seminar wrap-up. WRITING PROJECTS DUE.

APPENDIX B
"Great Theorems" from the NEH Seminar

1. **Hippocrates' quadrature of the lune** (ca. 440 B.C.)

> The oldest surviving mathematical proof, which raised issues of squaring the circle that would reverberate through mathematics for over two millennia. (Heath, *A History of Greek Mathematics,* Vol. 1. Dover Reprint, 1981, p. 183-185)

2. **The Pythagorean theorem from Euclid** (ca. 300 B.C.)

> The first book of the *Elements,* a treasure trove of synthetic geometry since its creation, culminates in Propositions I.47 and I.48, Euclid's original proofs of the great theorem of Pythagoras and its converse. (Heath, *The Thirteen Books of Euclid's Elements.* Vol. I, Dover Reprint, 1956, p. 349-350)

3. **Euclid and the infinitude of primes** (ca. 300 B.C.)

> A time-honored classic from Book IX of the *Elements* that may surprise those who have not read it in its original form. (Heath, *Ibid.,* Vol. 2, p. 412 or Dunham, "Euclid and the Infinitude of Primes," *Mathematics Teacher,* Jan., 1987, p. 16-17)

4. Archimedes' determination of circular area (ca. 225 B.C.)

In the hands of antiquity's greatest mathematician, the method of exhaustion and a "double reductio ad absurdum" argument reveal the secret of circular area. (Heath, *The Works of Archimedes*. Dover Reprint, 1953, p. 91-93)

5. Archimedes and spherical surface (ca. 225 B.C.)

A brilliant deduction from Archimedes' masterpiece *On the Sphere and the Cylinder* that remains astounding even after two thousand years. (*Ibid., p.* 16-44)

6. Heron's formula for triangular area (c. 75 A.D.)

This proof begins at the center of the triangle's inscribed circle and detours through a seemingly endless series of similarity arguments before dramatically establishing one of geometry's most peculiar results. (Heath, *A History of Greek Mathematics*, Vol. 2. Dover Reprint, 1981, p. 321-323)

7. Cardano's solution of the cubic (1545)

Cardano proves this algebraic landmark by dividing up a cube in a wonderfully ingenious fashion. (Cardano, *The Great Art*, M.I.T. Press, 1968, Chapter, XI or Struik, *A Source Book in Mathematics*. 1200-1800, Princeton U. Press, 1986, p. 63-69)

8. Newton's approximation of π (c. 1670)

Combine the generalized binomial theorem with the inverse method of fluxions and mix in a generous portion of Newtonian ingenuity to get a very efficient approximation of this famous constant. (Whiteside (Ed.), *The Mathematical Papers of Isaac Newton* Vol. 3. Cambridge, 1970, p. 223-227)

9. Newton's Method in its original form (c. 1667)

Newton: "I do not know whether this method...is widely known or not, but certainly in comparison with others it is both simple and suited to practice." (Whiteside (Ed.), *The Mathematical Papers of Isaac Newton*, Vol. 2. Cambridge, 1969, p. 219-223)

10. The Bernoullis and the harmonic series (1689)

The little known but quite clever proof that (in Jakob Bernoulli's words) "The sum of an infinite series whose final term vanishes perhaps is finite, perhaps infinite." (Struik,

Source Book in Mathematics: 1200-1800, p. 321-324)

11. Euler's evaluation of $\sum\limits_{n=1}^{\infty} \dfrac{1}{n^2}$ (1734)

In a stunning performance, the young Euler solves a problem that had stumped Leibniz, the Bernoullis, and everyone else who approached it. (*Euler, Opera Omnia*, Vol. 14. Leipzig, 1925, p. 83-85)

12. Euler's proof of the little Fermat theorem (1736)

Turning to number theory, Euler uses an inductive argument to establish the result stated by Fermat a century before. (Euler, *Opera Omnia*, Vol. 2. Leipzig, p. 65-74 or Dunham, *Journey Through Genius*, Wiley, 1990, Chapter 10)

13. Euler's refutation of a conjecture of Fermat (1732)

How Euler discovered that, contrary to Fermat's belief, $2^{2^n} + 1$ need not always be prime. (*Ibid.*)

14. Euler and the fundamental theorem of algebra (1749)

Although not entirely successful, Euler's deft fusion of algebra and analysis establishes that every real fourth or fifth-degree polynomial can be factored into real linear and/or real quadratic factors. (Euler, *Opera Omnia*, Vol. 6, p. 78-147 or Struik's *Source Book in Mathematics: 1200-1800 p.* 99-102)

15. Gauss' construction of the regular heptadecagon (1796)

A hitherto unsuspected compass-and-straightedge construction is revealed under the light of complex numbers in one of the most remarkable discoveries in the entire history of mathematics. (Gauss, *Disquisitiones Arithmeticae*. Yale U. Press, 1966, Articles 365, 366 or *Oystein Ore, Number Theory and its History*, Dover Reprint, 1976, p. 346-358)

16. Cantor's first proof of the uncountability of the continuum (1874)

In a paper of profound consequences, Cantor uses the nested interval theorem to prove that no interval is denumerable. (Cantor, *Gesammelte Abhandlungen*, Berlin, 1962, p . 115-118)

17. Cantor's second proof of uncountability of the continuum (1891)

Of his 1891 diagonalization argument, Cantor observes, "This proof seems remarkable not only because of its great simplicity but also because the principle which it follows can be extended directly to general cases." (Dauben, Georg Cantor: *His Mathematics and Philosophy of the Infinite.* Harvard U. Press, 1979, p. 165-167)

18 . Cantor's theorem (1891)

A mind-boggling result proving "...that in place of any given set L another set M can be placed which is of greater power than L." (*Ibid.*)

19. The Schröder-Bernstein Theorem (1898)

This surprisingly difficult argument shows that in the realm of the infinite the obvious can mask a profound subtlety. (Kelley, *General Topology,* Van Nostrand, 1955, p. 28-29)

Teaching Mathematics Humanistically:
A New Look at an Old Friend

Elena Anne Marchisotto
California State University, Northridge
Northridge, California

Introduction

The sense of excitement one gets from the study of mathematics derives in large measure from the pleasure of making connections in mathematics—a pleasure most mathematics students neither experience nor imagine. The problem, then, for mathematics teachers is to present courses that instill that excitement in the very design of the class.

How can we design courses that instill in students a sense of excitement about making connections in mathematics? One approach is to introduce classroom discussion via topics familiar to students, and to expand the dialogue with historical references and activities that lead to a discovery of new directions or themes in mathematics. The familiar topic then serves as an anchor for students—it gives them confidence to begin the exploration process. History gives context to the investigation.

One such anchor is the Pythagorean theorem, an "old friend" from high school geometry. At the post-secondary level, it is usually a topic in mathematics courses for teachers or non-majors. Its discussion, however, is frequently restricted to an exposition of the

Selected readings demonstrate a mathematician's need to invent proofs as the result of creative impulses and intellectual challenge.

formula and its applications—offering little for students to appreciate beyond the statement of the theorem and some limited possibilities for its use. Students who study the Pythagorean theorem are not always exposed to its rich history or to the activity and excitement it has generated for centuries. It is possible, however, to design a class, using the Pythagorean theorem as an anchor, that enlarges the students' understanding of the theorem, and encourages them to explore other avenues in mathematics.

In this article, I describe a course that not only enables students to appreciate the history and cultural significance of the Pythagorean theorem, but also introduces them to specific themes in the evolution of mathematics. The thesis—that activities which historically and currently involve the Pythagorean theorem exemplify events that characterize the evolution of mathematics—provides the rationale for the course and inspires seven themes:

I. Results in mathematics are often accepted as true before they can be proved.

II. One proof is rarely enough.

III. What motivates mathematicians to do proofs.

IV. Knowledge gives rise to new knowledge; problems generate new problems.

V. A change of context often alters an accepted result.

VI. A discovery in one field of mathematics can precipitate major crises in the mathematical world and beyond.

VII. New branches of mathematics are often the consequence of investigating familiar facts.

What follows is an outline of how I used these themes to teach a general education course for 35 liberal arts majors at California State University. An annotated list[1] of books, journal articles, and videotapes that served as resources for classroom discussion and projects is included in the appendix.

Theme I— Results are Often Accepted as True Before They Can Be Proved.

As early as 1700 B.C., ancient civilizations may have known of the relationship of integers that satisfy the formula of what later came to be known as the Pythagorean theorem. Although historical evidence suggests that a proof to deduce this special property of right triangles was not given until the 6th century B.C., the ancient civilizations of Babylonia, India, Greece, and China accepted the truth of the theorem for many years.

This historical survey prepares students to observe how instances of this theme continue in today's mathematical world. Students discover this by examining the Goldbach conjecture and "Fermat's last theorem"[2]. The Four Color theorem raises interesting questions about proofs by computer.

Group Projects and Assignments

Projects and assignments for this theme can be centered around the Pythagoreans' work with figurate numbers [22]. By arranging points in regular geometric patterns like triangles, squares, pentagons, etc., the Pythagoreans established many interesting facts about numbers simply by inspecting pictures. They were able, for example, to derive the formula for the sum of the first n odd integers by examining square numbers as follows:

Each square can be obtained from its predecessor by adding successive odd integers. From the square arrays, then, it is clear that the sum of the first n odd numbers is equal to n squared: $1+3+5+...+(2n-1) = n^2$.

Students quickly become intrigued with discovering patterns, obtaining sequences, and finding sums based on figurate numbers. They derive their own number facts and test their formulas, experimenting with different figurate numbers to get evidence for their statements. This can motivate a discussion of the differences between evidence and proof that eventually leads to the process of mathematical induction (which students want to learn in order to prove *their* formulas). As a result of these activities, students have: (1) a better understanding of how results can be accepted before rigorous mathematical proof is demonstrated; and (2) a concrete experience of the important distinction between conjecture and proof.

Theme II— One Proof is Rarely Enough.

In *The Mathematical Experience*, Davis and Hersh describe mathematics as the subject in which there are proofs. They ask: "Why do mathematicians and their students find it worthwhile to prove again and yet again the Pythagorean theorem?" (1981, p. 147).

In *every* volume of *The American Mathematical Monthly*, from 1886 to 1899, Yanney and Calderhead presented a series of papers under the title: "New and Old Proofs of the Pythagorean Theorem" [39]. Further, Loomis' *The Pythagorean Proposition* [28] contains 367 different proofs. Students survey these sources as well as proofs from more recent journals ([9], [11], [17], [19], [24], [36], [39]) to get a sense of what Davis and Hersh describe. They perceive how some proofs are an improvement of other proofs by comparing Proposition 47 in Book I of Euclid's *Elements* (which depends on a complicated diagram sometimes referred to as the Franciscan's cowl) to Bhaskara's dissection proof [12], which he appropriately captioned with only one word: "Behold."

Group Projects and Assignments

This theme can be used to encourage students to "act like mathematicians." The appendix includes proofs for a wide variety of skills levels ([2], [9], [11], [13], [16], [17], [19], [24], [28], [33], [36], [37], [39]).

Theme III— What Motivates Mathematicians To Do Proofs.

Selected readings (e.g., *Mathematics, A Cultural Approach* by Kline and "Proof" from *The Mathematical Experience*) demonstrate a mathematician's need to invent proofs as the result of creative impulses and intellectual challenge, and such readings show how the stimuli for new proofs often come from the nature of proof itself. Students learn how proof enables mathematicians to comprehend the structure of their discipline, reveals connections between topics, and empowers mathematicians to unify mathematical theories that lead to new discoveries.

Group Projects and Assignments

Euclid's superposition proofs show the need for rigorous proof in mathematics, and can alert students to the dangers of accepting visual evidence as proof.

Theme IV— Knowledge Gives Rise To New Knowledge; Problems Generate New Problems.

Proposition 31, Book VI of Euclid's *Elements* gives knowledge of a relationship between similar figures (i.e., that *any* similar shapes formed from the legs of a right triangle have areas whose sum is the area of the figure similar to those smaller shapes that are formed from the hypotenuse) that is not immediately evident from the original statement of the theorem. Generalizations of the Pythagorean theorem ([2], [12], [20]) include the Pappus extension to three-space (establishing a relationship between the volumes of three triangular prisms constructed via a vector formed by a vertex of the tetrahedron and the point of intersection of three planes intersecting the three prisms); and another three-space analogue of the Pythagorean theorem called de Gua's theorem (describing the relationship between the area of the base of a trirectangular tetrahedron[3] and the areas of its three other faces). The Pythagorean theorem has a special relationship to Euclid's parallel postulate ([25]), and it plays an important role in the development of the theory of relativity ([14]).

This theme easily leads to a discussion of: (1) how different fields in mathematics often "borrow" ideas from each other (e.g., topologists use group theory to design algorithms for constructing knots); and (2) how this "borrowing" phenomenon also extends to other sciences (e.g., geneticists use group theory to explain the behavior of DNA; financial analysts use probability in designing marketing strategies). The LINK sections of *The Mathematics Sampler: Topics for Liberal Arts* are a good source for these topics.

Group Projects and Assignments

The generation of Pythagorean triples still fascinates some contemporary mathematicians ([3], [6], [7], [18], [21], [22], [23], [26], [29], [30]). Students enthusiastically respond to the challenge of discovering common characteristics of triples with the goal of finding a generating form for them. (See "Connections in Mathematics: An Introduction to Fibonacci via Pythagoras" by Marchisotto and "Small Steps to a Student-Centered Classroom" by Castro et al.). The method of using Fibonacci numbers [7] to produce Pythagorean triples can give another historical perspective to the topic by illustrating how mathematics builds upon itself using newly found techniques to re-examine old problems. In addition this method raises interesting questions about discovery vs. invention of mathematics.

Theme V— A Change Of Context Often Alters An Accepted Result.

"What is Mathematics?" by Snapper responds to the misconception that mathematics is practiced in a vacuum. Students learn that mathematics is context sensitive by examining: Lines that always intersect in one geometry (Riemannian) but never cross in another (Lobachevskian); operations that maintain certain properties in one algebraic system (commutative multiplication of integers) but lose them in another (noncommutative multiplication of quaternions); proofs that are demonstrable in a computer world but inaccessible as yet without computers (the Four Color theorem); problems (trisection of an angle) that are

unsolvable under one set of constraints (Euclidean tools) but solvable under another (Euclidean tools and the conchoid).

Group Projects and Assignments

For this theme, students try to extend the Pythagorean theorem to three-space ([12], [31], [34]) and explore construction of triangles on the surface of a sphere to determine if the theorem holds. ([2], [13]). They can develop an appreciation for the restrictions of dimensionality, reading Abbott's *Flatland, A Romance of Many Dimensions*, and examining certain historical formulas that hold in two dimensions but not in three. Problems like the following present interesting challenges:

> The Hindu mathematician, Aryabhata, wrote early in the sixth century A.D. His work is a poem of 33 couplets called the *Ganita*. Following are translations of two of the couplets: (1) The area of a triangle is the product of the altitude and half the base; half of the product of this area and the height is the volume of the solid of six edges. (2) Half the circumference multiplied by half the diameter gives the area of the circle; this area multiplied by its own square root gives the volume of the sphere. Show that in each of the couplets, Aryabhata is correct in two dimensions but wrong in three. (Eves and Newsom, 1965, p. 21).

The Fundamental Theorem of Arithmetic provides another opportunity to explore this theme. The number system of Lagado (all integers of the form $3k+1$, where k is a non-negative integer) colorfully described by Stein in *Mathematics The Man-Made-Universe*, allows for two different factorizations of 100. Students, assuming unique factorization, uncover the failure of the theorem. This activity gives them insight into the importance of context in mathematics as well as some healthy skepticism about what often seems like "overwhelming evidence" for a conjecture.

Theme VI— A Discovery In One Field Can Precipitate Major Crises in the Mathematical World and Beyond.

The discovery of incommensurable lengths by Hippasus cleared the path for new investigations in mathematics, and opened mathematics to new self-consciousness that had its own rewards. The realization of the independence of the parallel postulate (See "The Euclid Myth," in *The Mathematical Experience*) destroyed the view of Euclidean geometry as the exemplar of truth, freed mathematics from a total reliance on spatial intuition, and paved the way for a flourishing of many new fields of geometry. Gödel's Incompleteness theorem, with its origins in the ancient Greek (liar) paradox of Epimenides, used mathematical reasoning to explore mathematical reasoning itself, and its results stunned the mathematical world.

Introducing these historical events helps to expand students' knowledge of mathematics. Perhaps most importantly, the references to crises and the responses they generated give students evidence of mathematics as a process and not a mere collection of facts or "eternal truth."

Group Projects and Assignments

This theme lends itself nicely to assigned readings and to oral or written responses, so class activities are primarily discussion groups. The crisis that resulted from an application of the Pythagorean theorem illustrates several characteristics of mathematics. A discussion of how the Pythagoreans were unable to acknowledge the existence of irrational numbers exposes the idea that mathematics is time-dependent, i.e., what was once considered true is later found not true (another

---

The independence of the parallel postulate destroyed the view of Euclidean geometry as the exemplar of truth.

---

example: all geometry is Euclidean). The fact that the Pythagoreans tried to suppress the existence of incommensurable lengths illustrates how some discoveries in mathematics are rejected or hidden and do not receive acceptance for many years. (Other examples are negative and imaginary numbers.)

Jordan and O'Malley's article "An Implication of the Pythagorean Theorem" [25] is used to motivate a discussion of how the Pythagorean theorem could have led to the formulation of Euclid's parallel postulate. Sagher's proof [32] of the irrationality of the square root of two (which he claims was accessible to Pythagoras) opens the door to Euclid's nice proof of this result. Students thus expand their repertoire of proofs, seeing again the reward of seeking simple proofs.

Theme VII— New Branches of Mathematics are Often the Consequence of Investigating Familiar Facts.

Investigations of the Pythagoreans provided the foundation for new branches of mathematics: The discovery of incommensurable lengths motivated the exact definition of the continuum; the paradoxes uncovered by the Pythagoreans provided for the

construction of the real number system; the Pythagorean theorem plays a central role in the creation of trigonometry and in the development of geometry of infinitely many dimensions [14]. The birth of new branches of mathematics like non-Euclidean geometries and non-traditional algebras resulted from the investigation of accepted "mathematical truths" (the parallel postulate of Euclid, and the commutativity of multiplication).

The idea that the Pythagorean theorem led (even remotely) to the creation of new mathematical fields is an important one for students, both with regard to their understanding how mathematics develops and with respect to how they themselves approach mathematics. Too often, students fail to thoroughly examine "the given" when they attempt to solve problems. The concept of "the familiar breeding with the unknown" encourages students to investigate what it is they do know when they begin the process of doing mathematics. (See *How to Solve It: A New Aspect of Mathematical Method* by Polya.)

Group Projects and Assignments

"It Started in Greece" [13] has nice illustrations of models of Riemannian and Lobachevskian geometries. It also makes the connection between the Pythagorean theorem and trigonometry, and can serve as a good source of discussion. Students can then work in groups to derive the law of cosines using the inductive and deductive procedures proposed by Bibby and French in their article "Pythagoras Extended" [5].

By emphasizing the idea that the familiar can be a source of new knowledge, this theme can be used to illustrate the multi-culturism of mathematics. Examining a decorative motif encountered all over the world—on baskets of the Salish Indians of British Colombia, plaited mats from Angola, textiles from Scandinavia, and game boards in Liberia, Gerdes [17] shows how this design motif can lead to a conjecture of the Pythagorean theorem and how the same discovery process also suggests new demonstrations of the theorem. Students are fascinated by the connection between the artifacts of different cultures and the Pythagorean theorem. This example can lead to other discussions of Frieze Patterns and symmetries [*For All Practical Purposes*] and can give further evidence of mathematics as part of world culture.

Conclusion

I concentrated on these themes in my mathematics class for liberal arts majors to try to give students some sense of what mathematics is and what mathematicians do, using the Pythagorean theorem—a familiar topic to most of them—as an anchor. Based on this experience,

I found that students wanted to discover more about the theorem, and about mathematics in general.

This kind of approach does not have to be limited to mathematics classes for liberal arts majors or to the themes here presented. One can use the Pythagorean theorem as a path to analytic geometry and Hilbert space. Other themes that characterize mathematics can be explored via the Pythagorean theorem: how partial solutions to problems contribute to the expansion of mathematical knowledge; or how mathematics builds upon itself in a way that other sciences do not. Any presentation of the Pythagorean theorem incorporating these themes enriches its exposition by providing historical insight into its discovery and by fostering greater understanding of the way knowledge expands in

---

Gödel's Incompleteness theorem, with its origins in the ancient Greek (liar) paradox of Epimenides, used mathematical reasoning to explore mathematical reasoning itself, and its results stunned the mathematical world.

---

the mathematical world. Such presentations are, in my opinion, humanistic and true to the spirit of Alvin White's challenge:

> The concepts and relationships of mathematics should be presented as the building blocks of this magnificent edifice created by the human imagination (1985, p. 850).

I have tried, with this "new look at an old friend", to rise to this challenge, and, based on my students' responses, I am convinced it is worth the effort. One student wrote:

> I now look at mathematics differently. I have gained an appreciation for proof and for the mathematicians themselves. Many times the class struggled to try to prove an idea, for example, Pythagorean triples, and in the end, we learned to admire the people who created the theory. Now, from my experiences in the class, I question ideas and I want to know the methods behind the theorems I used to take for granted.

Works Cited

Abbott, Edwin. *Flatland. A Romance of Many Dimensions*, Arion Press, San Francisco, 1980.

Berlinghoff, William and Grant, Kerry. *A Mathematics Sampler: Topics for the Liberal Arts*, Ardsley House Publishers, New York, 1988.

Castro, J., Gold, J., Marchisotto, E., Schilling, M., & Zeitlin, J., "Small Steps to a Student-Centered Classroom," *Primus*, Volume 1, No.3, September 1991, pp. 253-274.

Davis, Philip and Hersh, Reuben. *The Mathematical Experience*, Birkhauser, Boston, 1981.

Eves, Howard and Newsom, Carroll. *An Introduction to the Foundations and Fundamental Concepts of Mathematics*, Holt, Rinehart and Winston, New York, 1965.

Kline, Morris. *Mathematics. A Cultural Approach*, Addison-Wesley Publishing Co., Massachusetts, 1962.

Polya, George. *How To Solve It: A New Aspect of Mathematical Method*, Princeton Press, New Jersey, 1973.

Snapper, Ernst. "What Is Mathematics," *The American Mathematical Monthly*, August-September, 1979, pp. 551-557.

Stein, Sherman. *Mathematics The Man-Made Universe*, W.H. Freeman and Company, New York, 1976.

White, Alvin. "Beyond Behavioral Objectives," *The American Mathematical Monthly*, Volume 82, October 1985, pp. 849-851.

Notes

1. All numbers in brackets [ ] will refer to the books, journal articles, and videotapes cited in the annotated list of resources.

2. This conjecture can be considered a "stepchild" of the Pythagorean theorem. Venden Eynder [38] shows how the proof for the case p = 4 depends on Pythagorean triples, integers that satisfy the Pythagorean theorem.

3. A tetrahedron having a trihedral angle, all face angles of which are right angles, is a trirectangular tetrahedron. The trihedral angle is called the right angle of the tetrahedron and the face opposite this angle is called the base of the tetrahedron.

Appendix

Annotated List of Resources

[1] Anglin, W.S. "Using Pythagorean Triangles to Approximate Angles," *The American Mathematical Monthly*, Volume 95, Number 6, June/July 1988, pp. 540-541.

This article, which develops an "Approximation Theorem" to accomplish what the title suggests, assumes a knowledge of trigonometry and elementary number theory.

[2] Apostol, Tom. "The Theorem of Pythagoras," *MATHEMATICS!*, California Institute of Technology, 1988.

This is an award-winning 20 minute computer-animated videotape, with accompanying workbook, on history, proofs, and applications of the theorem.

[3] Arpaia, P.J. "A Generating Property of Pythagorean Triples," *The Mathematics Magazine*, Volume 44, Jan./Feb. 1971, pp. 26-27.

Based on the generating property of pairs of Pythagorean triples given by Courant and Robbins in *What Is Mathematics?*, this note establishes a generating property of any Pythagorean triple.

[4] Bankoff, Leon and Trigg, Charles. "The Ubiquitous 3:4:5 Triangle," *The Mathematics Magazine*, Volume 47, March 1974, pp. 61-70.

The authors illustrate 3:4:5 right triangles imbedded in a variety of configurations and show some under-lying patterns which generate these triangles. They claim: "The ancient Pythagoreans would have been delighted to have this evidence to support their number mysticism."

[5] Bibby, Neil and French, Doug. "Pythagoras extended," *The Mathematical Gazette*, Volume 72, Number 461, October 1988, pp. 184-188.

This article generalizes the Euclidean dissection proof to "make the cosine rule in some way geometrically obvious."

[6] Bergum, Gerald and Yocom, Ken. "Tchebysheff Polynomials and Primitive Pythagorean Triples," *Two Year College Mathematics Readings*, The Mathematical Association of America, Washington D.C., 1981.

This note illustrates and proves how to produce primitive integer-sides of a right triangle with hypotenuse c^n when the set of integers {a,b,c} is primitive.

[7] Boulger, William. "Pythagoras Meets Fibonacci," *Mathematics Teacher*, Volume 82, Number 4, April 1989, pp. 277-281.

Boulger makes a nice connection between Fibonacci numbers and Pythagorean triples and proves it. He also illustrates a relationship between Fibonacci numbers and the golden ratio.

[8] Bronowski, Jacob. "The Music of the Spheres," *Mathematics. People, Problems, Results. Volume I* [Campbell, Douglas and Higgens, John (Editors)], Wadsworth International, Belmont, California, 1984.

This chapter is excerpted from the book *The Ascent of Man*. Bronowski's discussion of the Pythagorean theorem (as with other topics treated in the essay) suggest how the idea grew, spread, and generated other ideas.

[9] Caners, Leonard. "Pythagorean Principle and Calculus," *The Mathematics Magazine*, Volume 28, May/June 1955, p. 256.

Using calculus, the author gives a proof of the Pythagorean theorem. A comment on this proof is given on page 40 of *The Mathematics Magazine*, Volume 29, 1955 by H.W. Becker who refers to Loomis' claim that no independent proof of the Pythagorean theorem can be based on trigonometry, analytic geometry, or calculus. Yet another comment (on Becker's comment) refuting this claim is given on page 204 of *The Mathematics Magazine*, Volume 29, 1956 by H.V. Craig.

[10] Duncan, Dewey. "Generalized Pythagorean Numbers," *The Mathematics Magazine*, Volume 10, 1936, pp. 209-211.

This note establishes general formulas that yield n numbers with the property that the sum of the square of n-1 of these numbers yields the square of the nth number.

[11] Eaves, James. "Pythagoras, His Theorems and Some Gadgets," *The Mathematics Magazine*, Volume 27, 1954, pp. 161-167.

This note presents several models which the author calls gadgets for proving the Pythagorean theorem. In each case, he illustrates the geometric figure with the necessary constructions, and gives the outline of a proof based on them. The article includes interesting historical information about the theorem.

[12] Eves, Howard. "The First Great Theorem," *Great Moments in Mathematics Before 1650*, The Mathematical Association of America, Washington D.C., 1983.

This chapter gives a brief history of the Pythagorean theorem, and very nicely illustrates several famous proofs. Among them are the proofs by Bhaskara, Dudeney, Euclid, President Garfield, Leonardo da Vinci, Wallis and of course Pythagoras.

[13] Freeman, W. H. (publisher). "It Started In Greece: Measurement," *For All Practical Purposes*, 1988.

This is a 30 minute videotape which discusses Pythagoras and gives proofs and applications of the Pythagorean theorem. This videotape supplements the mathematics textbook *For All Practical Purposes*.

[14] Fredrichs, K.O. *From Pythagoras to Einstein*, The Mathematical Association of America, Washington D.C., 1965.

This book discusses the Pythagorean theorem and some basic facts of vector geometry with a view towards illustrating the significance of these ideas in the special Theory of Relativity. It can serve as a good reference book and source of problems for instructors.

[15] Gandi, J.M. "An Infinity Descent Method to Prove Pythagorean Principle," *The Mathematics Magazine*, Volume 30, May/June 1957, p. 250.

Using trigonometric relationships which he claims can be derived without the use of the Pythagorean theorem and calculus, the author proves that $\sin^2 A + \cos^2 A = 1$.

[16] Garfunkel, Solomon (Editor). "Measurement," *For All Practical Purposes*, W.H. Freeman and Co., New York, 1988.

In the chapter entitled "Measurement," this text discusses the Pythagorean theorem. It includes two excellent illustrations of proofs for the theorem: the Byrne colored proof from his edition *The First Six Books of the Elements of Euclid in which Coloured Diagrams and Symbols are Used Instead of Letters for the Greater Ease of the Learner*, and a dissection proof for the students to try. In keeping with the book's title, the chapter includes applications and interesting problems relating to the Pythagorean theorem.

[17] Gerdes, Paulus. "A Widespread Decorative Motif and the Pythagorean Theorem," *For the Learning of Mathematics*, Volume 8, 1988, pp. 35-39.

The author gives an example of a decorative motif in designs from different cultures that motivates the introduction of the Pythagorean theorem in the classroom and generate different proofs of the theorem.

[18] Goodrich, Merton. "A Systematic Method of Finding Pythagorean Numbers," *The Mathematics Magazine*, Volume 19, 1945, pp. 395-397.

The author constructs a table to accomplish just what the title suggests.

[19] Hardy, Michael. "Behold! The Pythagorean Theorem via Mean Proportionals," *The College Mathematics Journal*, Volume 17, No.5, November 1986, p. 422.

The author gives a nice illustration of the theorem based on mean proportionals "almost" without words.

[20] Hazard, William. "Generalizations of the Theorem of Pythagoras and Euclid's Theorem of the Gnomon," *The American Mathematical Monthly*, Volume 36, January 1929, pp. 32-43.

This note generalizes the Hindu proof for the Pythagorean theorem to results regarding areas of parallelograms.

[21] Hildebrand, W.J. "Generalized Pythagorean Triples," *The College Mathematics Journal*, Vol. 16, No. 1, January 1985, pp. 48-52.

This article presents an algorithm to generate all possible integer triples (a,b,c) that are the sides of a triangle that contains angle C whose cosine is the rational number p/q. It also gives applications. The article assumes a knowledge of calculus and elementary number theory.

[22] Honsberger, Ross. "Pythagorean Arithmetic," *Ingenuity in Mathematics*, The Mathematical Association of America, Washington D.C., 1970.

This short essay describes the arithmetic methods of the Pythagoreans and develops a procedure for generating Pythagorean triples. It is interesting reading for high school and college students and a good source of problems for instructors.

[23] Horadam, A.F. "On Khazanov's Formulas," *The Mathematics Magazine*, Volume 36, Sept./Oct. 1963, pp. 219-220.

The author outlines Khazanov's method for finding Pythagorean triples, his own method for finding triples using a generalized Fibonacci sequence, and illustrates the connection between the two.

[24] Isaacs, Rufus. "Two Mathematical Papers Without Words," *The Mathematics Magazine*, Volume 48, 1975, p. 198.

On this one page is nicely illustrated the trisection of an angle and the Chinese proof of the Pythagorean theorem. Comments on these appear on pages 50 and 51 in Volume 49 of *The Mathematics Magazine*, 1976.

[25] Jordan, John and O'Malley, John. "An Implication of the Pythagorean Theorem," *The Mathematics Magazine*, Volume 43, Sept./Oct. 1970, pp. 186-189.

The author gives two proofs of the conjecture that the Pythagorean theorem implies Euclid's parallel postulate, or an equivalent proposition.

[26] Kalman, Dan. "Angling for Pythagorean Triples," *The College Mathematics Journal*, Volume 17, No.2, 1986, pp. 167-8.

The author generates Pythagorean triples from common fractions. A knowledge of trigonometry is assumed.

[27] Kramer, Edna. *The Nature and Growth of Modern Mathematics,* Princeton University Press, New Jersey, 1981.

This book traces the development of important mathematical concepts from their inception to their present formulation. It includes much historical information about the discoveries of the Pythagoreans and in the words of its author, answers the question: "Why should Pythagoras and his followers be credited with (or blamed for) some of the methodology of the 'new' mathematics?"

[28] Loomis, Elisha Scott. *The Pythagorean Proposition,* National Council of Teachers of Mathematics, Washington D.C., 1968.

This classic book, first published in 1940 gives 367 proofs of the Pythagorean Theorem. In the words of the author, it is written: "With the hope that this simple exposition of this historically renowned and mathematically fundamental proposition...may interest many minds and prove helpful and suggestive."

[29] Nishi, Akihiro. "A Method of Obtaining Pythagorean Triples," *The American Mathematical Monthly,* Vol. 94, No.4, 1987, pp. 869-872.

This article, which develops what the title indicates, assumes a knowledge of elementary number theory.

[30] Ore, Oystein. "The Pythagorean Problem," *Invitation to Number Theory,* The Mathematical Association of America, Washington D.C., 1967.

This chapter introduces Pythagorean triples and develops the formula for generating them. It concludes with problems related to Pythagorean triangles. Mathematical skill at the level of intermediate algebra is sufficient to comprehend this presentation.

[31] Polya, George. *Mathematical Discovery,* John Wiley & Sons, New York, 1981.

In a chapter entitled "The Cartesian Pattern," Polya gives several analogues to the Pythagorean theorem in solid geometry.

[32] Sagher, Yoram. "What Pythagoras Could Have Done," *The American Mathematical Monthly,* Vol. 95, No. 2, p. 117.

This note gives a nice proof that the square root of 2 is irrational, which, in the words of the author: "was fully accessible to Pythagoras . . . and is also accessible to high school students after one year of algebra."

[33] Swetz, Frank. "The Evolution of Mathematics in Ancient China," *Mathematics, People, Problems,*

Results. Volume I [Campbell, Douglas and Higgins, John (Editors)], Wadsworth International, Belmont, California, 1984.

The chapter gives the oldest known demonstration of a proof of the Pythagorean theorem from *Chou pei,* within the context of the development of mathematics in China.

[34] Thorton, Carol A. "A New Look, Pythagoras," *The Mathematics Teacher,* Volume 74, No. 2, February 1981, pp. 98-100.

The author illustrates how to extend the Pythagorean theorem into 3-space.

[35] Tirman, Alvin. "Geometric Parametrization of Pythagorean Triples," *The College Mathematics Journal,* Volume 17, #2, January 1986, p. 168.

This nicely illustrated parametrization confirms that the acute angles in the triangle formed by the median and altitude to the hypotenuse of an integer-sided right triangle are Pythagorean.

[36] Torczynski, J. R. "Dimensional Analysis and Calculus Identity," *The American Mathematical Monthly,* Volume 95, #8, October 1988, pp. 746-754.

In a section entitled "Dimensional Analysis and Geometry" the author gives an elegant proof of the Pythagorean theorem which depends solely on dimensional analysis.

[37] van der Waerden, B. L. *Geometry and Algebra in Ancient Civilizations,* Springer-Verlag, Berlin, 1983.

The first chapter of this book (pages 1-35) is concerned with Pythagorean triangles. It consists of three parts: documentation of written sources, archaelogical evidence, and examples of proofs and applications. This is a good reference book for university students and for instructors.

[38] Vanden Eynder, Charles. "Fermat's Last Theorem: 1637-1988," *Mathematics Teacher,* Volume 82, Number 8, November 1989.

Vanden Eynder gives a brief history of Fermat's last theorem. In his discussion, he shows how the proof for the case n = 4 depends on Pythagorean triples.

[39] Yanney, B.F. and Calderhead, James. "New and Old Proofs of the Pythagorean Theorem" [a continuing series], *The American Mathematical Monthly,* Volume 3, 1896; Volume 4, 1897; Volume 5, 1898; Volume 6, 1899.

This series gave several proofs of the Pythagorean theorem in each issue.

A Triptych of Pythagorean Triples

Peter Flusser
Kansas Wesleyan University
Salina, Kansas

1. About This Triptych.

The study of ancient mathematical works, especially of original works, helps us to recognize the human side of mathematics. Such works enable us to see how human beings have struggled to attain new knowledge while at the same time developing the necessary tools for attaining new knowledge. Their study of such works gives us insight into the circumstances, the society, and the intellectual climate in which their creators lived. Such study may upset many of our preconceived notions. For example, we all share more or less the same mental picture of the real numbers. This picture is so much a part of our intellectual baggage that it is hard for us to realize that it was not shared by all people at all times in history. We need to know what mental pictures numbers created in the minds of Euclid and his colleagues at the Museum in Alexandria, and why they

> **The study of ancient mathematical works enables us to see how human beings have struggled to attain new knowledge while at the same time developing the necessary tools for attaining new knowledge.**

found it necessary to distinguish between number and quantity. We need to learn how our mental pictures of the real numbers evolved over the centuries. We need to be aware of the contributions of the masters; we need to study not only their achievements but also their mistakes, and we need to learn how their mistakes often led to even greater achievements. All of this is part of being a humanistic mathematician.

But the works of the old masters, even if they deal with elementary topics, are hard for us to understand. The intellectual climate in which these works were created is different from ours. Various tools which simplify the study of these topics had not yet been invented. Most importantly, the students of the masters who would

eventually clarify many of the obscure points had not yet been born. It is the purpose of this essay to help you read what Euclid, Diophantus, and Euler—three mathematical giants living many centuries apart—had to say about *Pythagorean triples*, a subject of sufficient importance, beauty and simplicity to be of interest to a wide audience. I show you excerpts of their works together with suitable commentary to make the reading easier. A triptych is a picture consisting of three scenes in three separate compartments usually shown side by side. I would like you to think of the writings of Euclid, Diophantus and Euler quoted in this paper as the important part: the scenes themselves. My commentary is the frame that holds them together.

Pythagoras's Theorem tells us much more than that the square on the hypotenuse of a right triangle has an area equal to the sum of the areas of the squares on the other two sides. It enables us to compute the length of one side of a right triangle, provided that the lengths of the other two sides are known. In so doing, this theorem introduces us to numbers that cannot be obtained from the natural numbers by the standard processes of addition, subtraction, multiplication and division. For the first time in the history of mathematics we meet irrational numbers: even though two sides of a right triangle may have integer lengths, the length of the third side need not be a rational number. We are thus naturally led to the following problems:

> **Find (all) right triangles with integer (rational) sides.**

or: **Find (all) triples of natural (rational) numbers (a,b,c) such that**
$$a^2 + b^2 = c^2$$

We call triples of natural numbers (a,b,c) such that $a^2 + b^2 = c^2$ *Pythagorean triples*. If a, b and c are pairwise relatively prime we call the triple a *Primitive Pythagorean triple*.

We will study Euclid, Diophantus and Euler with 20-20 hindsight. We won't be so much interested in their results per se as in the manner in which they obtained these results. We will assume that the following facts are well known; if you wish to refresh your memory, a proof can be found in Courant and Robbins (1941, p. 40 ff.)

Proposition 1. Given the set of all primitive Pythagorean triples, all other triples can be obtained by multiplying these by a suitable constant.

Proposition 2. If p and q are relatively prime integers of opposite parity with p > q, then, if a = 2pq, b = $p^2 - q^2$ and c = $p^2 + q^2$, (a,b,c) form a primitive Pythagorean triple.

Proposition 3. Given a primitive Pythagorean triple (a,b,c), there exist relatively prime integers of opposite parity, p>q, such that a = 2pq, b = p^2-q^2 and c = $p^2 + q^2$

2. Pythagorean Triples in Ancient Babylon.

According to Herodotus, the Father of History, geometry was invented by the ancient Egyptians who needed a tool to calculate how much tax was owed by land-owners after the periodically flooding Nile had either washed away some of the old or deposited some new soil on their property. Somehow the idea that the Theorem of Pythagoras was discovered by a surveyor working for the Egyptian IRS does not appeal to me. I prefer to think that it was discovered by an artist who contemplated a design such as the one in Figure 1.

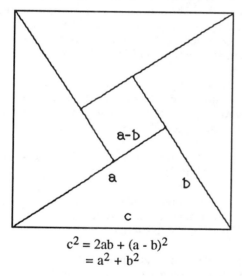

$$c^2 = 2ab + (a - b)^2$$
$$= a^2 + b^2$$

Figure 1. The Theorem of Pythagoras.

We know now that the Pythagorean Theorem was known more than a thousand years before Pythagoras was born. It may be that Pythagoras was the first to supply a geometric proof, but Figure 2 is conclusive evidence that the Babylonians must have had a thorough

understanding of that theorem. Let me explain.

The tablet shown in Figure 2 contains four columns of 15 numbers each. Starting from the right, the first column contains the numbers 1 through 15. The second column contains lengths of hypotenuses of right triangles (C), the third the lengths of one of the sides (B), and the fourth column the squares of the ratios of the hypotenuse to the third side $((C/A)^2)$. The numbers in the second and third columns are integers, those in the fourth column are terminating sexagesimal fractions. The largest triple represented is (13,500, 12,709, 18,541); it is, incidentally, a primitive triple.

[1,59,0,]15	1,59	2,49	1
[1,56,56,]58,14,50,6,15	56,7	3,12,1	2
[1,55,7,]41,15,33,45	1,16,41	1,50,49	3
[1,]5[3,1]0,29,32,52,16	3,31,49	5,9,1	4
[1,]48,54,1,40	1,5	1,37	5
[1,]47,6,41,40	5,19	8,1	6
[1,]43,11,56,28,26,40	38,11	59,1	7
[1,]41,33,59,3,45	13,19	20,49	8
[1,]38,33,36,36	9,1	12,49	9
1,35,10,2,28,27,24,26,40	1,22,41	2,16,1	10
1,33,45	45	1,15	11
1,29,21,54,2,15	27,59	48,49	12
[1,]27,0,3,45	7,12,1	4,49	13
1,25,48,51,35,6,40	29,31	53,49	14
[1,]23,13,46,40	56	53	15

Neugebauer's translation of Plimpton 322. Illegible digits are enclosed in square brackets.

1,59,0,15	1,59	2,49	1
1,56,56,58,14,50,6,15	56,7	1,20,25	2
1,55,7,41,15,33,45	1,16,41	1,50,49	3
1,53,10,29,32,52,16	3,31,49	5,9,1	4
1,48,54,1,40	1,5	1,37	5
1,47,6,41,40	5,19	8,1	6
1,43,11,56,28,26,40	38,11	59,1	7
1,41,33,59,3,45	13,19	20,49	8
1,38,33,36,36	8,1	12,49	9
1,35,10,2,28,27,24,26,40	1,22,41	2,16,1	10
1,33,45	45,0	1,15,0	11
1,29,21,54,2,15	27,59	48,49	12
1,27,0,3,45	2,41	4,49	13
1,25,48,51,35,6,40	29,31	53,49	14
1,23,13,46,40	56	1,46	15

Boyer's correction based on Neugebauer's comments. The entry in column 1, line 8 should read 1,41,33,45,14,3,45

Figure 2. (facing page) Plimpton 322 (above) Translations of Plimpton 322

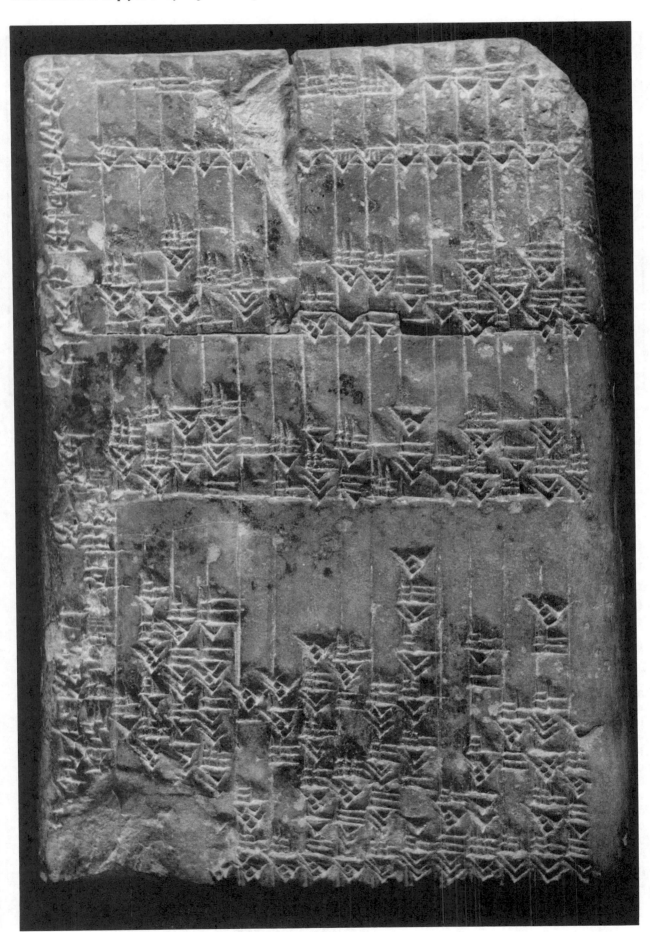

Plimpton Cuneiform Tablet #322, Photo courtesy of George A. Plimpton Collection, Rare Book and Manuscript Library, Columbia University

P	Q	A	$(C/A)^2$	B	C	N
12	5	120	1.983402777777	119	169	1
64	27	3456	1.949158552088	3367	4825	2
75	32	4800	1.91880212673	4601	6649	3
125	54	13500	1.886247906721	12709	18541	4
9	4	72	1.815007716049	65	97	5
20	9	360	1.785192901234	319	481	6
54	25	2700	1.719983676269	2291	3541	7
32	15	960	1.692709418402	799	1249	8
25	12	600	1.642669444444	481	769	9
81	40	6480	1.58612256611	4961	8161	10
60	30	3600	1.5625	2700	4500	11
48	25	2400	1.489416840277	1679	2929	12
15	8	240	1.450017361111	161	289	13
50	27	2700	1.430238820301	1771	3229	14
9	5	90	1.387160493827	56	106	15

$$A^2 + B^2 = C^2$$

$$A = 2PQ$$
$$B = P^2 - Q^2$$
$$C = P^2 + Q^2$$

Table 1: Plimpton 322 Corrected, Translated and Augmented.

The tablet of Figure 2 is No. 322 of the Plimpton Collection of Columbia University. It dates somewhere between 1900 and 1600 BC. It is broken on the left side. There is modern glue on that side, so the damage

The Babylonians were not interested in mathematics solely for utilitarian reasons. I believe that the creator of Plimpton 322 did his work for the sheer joy that the beauty of his results gave him. I believe that he truly was a humanistic mathematician.

must have occurred after the tablet was excavated. For reasonable conjectures about why the Babylonians were interested in these numbers see Boyer (1989). Boyer maintains that the Babylonians must have known Pythagoras's Theorem, and that calculations such as the ones on Plimpton 322 represented the beginnings of trigonometry of the right triangle. Neugebauer (1969) gives a detailed description of Plimpton 322. He conjectures that there are missing columns of numbers, one of which must have contained the lengths of the third sides (A) of the right triangles in question. He also states that at least the second of our three propositions must have been known to the ancient Babylonians and used to compute the tables on

Plimpton 322. Neugebauer maintains that it is impossible to find the triple (6480, 4961, 8161) without knowing that the corresponding values of p and q are 81 and 40. He therefore suggests that Plimpton 322 contained two more columns with the corresponding P's and Q's with A = 2PQ forming the third column from the left.

If Boyer and Neugebauer guessed right, then Table 1 is a modern version of what Plimpton 322 would have looked like. Boyer suggests that there might have been one more column containing the ratios $(B/A)^2$ if all but (possibly) one error had been corrected. In line 8, the ratio $(C/A)^2 = 1.69270941841$. In base 60 this number is 1,41.33.45.14.3.45, and not 1,42.33.59.3.45. I can't be one hundred percent sure, but I strongly suspect that this is a modern typographical error.

Even a cursory examination of Table 1 reveals several remarkable facts. Except in lines 11 and 15 all P's and Q's are relatively prime and of opposite parity; hence all triples except those appearing in those lines are primitive. In line 11 we find the famous (4,3,5) triple in disguise. Line 15 is even more interesting, since the P and Q that appear here are relatively prime. But they are both odd. To get the primitive triple (45,28,53) which is proportional to the triple (90,56,106) which appears on the tablet, the creator of Plimpton 322 would have had to choose P = 7 and Q = 2, but side A would then have been $P^2 - Q^2$ instead of 2PQ because
$$(7^2 + 2^2)/(2 \times 7 \times 2) = 53/28$$
is not a terminating sexagesimal, as all ratios that appear on Plimpton 322 are.

The numbers in the fourth column from the right are the squares of the secant of the angle opposite to the side which I call B in Table 1. We can tell that all these numbers are terminating sexagesimal fractions because P and Q are divisible by 2, 3 and 5 only and hence the same is true for the numbers representing the side A. Why were the numbers p and q chosen so that these ratios decrease to 1? If we could answer this question, we might be able to guess why the Babylonians constructed Plimpton 322 in the first place. The existence of that tablet strongly suggests one thing, however. I believe that the Babylonians were not interested in mathematics solely for utilitarian reasons. I believe that the creator of Plimpton 322 did his work for the sheer joy that the beauty of his results gave him. I believe that he truly was a humanistic mathematician.

3. Pythagoras and Pythagorean Triples.

The Plimpton tablet was interpreted only after we were well into the middle thirties of this century. Until then the earliest knowledge of a treatment of Pythagorean triples had come to us indirectly through Proclus (410-485 A.D.), the famous commentator on Euclid's works. Proclus has preserved for us Pythagoras's method for finding Pythagorean triples (Heath 1956, Vol 1 pg 356). Zhmud (1989) shows how Pythagoras might have used figured numbers, that is numbers represented by suitably arranged dots, for that purpose.

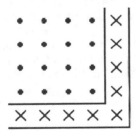

Figure 3. Pythagoras's method for generating triples.

Consider the square number 16 enclosed by the shorter of the two pairs of perpendicular lines in Figure 3. The figured number consisting of the nine crosses surrounding that number is called a gnomon. Adding a gnomon to a square yields the next square—in our case the square number $16 + 9 = 25$. If, as in our example, we choose a gnomon which is itself a square, we obtain a square which is the sum of two squares.

Now such a construction can always be accomplished as follows: Let the gnomon be the number m^2, and let the side of the preceding square be a. From Figure 3 we see that $m^2 = 2a+1$, hence $a = (m^2 - 1)/2$. Let b be the side

of the succeeding square; then $b = a + 1 = (m^2 + 1)/2$. This imposes the only condition on m: it must be odd. Thus for any odd number m, Pythagoras obtains the triple $(m, (m^2 - 1)/2, (m + 1)/2)$.

A note on references: A referee remarked that, "The year of quoted publications next to references of ancient utterings is a bit confusing." That is correct. Unfortunately neither I nor, I suspect, many readers have access to ancient manuscripts or early printed versions of the works of Euclid, Diophantus, and Euler. I have therefore referred to modern editions of these works, editions to which most people do have access and from which I have quoted so extensively. Thus, for example, all references to Euclid are to Euclid (1956), Thomas L. Heath's authoritative edition which is still available from Dover Publications. When quoting one of Heath's remarks, I refer to that work as Heath (1956).

4. Greek Mathematics at the Time of Euclid.

We are now ready to examine the first masterwork in our triptych, Euclid's Proposition II. 6 and Lemma 1 to Proposition 29 in Book X. Before we offer a reading from Euclid, a few remarks about the nature of Greek mathematics are needed.

The Babylonians distinguished between regular and irregular numbers. The former are natural numbers whose only prime factors are 2,3 and 5; hence their reciprocals are terminating sexagesimal fractions. The reciprocals of irregular numbers are repeating sexagesimal fractions. Note that of the first ten natural numbers only one, namely 7, is irregular. The Babylonians constructed many tables of reciprocals as aids to computation; in many of these one finds comments to the effect that irregular numbers have no reciprocals. It is thus reasonable to suppose that the Babylonians considered both irrational numbers and repeating sexagesimals as undesirable characters and therefore to be avoided. So they might not have bothered to make fine distinctions between irrational numbers and repeating sexagesimals.

The Greeks, on the other hand, were very much aware of the distinction between what we call today rational and irrational numbers, and the existence of irrational numbers bothered them a great deal. It upset one of their favorite theories, one proposed by Pythagoras and his disciples, that "all is number." As the very name suggests, they considered irrationals to be unreasonable, crazy, mad. Irrationals had to be expelled from the realm of number. And so, as a result of the unwelcome discovery of quantities that are not ratios of whole numbers, the Greeks created a theory in which they distinguished between three types of mathematical objects: numbers, ratios, and quantities or magnitudes.

For the Greeks, numbers were what we call today natural numbers. Negative numbers did not exist for them; neither did complex numbers.

The concept of magnitude covered a multitude of mathematical objects. Perhaps the easiest definition to understand is that given by Euler (1984):

> Whatever is capable of increase or diminution is called *magnitude*, or *quantity*. A sum of money is therefore a quantity, since we may increase or diminish it. It is the same with a weight and other things of this nature.

According to the Greeks, there were magnitudes of dimension zero (angles), one (line segments), two (plane figures), and three (solid bodies). It was not the length of a line segment or the area of a plane figure, or the volume of a solid, that was a magnitude; it was the line segment, the plane figure or the solid itself. Rigid motion in space was a permissible operation on magnitudes; thus the Greeks would say that the line segment AB was *equal* to the segment CD if AB could be made to coincide with CD by means of a rigid motion. For plane or solid figures *equal* meant what we would call today equal in area, or volume, respectively. If two plane or solid figures could be made to coincide by means of a rigid motion, they were said to be *equal in all respects*, or *congruent*.

The Greek concept of ratio is the hardest to explain to the modern reader. Euclid defined ratios (Definition 3, Book V) as follows:

> A *ratio* is a sort of relation in respect of size between two magnitudes of the same kind.

This is indeed a hard definition to understand. I suppose no modern editor would allow words such as "a sort of relation" to be used in a formal definition. For us a ratio is the result obtained when one (complex) number divides another. In this sense it is like a product which is the result obtained when one number multiplies another. Euclid would have agreed with our definition of multiplication. He used to say: "Two numbers, by multiplying one another, make some number." But he never said: "One number, by dividing another, makes some number" because, in general, when some number, like 4, divided another, like 6, it did not make a number. Somehow, after the division process, the dividend and the divisor managed to retain a part of their identities, and their ratio represented a way in which their "sizes" related to each other. Even as late as the early 20th century attempts were made to clarify this idea by distinguishing the ratio that a has to b (in symbols: a:b) from the quotient obtained when a is divided by b (in symbols: a/b). While it would not be wrong to identify the modern concept of "rational

number" with the Greek concept of "ratio of two numbers", it should be borne in mind that the idea of a ratio was far richer for the Greeks than the notion of number. Ratio was a property shared by numbers as well as pairs of magnitudes and even pairs of ratios. Equality of ratios (expressed symbolically: a:b::c:d) was

---

As the very name suggests, they considered irrationals to be unreasonable, crazy, mad. Irrationals had to be expelled from the realm of number.

---

called a proportion, a concept I will discuss later in more detail. These overtones were lost over the centuries as number, ratio, and magnitude were first amalgamated under the concept of positive real number and as this new number concept was eventually extended to the system of complex numbers.

It is hard for us to get used to such a line of thought. We do distinguish between natural numbers, integers, rationals, reals and complex numbers, but what is important for us is the fact that they all belong to the same species—in other words, that they are all elements of the same set, the set of numbers. What was important to the Greeks was the fact that they are all different; thus Euclid proves many theorems, such as the distributive law, twice: once for numbers (in Book VII) and once for magnitudes (in Book II). It is fascinating to watch Euclid do algebra geometrically. To give you a flavor of this modus operandi, I shall first quote Euclid II.6, a proposition which is needed in the proof of Lemma 1 to Proposition X.29.

5. Euclid Excerpt (II.6)

PROPOSITION II.6
If a straight line be bisected and a straight line be added to it in a straight line, the rectangle contained by the whole with the added straight line and the added straight line together with the square on the half is equal to the square on the straight line made up of the half and the added straight line.

For let a straight line AB be bisected at the point C, and let a straight line BD be added to it in a straight line;
    I say that the rectangle contained by AD,DB together with the square on CB is equal to the square on CD.

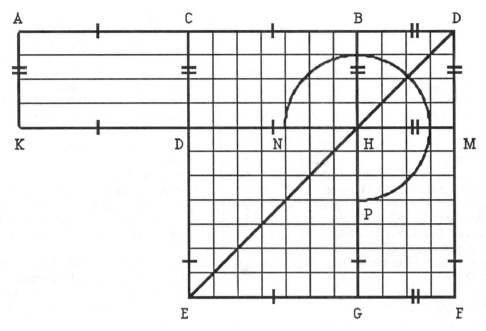

Figure 4.

For let the square CEFD be described on CD, [I.46] and let DE be joined; through the point B let BG be drawn parallel to either EC or DF, through the point H let KM be drawn parallel to either AB or EF, and further through A let AK be drawn parallel to either CL or DM.   [I.31]
Then, since AC is equal to CB, AL is also equal to CH,   [I.36]
But CH is equal to HF.   [1.43]
Therefore AL is also equal to HF.

Let CM be added to each; therefore the whole AM is equal to the gnomon NOP. But AM is the rectangle AD, DB, for DM is equal to DB; therefore the gnomon NOP is also equal to the rectangle AD, DB. Let LG, which is equal to the square on BC, be added to each; therefore the rectangle contained by AD, DB together with the square on CB is equal to the gnomon NOP and LG. But the gnomon NOP and LG are the whole square CEFD, which is described on CD; therefore the rectangle contained by AD, DB together with the square on CB is equal to the square on CD.
Therefore etc.   Q. E. D.

6.   Euclid II.6. Discussion.

What does this mean? Before you read any further, review the statement of the proposition and see whether you can figure out which algebraic identity is expressed here in geometric terms.

I will follow the Greek custom of thinking of a line segment as a magnitude. Thus I will write AB = a when I really mean: "the length of the segment AB is

equal to a." Neither will you be confused if you think of the expression AB = CD as an abbreviation of the more correct m(AB) = m(CD). (The measure of AB is equal to the measure of CD.) I attended high school in the days prior to the introduction of the "New Math," and so this (wrong?) terminology seems natural to me.

So let AB = 2a and BD = b. Then AC = CB = a, "the rectangle contained by the whole with the added straight line and the added straight line is" [in other words has area] $(2a + b)b$, "the square on the half is" [has area] a^2 and so "the rectangle contained by the whole with the added straight line and the added straight line together with the square on the half" is [has area] $(2a + b)b + a^2$. This expression is equal to (the area of) "the square on the straight line made up of the half and the added straight line", in other words $(a + b)^2$. So this whole complicated statement expresses the trivial fact that
$$(2a + b)b + a^2 = (a + b)^2.$$

Now the geometric formulation only seems complicated to us because we are not used to it. I don't think that Euclid's students at the Museum (University) of Alexandria, Egypt, found this theorem as difficult as we do. A look at Figure 4 would have convinced these students: the theorem merely says that the area of the T-shaped region ADMHGELK is equal to that of the square region CDFE. And that follows from the fact that ACLK and HMFG are congruent rectangles.

Now let us examine Euclid's proof. As in all his theorems, Euclid goes from the general to the particular: he names the given straight lines AB and BD, constructs the bisector C of AB, and rephrases the theorem in terms of this data. The general statement follows from the particular one, because the particular proof can be repeated for any other special case.

Euclid then constructs Figure 4 one step at a time, referring to Propositions in Book I to establish the fact that each construction is permissible. The next few sentences may be a bit confusing. Read them as follows:

Since the segment AC = the segment CB, the (area of) rectangle A(C)L(K) = the (area of) rectangle C(B)H(L) = the (area of) rectangle H(M)F(G). Let the (area of) rectangle C(D)M(L) be added to each, then the (area of) rectangle A(D)M(K) = [AD][DB] = (the area of) the L-shaped figure CDFGHL (gnomon NOP) and so (the area of) the L-shaped figure CDFGHL (gnomon NOP) is equal to (the area of) the rectangle contained by the segments AD and DB.

Now let the rectangle L(H)G(E) whose area is [BC]^2 be added to each (side of this equation), then the (area of) the rectangle contained by AD and DB plus (that of) the square on CD is equal in area to the gnomon NOP plus the rectangle L(H)G(E) which actually *is* the square C(D)F(E);

> "... therefore the rectangle contained by AD, BD together with the square on CB is equal to the square on CD.
> Therefore etc.   Q. E. D."

The "etc" stands for the repetition of the original statement of the Proposition.

Note again the way Euclid uses the word "equal." Two line segments are equal if they have the same length; that makes them congruent also. Two plane figures need not be of the same kind to be equal; all that matters is that they have the same area. Thus a rectangle can be equal to a gnomon.

7. Euclid Lemma 1 to X.29 Excerpt.

We are now ready to study Lemma 1 to Proposition X.29.
        LEMMA 1.

To find two square numbers such that their sum is also a square.

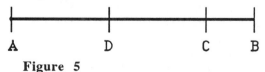

Figure 5

        Let two numbers AB, BC be set out, and let them be either both even or both odd.
        Then since, whether an even number is subtracted from an even number, or an odd number

from an odd number, the remainder is even, [IX. 24, 26] therefore the remainder AC is even.
        Let AC be bisected at D.
        Let AB, BC also be either similar plane numbers, or square numbers, which are themselves also similar plane numbers.
        Now the product of AB, BC together with the square on CD is equal to the square on BD [II.6]
        And the product of AB, BC is square, inasmuch as it was proved that, if two similar plane numbers by multiplying one another make some number, the product is square. [IX.1]
        Therefore two square numbers, the product of AB, BC, and the square on CD, have been found which, when added together, make the square on BD.
        And it is manifest that two square numbers, the square on BD and the square on CD, have again been found such that their difference, the product of AB, BC, is a square, whenever AB, BC are similar plane numbers.
        But when they are not similar plane numbers, two square numbers, the square on BD and the square on DC, have been found such that their difference, the product of AB, BC, is not a square.   Q. E. D.

8. Euclid Lemma 1 to X.29 Discussion.

To find a Pythagorean triple Euclid proceeds as follows: He chooses two similar plane numbers of the same parity AB and BC. Similar plane numbers are defined in Definitions 20 and 21 of Book VII.

> *Definition VII.20*: Numbers are proportional when the first is the same multiple, or the same part, or the same parts of the second as the third is of the fourth.
> *Definition VII.21*: Similar plane and solid numbers are those which have their sides proportional.

After giving a relatively unsatisfactory definition of ratio, Euclid tells us exactly what the expression a:b::c:d means; he defines the concept "equality of ratios" precisely. Recall that a, b, c and d are natural numbers. The natural number a is a *multiple* of b if there exists a natural number k such that a = kb; it is a *part* of b, or equivalently, *measures* b if b is a multiple of a. Thus if a = kb (respectively b = ka), then a,b,c and d are proportional if c = kd (respectively, d = kc). Finally, the number a is *parts* of the number b if neither measures the other. Since a and b are natural numbers, they are *commensurable*; that is, some part of a measures b. Specifically, the unit u can be taken as that part of a that measures b. Thus the expression "a is the same parts of b as c is of d" means that the ratio of the number of times u goes into a to the number of times u goes into b is equal to the ratio of the number of times u goes into c to the number of times u goes into d. Hence in all three cases considered here, we find that Euclid's definition of proportionality is equivalent to the one we are used to, namely: a, b, c and d are proportional if $a/b = c/d$.

Definition VII.21 expresses in arithmetic terms the geometric idea that similar rectangles and rectangular parallelpipeds have the same shape by virtue of the fact that their sides are proportional. Plane numbers are numbers that are products of two factors. I was going to add "neither of which is 1", but there is nothing to keep us from constructing a rectangle of sides 3 and 1, and hence to think of 3 as a plane number. No wonder

---

This schism between pure and applied mathematics is detrimental to the health of our subject; progress is achieved when all areas of mathematics are allowed to cross-fertilize each other.

---

the Greek concepts of plane and solid numbers did not survive to modern times. Solid numbers are products of three factors. The simplest and most general way to construct two similar plane numbers ab and cd is to choose arbitrary numbers m, n, p and q and to set a = mp, b = np, c = mq and d = nq. Then $a/c = p/q = b/d$ and so $ab = mnp^2$ and $cd = mnq^2$ are similar plane numbers. Notice (1) that two square numbers are automatically similar (take m = n and p = q = 1) and (2) that the product of two plane numbers is a square if and only if they are similar. Euclid takes the first statement for granted; he proves the second in Propositions IX.1 and IX.2.

> *Proposition IX.1: If two similar plane numbers by multiplying one another make some number, the product will be square.*

Modern Proof: If a = mp, b = np, c = mq and d = nq then ab and cd are similar plane numbers and $(ab)(cd) = m^2n^2p^2q^2$.

And conversely:

> *Proposition IX.2: If two numbers by multiplying one another make a square number they are similar plane numbers.*

Euclid's proof modernized: Let xy be a square. Then $z = \sqrt{xy}$ is a number and $x/z = z/y$. Let a/b be the fraction

$x/z = z/y$ in lowest terms, that is let a and b be relatively prime and let $a/b = x/z = z/y$. Then there exist numbers m and n such that x = ma, z = mb and z = na and y = nb. Hence x and y are plane numbers. Moreover $a/b = m/n$ and so x and y are similar.

With these preliminaries understood, Euclid proceeds with the proof of his Lemma. Having chosen two similar plane numbers AC and BC of the same parity, he arranges them as in Figure 5.

Now AC = AB - BC, and since AB and BC are either both even or both odd, AD = AC/2 is a number. By Proposition II.6 $(AB)(CD) + (DC)^2 = (DB)^2$, and since (AB)(CD), being the product of two similar plane numbers, is a square, we have found two square numbers whose sum is a square.

Further, if AB and BC are similar plane numbers, then squares have been found whose difference is a square, but, if AB and BC are not similar plane numbers, then squares have been found whose difference is not a square.

Algebraic interpretation: Let $x = mnp^2$, $y = mnq^2$ be of the same parity so that (x-y)/2 and (x+y)/2 are (whole) numbers. Then if $a = \sqrt{xy} = mnpq$, b = (x - y)/2, and c = (x + y)/2, (a,b,c) is a Pythagorean triple. If x and y are not similar plane numbers, then xy is not a square and this construction does not work. Thus, even though Euclid neither proves nor asserts this, it can easily be shown that his method generates all Pythagorean triples. Euclid could have simplified his construction by working only with squares of the same parity rather than with similar plane numbers.

In spite of the geometric language, we have really been doing arithmetic and algebra. This is an important lesson we learn from reading Euclid. The *Elements* is not a geometry text; it is a text that covers every aspect of elementary mathematics known in Euclid's day. Well into the modern era the title "geometer" was bestowed not just on experts in geometry but on all theoretical mathematicians and even on physicists—people who derived abstract results from basic principles. Those who computed numerical answers to practical problems were sometimes called abacists and sometimes algorizmists. Even today, in certain high schools geometry is thought of as the class in which one proves things, while algebra is the class where one calculates. Graduates of such schools receive the shock of their lives in college when they are asked to use the field axioms to prove that (-1)(-1) = 1. This schism between pure and applied mathematics is detrimental to the health of our subject; progress is achieved when all areas of mathematics are allowed to cross-fertilize each other.

9. Diophantus.

The second mathematician whose work is featured in this triptych is Diophantus of Alexandria. What little we know of his life is told in Heath (1964). Diophantus has been called the "father of algebra" even though his most famous work, the *Arithmetica*, is neither an algebra nor an arithmetic text. This book is quite different from the other classic Greek texts, and, for that matter, from classic Greek mathematics. Boyer (1989) says: "It represents a new branch and makes use of a different approach. Being divorced from the geometrical approach it resembles Babylonian mathematics ... (but it) is almost entirely devoted to the exact solution of equations both determinate and indeterminate."

The *Arithmetica* consists of a number of problems which fall under the category of number theory even though they differ from number theory in one important respect. Number theory is the study of integers; a classical number theoretical problem, which is an extension of ours and which appears in the *Arithmetica* (Book IV, Sesiano 1982, pg. 89) is the following: "Find *integers* x, y and z such that $x^2 + y^2 = z^3$." The *Arithmetica* contains many similar problems, except that in most cases the solutions sought are rational numbers.

The reason that Diophantus is called the "father of algebra" is that he is the first mathematician we know of who used a symbol to represent the unknown in an equation, just as we use the letter x. (He also introduced other notation which eventually developed into our modern algebraic notation.) Diophantus did not use a letter of the alphabet; instead he used a symbol of his own invention that looked much like the

---

Just as the early Greek playwrights allowed only one character to be on stage at a time, Diophantus allowed only one unknown quantity in his problems at one time.

Greek letter s. This symbol was later referred to as "thing", and later writers used the word *kos* (i.e. "thing") to represent the unknown in an equation. During the Middle Ages algebra was often called "the art of kos."

Just as the early Greek playwrights allowed only one character to be on stage at a time, Diophantus allowed only one unknown quantity in his problems at one

time. After all he had only one symbol for the unknown. Unlike Dr. Seuss's charming Cat in the Hat, Diophantus did not have a "thing-one" and a "thing-two" to help him with his chores.

The *Arithmetica* was not intended as a text as were the other two works in our triptych. The problems seem independent of each other, the method of solution arbitrary, and no attempt is made to find all solutions of indeterminate problems. But enough material was provided so that great mathematicians like Fermat were able to find general solutions. Lesser mathematicians had more difficulty with the *Arithmetica*. One medieval manuscript is said to have the following marginal note written by the copyist: "Thy soul, O Diophantos, to Satanas for the difficulty of thy problems and this one in particular."

The second of our three scenes is Problem 8 in Book II. I have translated Arthur Czwalina's German version into English (Diophantos 1952) because, for reasons of clarity, Heath's English translation contains two different symbols for two different unknowns. Even though Heath explains in a footnote that Diophantus did not use a second symbol, the German translation seems to me to be closer in spirit to Diophantus' work.

10. Diophantus Excerpt Problem 8 Book II.

TO EXPRESS A GIVEN SQUARE AS THE SUM OF TWO SQUARES.

Let the given square be 16. To express it as the sum of two squares.
Let the first term be x^2, then the second is $16 - x^2$. This term should be a square. I form the square of the difference of an arbitrary multiple of x and the side of 16 which is 4. For example I form the square of 2x - 4. This is $4x^2 - 16x + 16$. This expression I set equal to $16 - x^2$. I add x^2 to both sides and subtract 16. Thus I obtain $5x^2 = 16x$ or x = 16/5.

The first square is thus 256/25, the other 144/25. The sum of these numbers is 16, and each summand is a square.

11. Diophantus Discussion.

Note that the solution to this problem is in rationals and not in integers. Under these conditions any square can be expressed, in many ways, as the sum of two squares.

Note also the constraints under which Diophantus is

working by allowing himself to use symbols for only a single unknown. A modern mathematician would first replace 16 by z^2, and then 2 (the arbitrary multiple of x) by m, as Heath does. The problem and solution would then read as follows:

TO EXPRESS A GIVEN SQUARE AS THE SUM OF TWO SQUARES.

Let the given square be z^2. To express it as the sum of two squares: Let the first term be x^2; then the second is $z^2 - x^2$. This term should be a square. I form the square of the difference of an arbitrary multiple of x and the side of z^2 which is z. (For example) I form the square of mx - z. This is $m^2x^2 - 2mxz + z^2$. This expression I set equal to $z^2 - x^2$. Solving the equation:
$$(mx - z)^2 = z^2 - x^2$$
I obtain:
$$x = 2mz/(m^2 + 1), \quad y = (m^2 - 1)z/(m^2 + 1)$$
from which $x^2 + y^2 = z^2$ follows easily.

Thus, given z, Diophantus finds the "rational" Pythagorean triple $(x,y,z) = (2mz/(m^2 + 1), (m^2 - 1)z/(m^2 + 1), z)$. Clearing fractions, we get the Pythagorean triple $(2mz, (m^2-1)z, (m^2+1)z)$. If m is odd, the greatest common divisor of $2mz$, $(m^2 - 1)z$ and $(m^2 + 1)z$ is 2z, else it is z. Dividing these numbers by 2z or z as needed to get relatively prime integers, we get Pythagoras's original solution, a primitive Pythagorean triple like (4,3,5) or (12, 5, 13) subject to the condition that z = x + 1. Diophantus's algorithm finds all such Pythagorean triples.

12. Fermat.

The *Arithmetica* seems to attract famous comments to its margins. It was in the margin of his copy of the *Arithmetica*, on the page containing Problem 8 of Book II, that Fermat wrote:

> "On the other hand, it is impossible for a cube to be written as a sum of two cubes or a fourth power to be written as the sum of two fourth powers or, in general, for any number which is a power greater than the second to be written as a sum of two like powers. I have a truly marvelous demonstration of this proposition which this margin is too narrow to contain." (Edwards, 1977, pg 2)

In other words, if n > 2, then the equation $x^n + y^n = z^n$ has no solution in natural numbers.

Fermat's conjecture is one of the great open problems in mathematics. Clearly the study of Pythagorean triples led Fermat to consider this obvious generalization which then led him to make his conjecture. Even though all published material shows only how to generate Pythagorean triples, it seems

obvious from Fermat's other writings that he was aware of the truth of Propositions 1, 2, and 3 stated at the beginning of this paper. Fermat published one proof in his lifetime, a proof of the theorem that:

> The area of a right triangle (with integer sides) cannot be a square. (Edwards, 1977)

In the proof of this theorem Fermat uses Proposition 3. It is worth noting that a proof of Fermat's conjecture for n = 4 can be obtained easily from this theorem.

I mention Fermat because Fermat was the bridge between Diophantus and Euler, the last, but by no means least, mathematician represented in our triptych. Fermat was a lawyer and later a councillor who lived a quiet life in his home town of Toulouse during the years 1601 - 1665. Mathematics was his hobby. In 1621 Claude Gaspard de Bachet published Diophantus's *Arithmetica*, and Fermat studied it with great interest, making copious notes in the margins, among them the one quoted above. Most of these notes were assertions of great importance and amazing insight for which Fermat never published a proof. After Fermat's death, his son published a new edition of the *Arithmetica* and included his father's marginal notes. This book aroused little interest; mathematicians of the period were too busy investigating the implications of the newest mathematical theory: the calculus. It was not until he was already a well-established mathematician (some 60 years after Fermat's death) that Euler (1707 - 1783) took up the work where Fermat left off.

13. Euler.

Euler never published a book on number theory; his main achievements in that area were published posthumously under the title *Tractatus de doctrina numerorum* in 1849, ninety-nine years after they were ready for publication. During Euler's lifetime most of his number theoretic results were made known to his contemporaries through personal correspondence. His most important correspondence was with his friend Goldbach (who conjectured that every even number greater than 2 is the sum of two primes). Goldbach, who, according to a contemporary, "loved numbers", had not read Fermat, but he knew of him and was familiar with some of his conjectures which had become part of mathematical folklore.

Towards the end of 1729 Goldbach mentioned the following conjecture of Fermat's to Euler: "Every number of the form $2^{2^n} + 1$ is a prime." Euler disproved this conjecture by showing that:
$$2^{2^5} + 1 = 2^{32} + 1 \qquad = 4,294,967,297$$
$$= 641 \times 6,700,417.$$
But this was the only one of Fermat's conjectures

which turned out to be false. Euler then set himself the task of proving Fermat's remaining conjectures, and succeeded with all except the one we quoted above and which later came to be known as Fermat's Last Theorem. [Weil (1983)].

In 1770 there appeared Euler's *Vollstaendige Einfuehrung zur Algebra*, a book that appeared in English translation as *Elements of Algebra* around 1820; the fifth edition appearing in 1840. This edition was recently reissued by Springer-Verlag and is referred

---

Reading Euler is a special experience; besides being mathematical treasures his works are entertaining and suspenseful.

---

to here as Euler (1984). Euler's *Elements* is not just a mathematical gem; it is a monument to Euler's indomitable spirit. Euler lost the sight of his right eye when he was in his early thirties. Some thirty years later he developed glaucoma in his left eye and gradually became totally blind. But Euler did not allow this misfortune to keep him from his work; rather he prepared to cope with his handicap, which meant a life where formulas would have to be read to him and where his own works would have to be dictated. As part of necessary mental exercises he dictated this algebra text to one of his servants. Several sources reveal that this man had been a tailor's apprentice who entered Euler's employ while the latter lived in Berlin and then followed his master to Saint Petersburg when Euler reassumed his former position in the Russian capital. It is ironic that none of these sources reveal the secretary's name.

Euler begins his text by explaining the meaning of the symbols +, -, and x; in the later chapters he almost proves Fermat's Last Theorem for the case $n = 3$ using techniques which he had recently discovered and which are based on the arithmetic of complex numbers. This proof contains one unproved statement, which, though correct, is not obvious. Before tackling this last problem, Euler applies his newly discovered techniques to Pythagorean triples. It is this passage which we will study in detail.

Euler generalizes the problem somewhat; instead of studying the conditions under which an expression such as $x^2 + y^2$ is a square, he devotes Chapter XII of the second part of his book to the topic:

Of the Transformation of the Formula $ax^2 + cy^2$
into Squares and higher Powers.

Euler then refers to previous work where he has shown:

that it is frequently impossible to reduce numbers of the form $ax^2 + cy^2$ to squares; but whenever it is possible, we may transform the formula into another, in which $a = 1$.

We now quote paragraphs 182 to 185 from Euler (1984).

14. Euler Excerpt.

182. Let, therefore, the formula $x^2 + cy^2$ be proposed, and let it be required to make it a square. As it is composed of the factors $(x + y\sqrt{-c}) * (x - y\sqrt{-c})$, these factors must either be squares, or squares multiplied by the same number. For, if the product of two numbers, for example, pq, must be a square, we must have $p = r^2$, and $q = s^2$; that is to say, each factor is of itself a square; or $p = mr^2$, and $q = ms^2$; and therefore these factors are squares multiplied both by the same number. For which reason, let us make $x + y\sqrt{-c} = m(p + q\sqrt{-c})^2$; it will follow that $x - y\sqrt{-c} = m(p - q\sqrt{-c})^2$, and we shall have $x^2 + cy^2 = m^2(p^2 + cq^2)^2$, which is a square. Farther, in order to determine x and y, we have the equations:

$x + y\sqrt{-c} = mp^2 + 2mpq\sqrt{-c} - mcq^2$, and

$x - y\sqrt{-c} = mp^2 - 2mpq\sqrt{-c} - mcq^2$; in which x is necessarily equal to the rational part, and $y\sqrt{-c}$ to the irrational part; so that $x = mp^2 - mcq^2$, and

$y\sqrt{-c} = 2mpq\sqrt{-c}$, or $y = 2mpq$; and these are the values of x and y that will transform the expression $x^2 + cy^2$ into a square, $m^2(p^2 + cq^2)^2$ the root of which is $mp^2 + mcq^2$.

183. If the numbers x and y have not a common divisor, we must make $m = 1$. Then, in order that $x^2 + cy^2$ may become a square, it will be sufficient to make $x = p^2 - cq^2$ and $y = 2pq$, which will render the formula equal to the square $(p^2 + cq^2)^2$.

Or, instead of making $x = p^2 - cq^2$ we may also suppose $x = cq^2 - p^2$, since the square x^2 is still left the same.

Besides, the same formula having been already found by methods altogether different, there can be no doubt with regard to the accuracy of the method which we have now employed. In fact, if we wish to make $x^2 + cy^2$ a square, we suppose, by the former method, the root to be $x + (py)/q$, and find $x^2 + cy^2 = x^2 + (2pxy)/q + (p^2y^2)/q^2$. Expunge the x^2, divide the other terms by y, multiply by q^2 and we shall have $cq^2y = 2pqx + p^2y$; or $cq^2y - p^2y = 2pqx$. Lastly, dividing by 2pq, and also by y, there results $(x/y) = (cq^2 - p^2)/2pq$. Now, as x and y, as well as p and q, are to have no common divisor, we must make x equal to the numerator, and y equal to the denominator, and hence we shall obtain the same results as we have already found, namely, $x = cq^2 - p^2$, and $y = 2pq$.

184. This solution will hold good, whether the number c be positive or negative; but, farther, if this number itself had factors, as, for instance, the formula $x^2 + acy^2$, we should not only have the preceding solution, which gives $x = acq^2 - p^2$, and $y = 2pq$, but this also, namely, $x = cq^2 - ap^2$, and $y = 2pq$; for, in this last case, we have, as in the other,
$x^2 + acq^2 = c^2q^4 + 2acp^2q^2 + a^2p^4 = (cq^2 + ap^2)^2$;
which takes place also when we make $x = ap^2 - cq^2$, because the square x^2 remains the same.

    This new solution is also obtained from the last method, in the following manner: If we make
$x + y\sqrt{-ac} = (p\sqrt{-a} + q\sqrt{-c})^2$, and
$x - y\sqrt{-ac} = (p\sqrt{-a} - q\sqrt{-c})^2$,
we shall have $x^2 + acy^2 = (ap^2 + cq^2)^2$, and, consequently, equal to a square. Further, because
$x + y\sqrt{-ac} = ap^2 + 2pq\sqrt{-ac} - cq^2$, and
$x - y\sqrt{-ac} = ap^2 - 2pq\sqrt{-ac} - cq^2$, we find
$x = ap^2 - cq^2$ and $y = 2pq$.
    It is farther evident, that if the number ac be resolvible into two factors, in a greater number of ways, we may also find a greater number of solutions.

185. Let us illustrate this by means of some determinate formula; and, first, if the formula $x^2 + y^2$ must become a square, we have ac = 1; so that $x = p^2 - q^2$, and $y = 2pq$; whence it follows that $x^2 + y^2 = (p^2 + q^2)^2$.
    If we would have $x^2 + y^2 = '$; we have ac = -1; so that we shall take $x = p^2 + q^2$, and $y = 2pq$, and there will result $x^2 - y^2 = (p^2 - q^2)^2 = '$.
    If we would have the formula $x^2 + 2y^2 = '$, we have ac = 2; let us therefore take $x = p^2 - 2q^2$ or $x = 2p^2 - q^2$, and $y = 2pq$, and we shall have $x^2 + 2y^2 = (p^2 + q^2)^2$ or $x^2 + 2y^2 = (2p^2 + q^2)^2$.
    If, in the fourth place, we would have $x^2 - 2y^2 = '$, in which ac = -2, we shall have $x = p^2 + 2q^2$, and $y = 2pq$; therefore $x^2 - 2y^2 = (p^2 - 2q^2)^2$.
    Lastly, let us make $x^2 + 6y^2 = '$. Here we shall have ac = 6; and, consequently, either a = 1, and c = 6, or a = 2, and c = 3. In the first case, $x = p^2 - 6q^2$, and $y = 2pq$; so that $x^2 + 6y^2 = (p^2 + 6q^2)^2$; in the second, $x = 2p^2 - 3q^2$, and $y = 2pq$; whence $x^2 + 6y^2 = (2p^2 + 3q^2)^2$

15. Euler's Possible Intentions.

Let us examine two interpretations of this selection. Consider first a statement I'll call "Theorem" EF:

"THEOREM" EF: If c is a prime or ±1, then the expression $x^2 + cy^2$ is a square if and only if there exist integers m, p and q such that $x = m(p^2 - cq^2)$ and $y = 2mpq$. If x and y are relatively prime, then m = 1.

Unfortunately the "necessary" part of this "theorem" is false. Let c = 5, x = 2 and y = 3. Then $x^2 + cy^2 = 2^2 + (5)3^2 = 49 = 7^2$, and, since 2 and 3 are relatively prime, m = 1. But it is quite clear that the equations:
$$p^2 - 5q^2 = 2 \qquad 2pq = 3$$
have no solution in integers (or even rational solutions for that matter).

Let us give the "sufficient" part of "Theorem" EF the name Theorem E:

THEOREM E: The expression $x^2 + cy^2$ is a square if there exist integers m, p and q such that $x = m(p^2 - cq^2)$ and $y = 2mpq$. If x and y are relatively prime, then m = 1.

It is a trivial matter to check that this less powerful theorem is true. The question then becomes: Did Euler think that "Theorem" EF is true, or did he only mean to assert Theorem E?

This question is hard to answer. I will not answer it. Instead I will list reasons on both sides of the question, and let you make up your mind one way or the other or, if you prefer, let you suspend judgment.

16. The Two Alternatives.

CASE E: EULER ASSERTS ONLY THAT THEOREM E IS TRUE.

1. Euler never states "Theorem" EF as explicitly as I have done. The argument leading to the statements made at the end of paragraph 182 are heuristic arguments; Euler is merely trying to expose for the benefit of the reader the thought processes that led him to the given conclusion. Such a generous sharing of insights is very typical of Euler. If Emil Fellman (1983) will allow me to translate freely a short passage from his biographical sketch of Euler:

> Euler was very modest; he never engaged in priority arguments. On the contrary, he frequently and most generously gave to others new discoveries he had made and insights he had gained. In his publications he never hides his techniques or his original train of thought, but by sharing all his

insights with the reader he often leads her to the very brink of something new and then leaves her the joy of discovery. Reading Euler is a special experience; besides being mathematical treasures his works are entertaining and suspenseful.

2. Euler is too great a mathematician to confuse a theorem with its converse. And besides, in the examples he gives in paragraph 185, he never asserts that the solutions he finds are the only ones.

3. Euler seems to be aware of the fact that when c = -5, his new techniques might get him into trouble. In Paragraph 195 he analyzes the following:

> QUESTION 4: Required squares in integer numbers, the double of which, diminished by 5, may be a cube; or it is required that $2x^2 - 5$ be a cube.
>
> If we begin by seeking the satisfactory cases for the formula $2x^2 - 5y^2$, we have, in the 188th Article, a = 2 and c = -5; whence $x = 2p^3 + 15pq^2$ and $y = 6p^2q + 5q^3$: so that, in this case, we must have y = ± 1; consequently: $6p^2q + 5q^3 = q(6p^2 + 5q^2) = ± 1$; and as this cannot be, either in integer numbers, or even in fractions, the case becomes very remarkable, because there is, notwithstanding, a satisfactory value of x; namely x = 4; which gives $2x^2 - 5 = 27$, or equal to the cube of 3. It will be of importance to investigate the cause of this peculiarity.

In paragraph 188 Euler had used arguments similar to those in paragraph 182 to derive (necessary and?) sufficient conditions for an expression of the form $ax^2 + cy^2$ to be a cube so that this passage indicates that Euler has discovered that his conditions are sufficient but not necessarily necessary.

CASE EF: EULER THOUGHT "THEOREM" EF IS TRUE.

1. Euler's argument is the classic argument used in solving equations: "If equation EQ has a solution then, according to Algorithm A which I use to solve EQ, that solution must belong to the set S. By direct substitution I find that the subset T of S contains all solutions of EQ. Hence x is a solution of EQ if and only if x belongs to T."

2. "Theorem" EF is true and Euler's proof is valid for several values of c. There are exactly nine positive values of c for which the theorem holds; these are c = 1, 2, 3, 7, 11, 19, 43, 67 and 163. There are exactly 38 values of c in the range -100 < c ≤ -2 for which Euler's proof holds; it is still an open question whether Euler's proof holds for an infinite number of negative values of c. (Stark 1978.) Thus, as noted at the beginning of

this essay, "Theorem" EF is valid for Pythagorean triples. Knowing this, and knowing also, as a result of extensive preliminary calculations, that "Theorem" EF holds for many values of c, Euler must have assumed that "Theorem" EF holds for all values of c; else he would not have brought to bear such beautiful and sophisticated new techniques to this problem.

3. Consider the following Theorem which is obvious when p and q are natural numbers and which we have already encountered as Euclid's Proposition IX.2.

THEOREM I. If the product pq is a square then either p an q are both squares or $p = mr^2$ and $q = ms^2$, that is p and q are squares multiplied by the same number.

Euler applies this theorem to the case when $p = x + y\sqrt{-c}$ and $q = x - y\sqrt{-c}$. Without being aware of it, Euler proved that, if such an extension of Theorem I holds for some c, then "Theorem" EF holds for that c. Such an extension of Theorem I is valid for many values of c, including the ones listed in #2 above, but it is not valid for other values of c such as c = 5. Apparently Euler was not aware of this fact.

4. Some scholars [see e.g. Edwards (1977)] suggest that Euler believed that the situation he describes in paragraph 195 is a special aberration brought about by the minus sign and that further study will enable mathematicians to fix any minor flaws in his argument and vindicate his method.

5. Euler believed that "Theorem" EF is true, but was not quite convinced of the validity of his method of proof. Hence in the second half of paragraph 183 he gives an alternate proof. Unfortunately there is an error in that proof. Euler says: "... there results $x/y = (cq^2 - p^2)/2pq$. Now, as x and y, as well as p and q are to have no common divisor, we must make x equal to the numerator, and y equal to the denominator, and hence we shall obtain the same results as we have already found, namely $x = cq^2 - p^2$, and y = 2pq." But the fact that p and q are relatively prime does not imply that $cq^2 - p^2$ and 2pq are relatively prime. (Try c = 6, p = 2 and q = 3: then $cq^2 - p^2 = 50$ and 2pq = 12.) Hence Euler's conclusion does not follow.

Even for c = 1 Euler's argument does not hold. If p and q are relatively prime then the greatest common divisor of 2pq and $q^2 - p^2$ is 1 if p and q are of opposite parity and 2 if they are both odd. Euler assumes that x and y are relatively prime and then sets z = x + (p/q)y. He then has the right to assume that p and q are relatively prime, but he does not have the right to assume that one is even and one is odd. Thus he cannot conclude that $x/y = (q^2 - p^2)/2pq$ implies that $x = q^2 - p^2$ and y = 2pq. The terms of the triple $(2pq, q^2 - p^2, q^2 + p^2)$, have greatest common divisor 2, hence the triple is not primitive; dividing all terms by 2 yields a primitive triple which is not of the form Euler predicts.

6. Euler looks at special cases which will make $x^2 + cy^2$ a square. Specifically he looks at the following values of c: 1, -1, 2, -2 and 6. I mentioned earlier that his proof holds when c = 1, 2 and -2; the case c = -1 is just a rewrite of the case c = 1. When c = 6 the situation is more complicated. "Theorem" EF does not apply in this case because c is neither +1 nor a prime, but Euler's argument for composite coefficients fails at the point where he asserts that if the coefficient is the product of two constants a and c, then $x + y\sqrt{-c}$ can be expressed either as a square of the form $(p\sqrt{-a} + q\sqrt{-c})^2$ or as a square of the form: $(p + q\sqrt{-c})^2$. Nevertheless an extensive computer search failed to reveal a counterexample to Euler's assertion that if $x^2 + 6y^2$ is a square, then there exist integers p and q such that either $\pm x = p^2 - 6q^2$ and $y = 2pq$ or $\pm x = 2p^2 - 3q^2$ and $y = 2pq$. But if Euler had tried to illustrate his method by "solving the formula" $x^2 + 15y^2 = q$, he would have missed the solution x = 1, y = 1, because the equation 2pq = 1 has no solutions in integers. If Euler knew that in his own examples he had found all values of x and y that make the corresponding "formulas" squares, he might easily but mistakenly have concluded that his method yields all solutions in all cases.

17. Significance and Implications of Theorem I.

In order to find a way out of this dilemma we will study Theorem I in some detail. We will investigate why it holds for the integers and to what other integral domains it can be generalized. The proof of Theorem I is a very simple application of the Fundamental Theorem of Arithmetic. Now let d be a square free integer, that is one whose factorization into primes consists of distinct primes. (In terms of Euler's notation, which we have used up to now, d = -c.) Let Ω be the field of rational numbers. We define:

$$\Omega(\sqrt{d}) = (a + b\sqrt{d} : a, b \text{ in } \Omega)$$

the field of rationals with $\sqrt{d}$ adjoined. We define $G(\sqrt{d})$ to be the domain of integers in $\Omega(\sqrt{d})$. In case d = -1, $G(\sqrt{-1})$ = G is the domain of Gaussian integers, i.e. the set (a + bi: a, b integers). For other values of d, $G(\sqrt{d})$ may be more complicated. Theorem I can be extended to a domain $G(\sqrt{d})$ if and only if an analogue of the Fundamental Theorem of Arithmetic holds in $G(\sqrt{d})$. Euler may not have known this. After all, why should it not be possible to construct a proof of Theorem I which does not rely on the Fundamental Theorem of Arithmetic? It seems that Euclid has provided such a proof in his proof of Proposition IX.2.

Euclid never states the Fundamental Theorem of Arithmetic as explicitly as we do when we teach it to our students, but Propositions VII.30, VII.31 and VII.32 together do imply that theorem. Thus it happens that the assumption of the Fundamental Theorem of Arithmetic is well hidden in the proof of Proposition IX.2, and it is quite possible that Euler never studied that proof closely enough to catch that assumption. To see that the Fundamental Theorem of Arithmetic is indeed essential in the Proof of Theorem I consider the domain E of even integers, a domain in which the Fundamental Theorem does not hold. In E, 144 (= 12 x 12) is a square, but 144 = 18 x 8 also and 18 and 8 are not similar plane numbers. As a matter of fact, since 1 is not an element of E and 18 is a prime in E, 18 is not even a plane number in E.

The Fundamental Theorem of Arithmetic holds for the Gaussian integers (Pollard 1950), and therefore so do Theorem I and "Theorem" EF. As mentioned earlier, the Fundamental Theorem of Arithmetic holds for exactly nine negative values of d, namely if d belongs to the set { -1, -2, -3, -7, -11, -19, -43, -67, -163} and for possibly an infinite number of positive values of d. The Fundamental Theorem of Arithmetic does not hold in $G(\sqrt{-5})$. For example 3, 7, $1 + 2\sqrt{-5}$ and $1 - 2\sqrt{-5}$ are all distinct primes in $G(\sqrt{-5})$ (Pollard 1950) and $21 = (3)(7) = (1 + 2\sqrt{-5})(1 - 2\sqrt{-5})$. Theorem I does not hold in $G(\sqrt{-5})$, and neither does "Theorem" EF; specifically the converse of Theorem E does not hold.

18. Euler and Pythagorean Triples.

Let us now apply Euler's techniques to prove Propositions 2 and 3. (Proposition 1 is simple enough to be left as an exercise for the reader.) Clearly Propositions 2 and 3 are equivalent to the statement that "Theorem" EF holds in G, the domain of the Gaussian integers. We need two lemmas.

Lemma 1: In G, the domain of Gaussian integers, the Fundamental Theorem of Arithmetic and Theorem I both hold.

A proof of the validity of the Fundamental Theorem of Arithmetic in G can be found in Pollard (1950). It is an elementary exercise to show that Theorem I holds in any integral domain in which the Fundamental Theorem of Arithmetic holds.

With these preliminaries out of the way, Euler's argument becomes flawless. Here it is:

THEOREM: (x,y,z) is a primitive Pythagorean triple if and only if there exist relatively prime integers p and q of opposite parity such that $x = p^2 - q^2$, $y = 2pq$ and $z = p^2 + q^2$.

Proof: Assume that x and y are relatively prime integers and that $x^2 + y^2 = z^2$. Then $(x + iy) (x - iy) = z^2$ and since x and y are relatively prime, so are $x + iy$ and $x - iy$. Since Theorem I holds for the Gaussian integers we can find integers p and q such that

$$x + iy = (p + iq)^2 = p^2 - q^2 + 2pqi \text{ and hence}$$
$$x - iy = (p - iq)^2 = p^2 - q^2 - 2pqi$$

Equating real and imaginary parts we get:
$$x = p^2 - q^2, \text{ and } y = 2pq.$$
Thus we have shown that there exist integers p and q such that $x = p^2 - q^2$, $y = 2pq$ and hence $z = p^2 + q^2$. This prove Proposition 3. Proposition 2 is now verified by direct substitution. Finally, it is easily seen that if x, y and z are relatively prime then they are pairwise relatively prime and that this happens if and only if p and q are relatively prime and of opposite parity.

19. "Wabi".

Unfortunately the sheer beauty of this analysis cannot obviate the fact that Euler either attempted to characterize the case $x^2 + cy^2 = z^2$ and obtained a wrong result or that he merely attempted to state the weaker Theorem E. Either way, that is certainly a defect in Euler's work.

I suggest that this defect is what the Japanese call a "wabi". A "wabi" is an imperfection in an otherwise perfect piece of workmanship, a defect which turns that item into a work of art; Rheingold (1988). The Japanese recognize that a perfectly executed work is not

---

The Japanese recognize that a perfectly executed work is not really a work of art because it lacks the human touch.

---

really a work of art because it lacks the human touch. What humanizes such a piece is a minor imperfection because being human means being imperfect.

In my opinion, Euler's work is, partly because of its "wabi", the greatest of the three masterpieces in our triptych. This might appear to be a strange assertion by a mathematician about a mathematical work. After all, mathematics does not only strive for perfection, it expects it, demands it, and is satisfied with nothing less. Such at least was the wisdom of our fathers who maintained that the objects which mathematicians study live in an ideal Platonic heaven and are therefore ideally

perfect. We used to speak of the cold and austere beauty of mathematics and write poems to the effect that: "Euclid alone has looked on beauty bare...."

While most of us remain Platonists, we have also become humanists. We have come to recognize several new truths. We have found Plato's world of ideas: it is inside our heads. Thus we no longer pretend to work in a perfectly antiseptic, ideal environment proving, with absolute rigor, statements which are absolutely true. We realize that standards of rigor change over time. We do not know whether God eternally geometrizes: we do know, however, that in case God does, God's activity differs radically from ours. We see mathematics as a human activity, one that involves a lot of very hard work and anguish and "blood and sweat and tears" but one that is nevertheless supremely satisfying. When we "do" mathematics, we do the best we can not to make mistakes, when we referee our colleagues' works, we assiduously look for errors and weed out all we can find. But we no longer expect perfection; we let time temper, improve, and justify our work. Euler with his "wabi" teaches us these lessons.

You are now invited to look for "wabis" in Euclid and Diophantus. I'm sure you will find some, and not only in these excerpts; Euclid's "wabis" have recently been well documented in the literature. Still, as far as this triptych is concerned, Euler's "wabi" is really a "whopper" in comparison to those of the other two authors. Nevertheless Euler's work surpasses that of Euclid and Diophantus in one important respect. Euler's "wabi" raises many new questions, leads us to study new topics such as the structure of quadratic fields, and gives us greater insights into well known mathematical facts such as the Fundamental Theorem of Arithmetic and its consequences. It would seem that the more humanistic pieces of mathematics are also the more fruitful.

BIBLIOGRAPHY

Boyer, Carl & Uta Merzbach. *A History of Mathematics*. New York, John Wiley & Sons, (1989)

Courant, Richard & Herbert Robbins. *What is Mathematics?*, New York, Oxford University Press, (1941)

Czwalina, Arthur. *Arithmetik des Diophantos aus Alexandria* Goettingen, Vandenhoeck & Ruprecht (1952)

Diophantos: *Arithmetik*, Goettingen, Vandenhoeck & Ruprecht (1952)

Edwards, Harold M. *Fermat's Last Theorem*. New York, Springer Verlag, (1977)

Euclid, *The Elements*, New York, Dover, (1956)

Euler, Leonhard. *Elements of Algebra*, New York, Springer Verlag, (1984)

Fellman, Emil A. "Ein Essay ueber Leben und Werk" in *Leonhard Euler: Beitraege zu Leben und Werk, Gedankenband des Kantons Basel-Stadt.* J. J. Burckkardt, E. A. Fellman & W. Habicht, editors, Birkhaeuser Verlag, Basel, (1983)

Heath, Thomas L. *Euclid, the Thirteen Books of the Elements*, New York, Dover, (1956)

Heath, Thomas L. *Diophantus of Alexandria: A Study in the History of Greek Algebra*, New York, Dover, (1964)

Neugebauer, Otto. *The Exact Sciences in Antiquity*, New York, Dover, (1969)

Pollard, Harry. *The Theory of Algebraic Numbers*. Washington, D.C., The Mathematical Association of America, (1950)

Rheingold, Howard. *They Have a Word for It*, Los Angeles, Jeremy P. Tarcher, Inc, (1988)

Sesiano, Jacques. *Books IV to VII of Diophantus' Arithmetica in the Arabic Translation Attributed to Qusta Ibn Luga*, New York, Springer Verlag, (1982)

Stark, Harold M. *An Introduction to Number Theory*, Cambridge, The MIT Press, (1978)

Weil, Andre. L'Oeuvre Arithmetique d'Euler" in *Leonhard Euler: Beitraege zu Leben und Werk. Gedankenband des Kantons Basel-Stadt.* J. J. Burckkardt, E. A. Fellman & W. Habicht, editors, Birkhaeuser Verlag, Basel, (1983)

Zhmud, Leonid. "Pythagoras as a Mathematician", *Historia Mathematica*, Vol. 16 #3, August 1989

Huygens' Cycloidal Clock

Abe Shenitzer
York University
North York, Ontario, Canada

Galileo conjectured the (near-) isochronous property of the pendulum but relied on water clocks to measure time. The first pendulum clock was invented in 1656 by the great mathematician and physicist Huygens. Not satisfied with the accuracy of the pendulum clock, Huygens sought to devise a pendulum with period strictly independent of the amplitude. The path of the bob of such a pendulum is called a tautochrone. Huygens proved that the cycloid is a tautochrone, and that its evolute is a congruent cycloid. These insights provided the theoretical basis for clocks (built around 1700) in which cycloidal jaws forced the bob to move along a cycloidal path.

While the practical value of Huygens' insights proved insignificant (pendulum friction and air resistance overshadow the improvement resulting from replacing

an ordinary pendulum with a cycloidal pendulum), his reasoning in the proof that the cycloid is the *only* tautochrone "goes far beyond the differential and integral calculus . . .

What follows are three computations, marked (a), (b), and (c), respectively. (a) is a calculus-based proof of the tautochrone property of the cycloid. (b) is Huygens' proof of this property. (c) is an application of the setup in (b).

Huygens' proof (b) is brilliant but beyond the inventive capacity of ordinary mortals. The calculus proof (a) calls for a measure of skill in the handling of integrals but is essentially routine. A comparison of the two

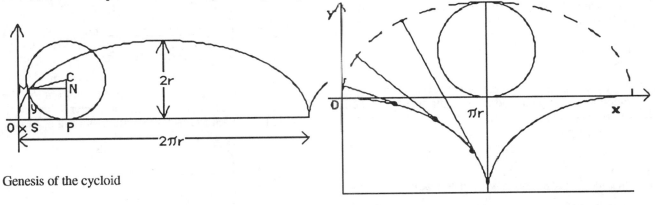

Genesis of the cycloid

The evolute of a cycloid is a cycloid

The cycloidal clock "in the raw"

The cycloidal clock "in finished form"

solutions is a dramatic illustration of the "routinizing" power of the calculus.

(a) *How to show that a cycloid is a tautochrone by the use of calculus*

For uniform motion (that is, motion with constant velocity), the time of travel t, the distance travelled s, and the velocity v are related by

$$v = \frac{s}{t}$$

Hence

$$t = \frac{s}{v}$$

The integral calculus generalizes this to nonuniform motion. If the velocity v depends on the distance s travelled, then the formula for the time t of travel is

(1) $$t = \int \frac{ds}{v} ,$$

the limits of integration being the initial and final distances, respectively.

Suppose the motion is along a curve given parametrically as

$$x = x(u) , \; y = y(u)$$

Then

$$ds = \sqrt{x'^2 + y'^2} \, du$$

where ' denotes differentiation with respect to the parameter. Setting this into the integral (1) for the time of travel gives

(1)' $$t = \int \frac{\sqrt{x'^2 + y'^2}}{v} \, du .$$

To evaluate this integral we need to know v as a function of the parameter. To do that we apply the law of conservation of energy. If a unit mass particle is at rest at the point

$$M = (x_0, y_0) = (x(u_0) , y(u_0))$$

(see Figure 1),

then its kinetic energy $\frac{1}{2} v^2$ at the point N = (x,y) is equal to the loss of potential energy between M and N. This loss is $g(y - y_0)$, so

(2) $$\frac{1}{2} v^2 (u) = g(y-y_0)$$

Now take for the curve a cycloid, given parametrically by

$$x = r (u - \sin u), \qquad y = r (1 - \cos u).$$

Then

$$x' = r (1 - \cos u), \;\; y' = r \sin u .$$

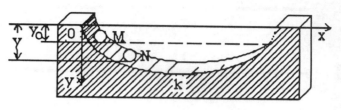

Figure 1

Hence

$$\sqrt{x'^2 + y'^2} = r\sqrt{2 - 2\cos u} \; ,$$

and, from (2),

$$v = \sqrt{2gr(\cos u_0 - \cos u)} \; .$$

Setting these into (1)' gives for the time of travel between M and N

$$t = \sqrt{\frac{r}{g}} \int_{u_0} \sqrt{\frac{1 - \cos u}{\cos u_0 - \cos u}} \, du$$

Using the identities

$$1 - \cos u = 2 \sin^2 \frac{u}{2} ,$$

$$\cos u = 2 \cos^2 u - 1 ,$$

we can rewrite the above as

$$t = \sqrt{\frac{r}{g}} \int_{u_0} \frac{\sin \frac{u}{2}}{\sqrt{\cos^2 \frac{u_0}{2} - \cos^2 \frac{u}{2}}} \, du .$$

Now take N to be the point at the bottom of the cycloid. This corresponds to the parameter value $u = \pi$, so the time taken to reach the bottom starting at M is

$$t = \sqrt{\frac{r}{g}} \int_{u_0}^{\pi} \frac{\sin \frac{u}{2}}{\sqrt{\cos^2 \frac{u_0}{2} - \cos^2 \frac{u}{2}}} \, du .$$

Introduce $\cos \frac{u_0}{2}$ as a new variable of integration. Then the limits are $p_0 = \cos \frac{u_0}{2}$ and $\cos \frac{\pi}{2} = 0$, so the time of travel is given by the integral

$$t = 2\sqrt{\frac{r}{g}} \int_{0}^{p_0} \frac{dp}{\sqrt{p_0^2 - p^2}} = \pi \sqrt{\frac{r}{g}}$$

This is a remarkable result! It shows that the time it takes for the particle to get from the release point M to the lowest point K on the cycloid is independent of M, that is, the cycloid is a tautochrone. Q.E.D.

(b) *How Huygens showed (more than 300 Years ago) that the cycloid is a tautochrone*

If two particles are released from the same height and move along two curves with velocities whose *vertical* components are the same, then they will reach ground level at the same time.

Huygens' idea is to replace the gravity-induced motion of a particle on a cycloid with a motion of a particle on a circle such that the vertical components of the velocities of the two particles are the same. The details follow.

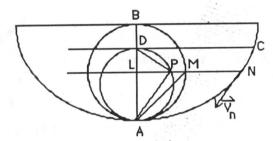

Figure 2

Consider the cycloid generated by the circle with diameter AB (Figure 2). The particle on the cycloid is released at C (Figure 2). N is a "typical" position of this particle. The corresponding particle moves on the semicircle DPA with diameter DA, where D is at the same height as C and A is the lowest point on the cycloid. P is a "typical" position of that particle. We recall that our aim is to impart to the particle on the semicircle a velocity whose vertical component is equal to the vertical component of the gravity-induced velocity of the particle on the cycloid.

We have: $|\vec{v}_n| = \sqrt{2g\,DL}$. A (nonobvious!) geometric fact of fundamental importance is that *the tangent to the cycloid at N is parallel to the chord AM* (this can be easily deduced by calculus from the parametric equations of the cycloid). This being so, the vertical component of $\vec{v}_n$ is

$\sqrt{2g \cdot DL} \cdot \sin \angle AML$, that is, $\sqrt{2g \cdot DL} \dfrac{AL}{AM}$.

Now we replace AM with quantities in the semicircle DPA. Since $LP^2 = DL \cdot AL$ implies that

$LP = \sqrt{DL \cdot AL}$, and $AM^2 = AL \cdot AB$ implies that $AM = \sqrt{AL} \cdot \sqrt{AB}$, it follows that the magnitude of the vertical component of $\vec{v}_n$ is

$$\sqrt{2g\,DL}\,\frac{AL}{AM} = \sqrt{2g}\,\frac{\sqrt{DL \cdot AL} \cdot \sqrt{AL}}{\sqrt{AL} \cdot \sqrt{AB}}$$

$$= \sqrt{\frac{2g}{AB}}\,LP \ .$$

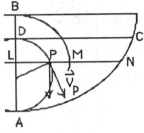

Figure 3

The vertical component of the (as yet undetermined) velocity $\vec{v}_p$ is $|\vec{v}_p| \cdot LP/\dfrac{AD}{2}$ (see Figure 3). Since we insist on the equality $2|\vec{v}_p| \cdot \dfrac{LP}{AD} = \sqrt{\dfrac{2g}{AB}}\,LP$, it follows that $|\vec{v}_p| = \dfrac{AD}{2}\,\sqrt{2g/AB}$. But then the time it takes the particle moving on the semicircle DPA to get from D to A is

$$\pi \cdot \frac{\dfrac{AD}{2}}{\sqrt{2g/AB} \cdot \dfrac{AD}{2}} = \pi / \sqrt{2g/AB} \ , \text{ which is}$$

independent of the elevation of D. It follows that *the time of descent of the particle moving under the action of gravity from C to A along the cycloidal arc CNA is independent of the elevation of C*, that is, that *the cycloid is a tautochrone* and its amplitude-independent quarter-period is $\pi/\sqrt{2g/AB}$. (N.B. Putting AB = 2r , where r is the radius of the circle generating the cycloid in our first computation, we see that the quarter-period value obtained by Huygens is also $\pi\sqrt{r/g}$.

Historical note. Since π was introduced in the 18th century (by Euler) it could not have been used by Huygens. Huygens' result was that

$$\frac{\text{CNA-time}}{\text{free-fall BA-time}} = \frac{\text{DPA-time}}{\text{free-fall BA-time}}$$

$$= \frac{\text{length of DPA}}{|\vec{v}_p|\,\sqrt{2AB/g}}$$

$$= \frac{\text{length of DPA}}{\dfrac{AD}{2}\sqrt{2g/AB}\,\sqrt{2AB/g}}$$

$$= \frac{\text{length of DPA}}{AD} \ .$$

This shows that the CNA-time, that is the time of descent of the particle on the cycloid from C to A , is independent of C. Q.E.D.

(c) *An application of the setup in (b)*

Let us introduce yet another particle into the Huygens scheme. This particle will be the projection—one might say the shadow—on AD of the particle traversing the semicircle DPA. The reason for our interest in this particle is that its velocity is directed vertically and equal to the vertical components of the velocities of the two previous particles. Since the speed of the particle on the semicircle DPA is $\dfrac{AD}{2}\sqrt{2g/AB} = \dfrac{AD}{2}\sqrt{g/r}$,

its angular velocity is $\omega = \sqrt{g/r}$. As we vary the release point C on the cycloid we obtain a family of semicircles DPA traversed with angular velocity ω. To each constant-speed motion on a semicircle of the family there corresponds a "shadow-motion" along the corresponding diameter DA (see Figure 2). When the shadow particle is at L its speed is $\sqrt{g/r} \cdot LP = \omega LP$. The shadow is executing a simple harmonic motion with half-period π/ω. If we introduce a coordinate system with center at the midpoint of DA and x-axis determined by AD, then the half-period of the shadow-motion is

$$\int_{-\rho}^{\rho}\frac{dx}{\omega LP} = \frac{1}{\omega}\int_{-\rho}^{\rho}\frac{dx}{\sqrt{\rho^2 - x^2}} \quad, \text{ where } \rho = AD/2. \text{ The}$$

equality $\dfrac{1}{\omega}\displaystyle\int_{-\rho}^{\rho}\dfrac{dx}{\sqrt{\rho^2 - x^2}} = \dfrac{\pi}{\omega}$ implies that

$$\int_{-\rho}^{\rho}\frac{dx}{\sqrt{\rho^2 - x^2}} = \pi,$$

regardless of the (positive) value of ρ. We have evaluated a nontrivial integral by giving it a physical interpretation!

While it is true that

$$\int_{-\rho}^{\rho}\frac{dx}{\sqrt{\rho^2 - x^2}} = [\text{arc sin } \tfrac{x}{\rho}]_{-\rho}^{\rho} = \pi,$$

it is nice to be able to predict this by interpreting

$\dfrac{1}{\omega}\displaystyle\int_{-\rho}^{\rho}\dfrac{dx}{\sqrt{\rho^2 - x^2}}$ as the (amplitude-independent!) half-period of a simple harmonic motion with frequency $\dfrac{\omega}{2\pi}$.

References

The computations in (a), originally based on 11, pp. 240-245 of A. A. Savelov, *Plane Curves*, Moscow, 1960 (in Russian), were made more explicit and transparent by Peter Lax.

(b) is adapted from Vol. 2, pp. 206-210 of *History of Mathematics*, ed. A. P. Yushkevich, Nauka, 1970 (in Russian).

NATIONAL UNIVERSITY LIBRARY   SAN DIEGO

8117